The Deep Learning Architect's Handbook

Build and deploy production-ready DL solutions leveraging
the latest Python techniques

Ee Kin Chin

‹packt›

BIRMINGHAM—MUMBAI

The Deep Learning Architect's Handbook

Group Product Manager: Ali Abidi
Book Project Manager: Shambhavi Mishra
Senior Editor: Rohit Singh
Technical Editor: Devanshi Ayare
Copy Editor: Safis Editing
Proofreader: Safis Editing
Indexer: Subalakshmi Govindhan
Production Designer: Ponraj Dhandapani
DevRel Marketing Executive: Vinishka Kalra

First published: December 2023

Production reference: 1301123

Published by Packt Publishing Ltd.
Grosvenor House
11 St Paul's Square
Birmingham
B3 1RB, UK.

ISBN 978-1-80324-379-5

www.packtpub.com

To my wife, Nina, the constant source of inspiration, support, and encouragement in my life. Without you, this book would have remained a dream.

– Ee Kin Chin

Contributors

About the author

Ee Kin Chin is a senior deep learning engineer at DataRobot. He led teams to develop advanced AI tools used by numerous organizations from diverse industries and provided consultation on many customer AI use cases. Previously, he worked on deep learning (DL) computer vision projects for smart vehicles and human sensing applications at Panasonic and offered AI solutions using edge cameras at a tech solutions provider. He was also a DL mentor for an online course. Holding a Bachelor of Engineering (honors) degree in electronics, with a major in telecommunications, and a proven track record of successful application of AI, Ee Kin's expertise includes embedded applications, practical deep learning, data science, and classical machine learning.

A huge shout-out to my fantastic friends, mentors, colleagues, family, book reviewers, and the open source community who've supported and motivated me during my professional career. Your shared knowledge, insights, and wisdom have been invaluable.

About the reviewers

Shivani Modi is a data scientist with expertise in machine learning, deep learning, and NLP, holding a master's degree from Columbia University. Her five years of professional experience spans IBM, SAP, and C3 AI, where she has excelled in deploying scalable AI models across various sectors. At Konko AI, Shivani spearheaded the development of tools to optimize LLM selection and deployment. Shivani's dedication to mentoring and talent development, coupled with her hands-on experience in leading complex projects, underscores her status as a thought leader in AI innovation. Her upcoming project aims to revolutionize how developers utilize LLMs, ensuring their secure and efficient implementation.

Ved Upadhyay is a seasoned data science and AI professional, bringing over seven years of hands-on experience in addressing enterprise-level challenges in deep learning. His expertise spans diverse industries, including retail, e-commerce, pharmaceuticals, agro-tech, and socio-tech, where he has successfully implemented AI solutions. Ved is currently working as a senior data scientist at Walmart, where he leads multiple initiatives focused on customer propensity and responsible AI. He earned his master's degree in data science from the University of Illinois Urbana-Champaign and has contributed as a deep learning researcher at IIIT Hyderabad.

Table of Contents

5

Understanding Autoencoders   101

6

Understanding Neural Network Transformers   113

7

Deep Neural Architecture Search   133

Part 2 – Multimodal Model Insights

11

Explaining Neural Network Predictions   239

12

Interpreting Neural Networks   259

13

Exploring Bias and Fairness   275

14

Analyzing Adversarial Performance   307

Part 3 – DLOps

15

Deploying Deep Learning Models to Production   331

19

Architecting LLM Solutions   449

Preface

As a deep learning practitioner and enthusiast, I have spent years working on various projects and learning from diverse sources such as Kaggle, GitHub, colleagues, and real-life use cases. I've realized that there is a significant gap in the availability of cohesive, end-to-end deep learning resources. Traditional **Massively Open Online Courses (MOOC)**, while helpful, often lack the practical knowledge and real-world insights that can only be gained through hands-on experience.

To bridge this gap, I've created The Deep Learning Architect Handbook, a comprehensive and practical guide that combines my unique experiences and insights. This book will help you navigate the complex landscape of deep learning, providing you with the knowledge and insights that would typically take years of hands-on experience to acquire, condensed into a resource that can be consumed in just days or weeks.

This book delves into various stages of the deep learning life cycle, from planning and data preparation to model deployment and governance. Throughout this journey, you'll encounter both foundational and advanced deep learning architectures, such as **Multi-Layer Perceptrons (MLPs)**, **Convolutional Neural Networks (CNNs)**, **Recurrent Neural Networks (RNNs)**, autoencoders, transformers, and cutting-edge methods, such as **Neural Architecture Search (NAS)**. Divided into three parts, this book covers foundational methods, model insights, and DLOps, exploring advanced topics such as NAS, adversarial performance, and **Large Language Model (LLM)** solutions. By the end of this book, you will be well-prepared to design, develop, and deploy effective deep learning solutions, unlocking their full potential and driving innovation across various applications.

I hope that this book will serve as a way for me to give back to the community, by sparking conversations, challenging assumptions, and inspiring new ideas and approaches in the field of deep learning. I invite you to join me on this journey, and I look forward to hearing your thoughts and feedback as we explore the captivating world of deep learning together. Please feel free to reach out to me via LinkedIn through www.linkedin.com/in/chineekin, Kaggle through https://www.kaggle.com/dicksonchin93, or other channels listed on my LinkedIn profile. Your unique experiences and perspectives will undoubtedly contribute to the ongoing evolution of this book and the deep learning community as a whole.

Who this book is for

This book is best suited for deep learning practitioners, data scientists, and machine learning developers who want to explore deep learning architectures to solve complex business problems. The audience of this book is professionals in the deep learning and AI space who are going to use the knowledge in their business use cases. Working knowledge of Python programming and a basic understanding of deep learning techniques is needed to get the most out of this book.

What this book covers

Chapter 1, Deep Learning Life Cycle, introduces the key stages of a deep learning project, focusing on planning and data preparation, and sets the stage for a comprehensive exploration of the deep learning life cycle throughout the book.

Chapter 2, Designing Deep Learning Architectures, dives into the foundational aspects of deep learning architectures, including MLPs, and discusses their role in advanced neural networks, as well as the importance of backpropagation and regularization.

Chapter 3, Understanding Convolutional Neural Networks, provides an in-depth look at CNNs, their applications in image processing, and various model families within the CNN domain.

Chapter 4, Understanding Recurrent Neural Networks, explores the structure and variations of RNNs and their ability to process sequential data effectively.

Chapter 5, Understanding Autoencoders, examines the fundamentals of autoencoders as a method for representation learning and their applications across different data modalities.

Chapter 6, Understanding Neural Network Transformers, delves into the versatile nature of transformers, capable of handling diverse data modalities without explicit data-specific biases, and their potential applications in various tasks and domains.

Chapter 7, Deep Neural Architecture Search, introduces the concept of NAS as a way to automate the design of advanced neural networks and discusses its applications and limitations in different scenarios.

Chapter 8, Exploring Supervised Deep Learning, covers various supervised learning problem types, techniques for implementing and training deep learning models, and practical implementations using popular deep learning frameworks.

Chapter 9, Exploring Unsupervised Deep Learning, discusses the contributions of deep learning to unsupervised learning, particularly highlighting the unsupervised pre-training method. Harnessing the vast amounts of freely available data on the internet, this approach improves model performance for downstream supervised tasks and paves the way toward general **Artificial Intelligence (AI)**.

Chapter 10, Exploring Model Evaluation Methods, provides an overview of model evaluation techniques, metric engineering, and strategies for optimizing against evaluation metrics.

Chapter 11, Explaining Neural Network Predictions, delves into the prediction explanation landscape, focusing on the integrated gradients technique and its practical applications for understanding neural network predictions.

Chapter 12, Interpreting Neural Networks, delves into the nuances of model understanding and showcases techniques for uncovering patterns detected by neurons. By exploring real images and generating images through optimization to activate specific neurons, you will gain valuable insights into the neural network's decision-making process.

Chapter 13, *Exploring Bias and Fairness*, addresses the critical issue of bias and fairness in machine learning models, discussing various types, metrics, and programmatic methods for detecting and mitigating bias.

Chapter 14, *Analyzing Adversarial Performance*, examines the importance of adversarial performance analysis in identifying vulnerabilities and weaknesses in machine learning models, along with practical examples and techniques for analysis.

Chapter 15, *Deploying Deep Learning Models in Production*, focuses on key components, requirements, and strategies for deploying deep learning models in production environments, including architectural choices, hardware infrastructure, and model packaging.

Chapter 16, *Governing Deep Learning Models*, explores the fundamental pillars of model governance, including model utilization, model monitoring, and model maintenance, while providing practical steps for monitoring deep learning models.

Chapter 17, *Managing Drift Effectively in a Dynamic Environment*, discusses the concept of drift and its impact on model performance, along with strategies for detecting, quantifying, and mitigating drift in deep learning models.

Chapter 18, *Exploring the DataRobot AI Platform*, showcases the benefits of AI platforms, specifically DataRobot, in streamlining and accelerating the deep learning life cycle, and highlights various features and capabilities of the platform.

Chapter 19, *Architecting LLM Solutions*, delves into LLMs and the potential applications, challenges, and strategies for creating effective, contextually aware solutions using LLMs.

To get the most out of this book

The code provided in the chapters has been tested on a computer with Python 3.10, Ubuntu 20.04 LTS 64-bit OS, 32 GB RAM, and an RTX 2080TI GPU for running deep learning models. Although the code has been tested on this specific setup, it may also work on other configurations; however, compatibility and performance are not guaranteed. Python dependencies are included in the `requirements.txt` file for easy installation in each chapter's respective GitHub folders. Additionally, some non-Python software might be required; their installation instructions will be mentioned at the beginning of each relevant tutorial. For these software installations, you need to refer to external manuals or guides to install them. Do keep in mind the potential differences in system configurations as you carry out the practical code sections in this book.

If you are using the digital version of this book, we advise you to type the code yourself or access the code from the book's GitHub repository (a link is available in the next section). Doing so will help you avoid any potential errors related to the copying and pasting of code.

Download the example code files

You can download the example code files for this book from GitHub at `https://github.com/PacktPublishing/The-Deep-Learning-Architect-Handbook`. If there's an update to the code, it will be updated in the GitHub repository.

We also have other code bundles from our rich catalog of books and videos available at `https://github.com/PacktPublishing/`. Check them out!

Conventions used

There are a number of text conventions used throughout this book.

`Code in text`: Indicates code words in text, database table names, folder names, filenames, file extensions, pathnames, dummy URLs, user input, and Twitter handles. Here is an example: "We will be using `pandas` for data manipulation and structuring, `matplotlib` and `seaborn` for plotting graphs, `tqdm` for visualizing iteration progress, and `lingua` for text language detection."

A block of code is set as follows:

```
import pandas as pd
import matplotlib.pyplot as plt
import seaborn as sns
from tqdm import tqdm
from lingua import Language, LanguageDetectorBuilder
tqdm.pandas()
```

Any command-line input or output is written as follows:

```
sudo systemctl start node_exporter
sudo systemctl start prometheus
```

Bold: Indicates a new term, an important word, or words that you see onscreen. For instance, words in menus or dialog boxes appear in **bold**. Here is an example: "We can set up the Prometheus link now by clicking on the three-line button on the top-left tab and clicking on the **Data Sources** tab under the **Administration** dropdown."

> **Tips or important notes**
> Appear like this.

Get in touch

Feedback from our readers is always welcome.

General feedback: If you have questions about any aspect of this book, email us at `customercare@packtpub.com` and mention the book title in the subject of your message.

Errata: Although we have taken every care to ensure the accuracy of our content, mistakes do happen. If you have found a mistake in this book, we would be grateful if you would report this to us. Please visit `www.packtpub.com/support/errata` and fill in the form.

Piracy: If you come across any illegal copies of our works in any form on the internet, we would be grateful if you would provide us with the location address or website name. Please contact us at `copyright@packt.com` with a link to the material.

If you are interested in becoming an author: If there is a topic that you have expertise in and you are interested in either writing or contributing to a book, please visit `authors.packtpub.com`.

Share Your Thoughts

Once you've read *The Deep Learning Architect's Handbook*, we'd love to hear your thoughts! Scan the QR code below to go straight to the Amazon review page for this book and share your feedback.

`https://packt.link/r/1-803-24379-1`

Your review is important to us and the tech community and will help us make sure we're delivering excellent quality content.

Download a free PDF copy of this book

Thanks for purchasing this book!

Do you like to read on the go but are unable to carry your print books everywhere?

Is your eBook purchase not compatible with the device of your choice?

Don't worry, now with every Packt book you get a DRM-free PDF version of that book at no cost.

Read anywhere, any place, on any device. Search, copy, and paste code from your favorite technical books directly into your application.

The perks don't stop there, you can get exclusive access to discounts, newsletters, and great free content in your inbox daily

Follow these simple steps to get the benefits:

1. Scan the QR code or visit the link below

https://packt.link/free-ebook/9781803243795

2. Submit your proof of purchase
3. That's it! We'll send your free PDF and other benefits to your email directly

Part 1 – Foundational Methods

In this part of the book, you will gain a comprehensive understanding of the foundational methods and techniques in deep learning architectures. Starting with the deep learning life cycle, you will explore various stages at a high level, from planning and data preparation to model development, insights, deployment, and governance. You will then dive into the intricacies of designing deep learning architectures such as MLPs, CNNs, RNNs, autoencoders, and transformers. Additionally, you will learn about the emerging method of neural architecture search and its impact on the field of deep learning.

Throughout this part, you will also delve into the practical aspects of supervised and unsupervised deep learning, covering topics such as binary classification, multiclassification, regression, and multitask learning, as well as unsupervised pre-training and representation learning. With a focus on real-world applications, this part provides valuable insights into the implementation of deep learning models using popular frameworks and programming languages.

By the end of this part, you will have a solid foundation in deep learning architectures, methods, and life cycles, which will enable you to continue your journey to face other challenges involved in crafting deep learning solutions.

This part contains the following chapters:

- *Chapter 1, Deep Learning Life Cycle*
- *Chapter 2, Designing Deep Learning Architectures*
- *Chapter 3, Understanding Convolutional Neural Networks*
- *Chapter 4, Understanding Recurrent Neural Networks*
- *Chapter 5, Understanding Autoencoders*
- *Chapter 6, Understanding Neural Network Transformers*
- *Chapter 7, Deep Neural Architecture Search*
- *Chapter 8, Exploring Supervised Deep Learning*
- *Chapter 9, Exploring Unsupervised Deep Learning*

1

Deep Learning Life Cycle

In this chapter, we will explore the intricacies of the deep learning life cycle. Sharing similar characteristics to the machine learning life cycle, the deep learning life cycle is a framework as much as it is a methodology that will allow a deep learning project idea to be insanely successful or to be completely scrapped when it is appropriate. We will grasp the reasons why the process is cyclical and understand some of the life cycle's initial processes on a deeper level. Additionally, we will go through some high-level sneak peeks of the later processes of the life cycle that will be explored at a deeper level in future chapters.

Comprehensively, this chapter will help you do the following:

- Understand the similarities and differences between the deep learning life cycle and its machine learning life cycle counterpart

- Understand where domain knowledge fits in a deep learning project

- Understand the few key steps in planning a deep learning project to make sure it can tangibly create real-world value

- Grasp some deep learning model development details at a high level

- Grasp the importance of model interpretation and the variety of deep learning interpretation techniques at a high level

- Explore high-level concepts of model deployments and their governance

- Learn to choose the necessary tools to carry out the processes in the deep learning life cycle

We'll cover this material in the following sections:

- Machine learning life cycle
- The construction strategy of a deep learning life cycle
- The data preparation stage
- Deep learning model development
- Delivering model insights
- Managing risks

Technical requirements

This chapter includes some practical implementations in the **Python** programming language. To complete it, you need to have a computer with the following libraries installed:

- `pandas`
- `matplotlib`
- `seaborn`
- `tqdm`
- `lingua`

The code files are available on GitHub: `https://github.com/PacktPublishing/The-Deep-Learning-Architect-Handbook/tree/main/CHAPTER_1`.

Understanding the machine learning life cycle

Deep learning is a subset of the wider machine learning category. The main characteristic that sets it apart from other machine learning algorithms is the foundational building block called **neural networks**. As deep learning has advanced tremendously since the early 2000s, it has made many previously unachievable feats possible through its machine learning counterparts. Specifically, deep learning has made breakthroughs in recognizing complex patterns that exist in complex and unstructured data such as text, images, videos, and audio. Some of the successful applications of deep learning today are face recognition with images, speech recognition from audio data, and language translation with textual data.

Machine learning, on the other hand, is a subset of the wider artificial intelligence category. Its algorithms, such as tree-based models and linear models, which are not considered to be deep learning models, still serve a wide range of use cases involving tabular data, which is the bulk of the data that's stored by small and big organizations alike. This tabular data may exist in multiple structured databases and can span from 1 to 10 years' worth of historical data that has the potential to be used for building predictive machine learning models. Some of the notable predictive applications for machine learning algorithms are fraud detection in the finance industry, product recommendations in e-commerce, and predictive maintenance in the manufacturing industry. *Figure 1.1* shows the relationships between deep learning, machine learning, and artificial intelligence for a clearer visual distinction between them:

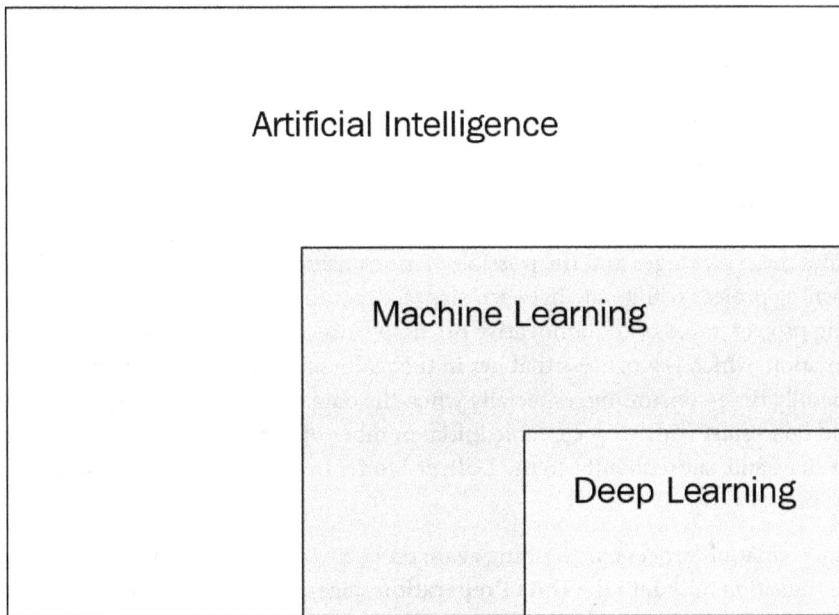

Figure 1.1 – Artificial intelligence relationships

Now that we know what deep learning and machine learning are in a nutshell, we are ready for a glimpse of the machine learning life cycle, as shown in *Figure 1.2*:

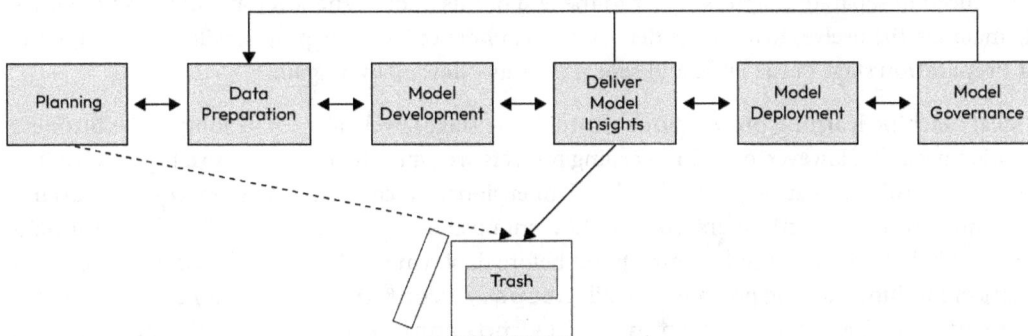

Figure 1.2 – Deep learning/machine learning life cycle

As advanced and complex the deep learning algorithm is compared to other machine learning algorithms, the guiding methodologies that are needed to ensure success in both domains are unequivocally the same. The machine learning life cycle involves six stages that interact with each other in different ways:

1. **Planning**
2. **Data Preparation**
3. **Model Development**
4. **Deliver Model Insights**
5. **Model Deployment**
6. **Model Governance**

Figure 1.2 shows these six stages and the possible stage transitions depicted with arrows. Typically, a machine learning project will iterate between stages, depending on the business requirements. In a deep learning project, most of the innovative predictive use cases require manual data collection and data annotation, which is a process that lies in the realm of the **Data Preparation** stage. As this process is generally time-consuming, especially when the data itself is not readily available, a go-to solution would be to start with an acceptable initial number of data and transition into the **Model Development** stage and, subsequently, to the **Deliver Model Insight** stage to make sure results from the ideas are sane.

After the initial validation process, depending again on business requirements, practitioners would then decide to transition back into the **Data Preparation** stage and continue to iterate through these stages cyclically in different data size milestones until results are satisfactory toward both the model development and business metrics. Once it gets approval from the necessary stakeholders, the project then goes into the **Model Deployment** stage, where the built machine learning model will be served to allow its predictions to be consumed. The final stage is **Model Governance**, where practitioners carry out tasks that manage the risk, performance, and reliability of the deployed machine learning model. Model deployment and model governance both deserve more in-depth discussion and will be introduced in separate chapters closer to the end of this book. Whenever any of the key metrics fail to maintain themselves to a certain determined confidence level, the project will fall back into the **Data Preparation** stage of the cycle and repeat the same flow all over again.

The ideal machine learning project flows through the stages cyclically for as long as the business application needs it. However, machine learning projects are typically susceptible to a high probability of failure. According to a survey conducted by Dimensional Research and Alegion, covering around 300 machine learning practitioners from 20 different business industries, 78% of machine learning projects get held back or delayed at some point before deployment. Additionally, Gartner predicted that 85% of machine learning projects will fail (https://venturebeat.com/2021/06/28/why-most-ai-implementations-fail-and-what-enterprises-can-do-to-beat-the-odds/). By expecting the unexpected, and anticipating failures before they happen, practitioners can likely circumvent potential failure factors early down the line in the planning stage. This also brings us to the trash icon bundled together in *Figure 1.2*. Proper projects with a good plan typically get discarded only at the **Deliver Model Insights** stage, when it's clear that the proposed model and project can't deliver satisfactory results.

Now that we've covered an overview of the machine learning life cycle, let's dive into each of the stages individually, broken down into sections, to help you discover the key tips and techniques that are needed the complete each stage successfully. These stages will be discussed in an abstract format and are not a concrete depiction of what you should ultimately be doing for your project since all projects are unique and strategies should always be evaluated on a case-by-case basis.

Strategizing the construction of a deep learning system

A deep learning model can only realize real-world value by being part of a system that performs some sort of operation. Bringing deep learning models from research papers to actual real-world usage is not an easy task. Thus, performing proper planning before conducting any project is a more reliable and structured way to achieve the desired goals. This section will discuss some considerations and strategies that will be beneficial when you start to plan your deep learning project toward success.

Starting the journey

Today, deep learning practitioners tend to focus a lot on the algorithmic model-building part of the process. It takes a considerable amount of mental strength to not get hooked on the hype of **state-of-the-art (SOTA)** research-focused techniques. With crazy techniques such as **pixtopix**, which is capable of generating high-resolution realistic color images from just sketches or image masks, and **natural language processing (NLP)** techniques such as **GPT-3**, a 175-billion parameters text generation model from OpenAI, and GPT-4, a multimodal text generation model that is a successor to GPT-3 and its sub-models, that are capable of generating practically anything you ask it to in a text format that ranges from text summarization to generating code, why wouldn't they?!

Jokes aside, to become a true deep learning architect, we need to come to a consensus that any successful machine learning or deep learning project starts with the business *problem* and not from the shiny new research paper you just read online complete with a public GitHub repository. The planning stage often involves many business executives who are not savvy about the details of machine learning algorithms and often, the same set of people wouldn't care about it at all. These algorithms are daunting for business-focused stakeholders to understand and, when added on top of the tough mental barriers of the adoption of artificial intelligence technologies itself, it doesn't make the project any more likely to be adopted.

Evaluating deep learning's worthiness

Deep learning shines the most in handling unstructured data. This includes image data, text data, audio data, and video data. This is largely due to the model's ability to automatically learn and extract complex, high-level features from the raw data. In the case of images and videos, deep learning models can capture spatial and temporal patterns, recognizing objects, scenes, and activities. With audio data, deep learning can understand the nuances of speech, noise, and various sound elements, making it possible to build applications such as speech recognition, voice assistants, and audio classification systems. For text data, deep learning models can capture the context, semantics, and syntax, enabling NLP tasks such as sentiment analysis, machine translation, and text summarization.

This means that if this data exists and is utilized by your company in its business processes, there may be an opportunity to solve a problem with the help of deep learning. However, never overcomplicate problems just so you can solve them with deep learning. Equating this to something more relatable, you wouldn't use a huge sledgehammer to get a nail into wood. It could work and you might get away with it, but you'd risk bending the nail or injuring yourself while using it.

Once a problem has been identified, evaluate the business value of solving it. Not all problems are born the same and they can be ranked based on their business impact, value, complexity, risks, costs, and suitability for deep learning. Generally, you'd be looking for high impact, high value, low complexity, low risks, low cost, and high suitability to deep learning. Trade-offs between these metrics are expected but simply put, make sure the problem you've discovered is worth solving at all with deep learning. A general rule of thumb is to always resort to a simpler solution for a problem, even if it ends up abandoning the usage of deep learning technologies. Simple approaches tend to be more reliable, less costly, less prone to risks, and faster to fruition.

Consider a problem where a solution is needed to remove background scenes in a video feed and leave only humans or necessary objects untouched so that a more suitable background scene can be overlaid as a background instead. This is a common problem in the professional filmmaking industry in all film genres today.

Semantic segmentation, which is the task of assigning a label to every pixel of an image in the width and height dimensions, is a method that is needed to solve such a problem. In this case, the task needs to assign labels that can help identify which pixels need to be removed. With the advent of many publicly available semantic segmentation datasets, deep learning has been able to advance considerably in the semantic segmentation field, allowing itself to achieve a very satisfactory fine-grained understanding of the world, enough so that it can be applied in the industry of autonomous driving and robot navigation most prominently. However, deep learning is not known to be 100% error-free and almost always has some error, even in the controlled evaluation dataset. In the case of human segmentation, for example, the model would likely result in the most errors in the fine hair areas. Most filmmakers aim for perfect depictions of their films and require that every single pixel gets removed appropriately without fail since a lot of money is spent on the time of the actors hired for the film. Additionally, a lot of time and money would be wasted in manually removing objects that could be otherwise simply removed if the scene had been shot with a green screen. This is an example of a case where we should not overcomplicate the problem. A green screen is all you need to solve the problem described: specifically, the rare chromakey green color. When green screens are prepped properly in the areas where the desired imagery will be overlaid digitally, image processing techniques alone can remove the pixels that are considered to be in the small light intensity range centered on the chromakey green color and achieve semantic segmentation effectively with a rule-based solution. The green screen is a simpler solution that is cost-effective, foolproof, and fast to set up.

That was a mouthful! Now, let's go through a simpler problem. Consider a problem where we want to automatically and digitally identify when it rains. In this use case, it is important to understand the actual requirements and goals of identifying the rain: is it sufficient to detect rain exactly when it happens? Or do we need to identify whether rain will happen in the near future? What will we use the information of rain events for? These questions will guide whether deep learning is required or not. We, as humans, know that rain can be predicted by visual input by either looking at the presence of raindrops falling or looking at cloud conditions. However, if the use case is sufficient to detect rain when it happens, and the goal of detecting rain is to determine when to water the plants, a simpler approach would be to use an electronic sensor to detect the presence of water or humidity. Only when you want to estimate whether it will rain in the future, let's say in 15 minutes, does deep learning make more sense to be applied as there are a lot of interactions between meteorological factors that can affect rainfall. Only by brainstorming each use case and analyzing all potential solutions, even outside of deep learning, can you make sure deep learning brings tangible business value compared to other solutions. Do not just apply deep learning because you want to.

At times, when value isn't clear when you're directly considering a use case, or when value is clear but you have no idea how to execute it, consider finding reference projects from companies in the same industry. Companies in the same industry have a high chance of wanting to optimize the same processes or solve the same pain points. Similar reference projects can serve as a guide to designing a deep learning system and can serve as proof that the use case being considered is worthy of the involvement of deep learning technologies. Of course, not everybody has access to details like this, but you'd be surprised what Google can tell you these days. Even if there isn't a similar project being carried out for direct reference, you would likely be able to pivot upon the other machine learning project references that already have a track record of bringing value to the same industry.

Admittedly, rejecting deep learning at times would be a hard pill to swallow considering that most practitioners get paid to implement deep learning solutions. However, dismissing it earlier will allow you to focus your time on more valuable problems that would be more useful to solve with deep learning and prevent the risk of undermining the potential of deep learning in cases where simpler solutions can outperform deep learning. Criteria for deep learning worthiness should be evaluated on a case-by-case basis and as a practitioner, the best advice to follow is to simply practice common sense. Spend a good amount of time going through the problem exploration and the worthiness evaluation process. The last thing you want is to spend a painstaking amount of time preparing data, building a deep learning model, and delivering very convincing model insights only to find out that the label you are trying to predict does not provide enough value for the business to invest further.

Defining success

Ever heard sentences like *"My deep learning model just got 99% accuracy on my validation dataset!"*? Data scientists often make the mistake of determining the success of a machine learning project just by using validation metrics they use to evaluate their machine learning models during the model development process. Model-building metrics such as accuracy, precision, or recall are important metrics to consider in a machine learning project but unless they add business values and connect to the business objectives in some way, they rarely mean anything. A project can achieve a good accuracy score but still fail to achieve the desired business goals. This can happen in cases when no proper success metrics have been defined early and subsequently cause a wrong label to be used in the data preparation and model development stages. Furthermore, even when the model metric positively impacts business processes directly, there is a chance that the achievement won't be communicated effectively to business stakeholders and the worst case not considered to be successful when reported as-is.

Success metrics, when defined early, act as the machine learning project's guardrails and ensure that the project goals are aligned with the business goals. One of the guardrails is that a success metric can help guide the choice of a proper label that can at inference time, tangibly improve the business processes or otherwise create value in the business. First, let's make sure we are aligned with what a label means, which is a value that you want the machine learning model to predict. The purpose of a machine learning model is to assign these labels automatically given some form of input data, and thus during the data preparation and model development stages, a label needs to be chosen to serve that purpose. Choosing the wrong label can be catastrophic to a deep learning project as sometimes, when data is not readily available, it means the project has to start all over again from the data preparation stage. Labels should always be indirectly or directly attributed to the success metric.

Success metrics, as the name suggests, can be plural, and range from time-based success definitions or milestones to the overall project success, and from intangible to tangible. It's good practice to generally brainstorm and document all the possible success criteria from a low level to a high level. Another best practice is to make sure to always define tangible success metrics alongside intangible metrics. Intangible metrics generate awareness, but tangible metrics make sure things are measurable and thus make them that much more attainable. A few examples of intangible and hard-to-measure metrics are as follows:

- Increasing customer satisfaction
- Increasing employee performance
- Improving shareholder outlook

Metrics are ways to measure something and are tied to goals to seal the deal. Goals themselves can be intangible, similar to the few examples listed previously, but so long as it is tied to tangible metrics, the project is off to a good start. When you have a clear goal, ask yourself in what way the goal can be proven to be achieved, demonstrated, or measured. A few examples of tangible success metrics for machine learning projects that could align with business goals are as follows:

- Increase the time customers spend, which can be a proxy for customer delight
- Increase company revenue, which can be a proxy for employee performance
- Increase the **click-through rate** (**CTR**), which can be a proxy for the effectiveness of targeted marketing campaigns

- Increase the **customer lifetime value** (**CLTV**), which can be a proxy for long-term customer satisfaction and loyalty

- Increase conversion rate, which can be a proxy for the success of promotional campaigns and website user experience

This concept is not new nor limited to just machine learning projects – just about any single project carried out for a company as every single real-world project needs to be aligned with the business goal. Many foundational project management techniques can be applied similarly to machine learning projects, and spending time gaining some project management skills out of the machine learning field would be beneficial and transferable to machine learning projects. Additionally, as machine learning is considered to be a software-based technology, software project management methodologies also apply.

A final concluding thought to take away is that machine learning systems are not about how advanced your machine learning models are, but instead about how humans and machine intelligence can work together to achieve a greater good and create value.

Planning resources

Deep learning often involves neural network architectures with a large set of parameters, otherwise called weights. These architecture's sizes can go from holding a few parameters up to holding hundreds of billions of parameters. For example, an OpenAI GPT-3 text generation model holds 175 billion neural network parameters, which amounts to around 350 GB in computer storage size. This means that to run GPT-3, you need a machine with a **random access memory** (**RAM**) size of at least 350 GB!

Deep learning model frameworks such as PyTorch and TensorFlow have been built to work with devices called **graphics processing units** (**GPUs**), which offer tremendous neural network model training and inference speedups. Off-the-shelf GPU devices commonly have a GPU RAM of 12 GB and are nowhere near the requirements needed to load a GPT-3 model in GPU mode. However, there are still methods to partition big models into multiple GPUs and run the model on GPUs. Additionally, some methods can allow for distributed GPU model training and inference to support larger data batch sizes at any one usage point. GPUs are not considered cheap devices and can cost anywhere from a few hundred bucks to hundreds of thousands from the most widely used GPU brand, Nvidia. With the rise of cryptocurrency technologies, the availability of GPUs is also reduced significantly due to people buying them immediately when they are in stock. All these emphasize the need to plan computing resources for training and inferencing deep learning models beforehand.

It is important to align your model development and deployment needs to your computing resource allocation early in the project. Start by gauging the range of sizes of deep learning architectures that are suitable for the task at hand either by browsing research papers or websites that provide a good summary of techniques, and setting aside computing resources for the model development process.

> **Tip**
>
> `paperswithcode.com` provides summaries of a wide variety of techniques grouped by a wide variety of tasks!

When computing resources are not readily available, make sure you always make purchase plans early, especially if it involves GPUs. But what if a physical machine is not desired? An alternative to using computing resources is to use paid cloud computing resource providers you can access online easily from anywhere in the world. During the model development stage, one of the benefits of having more GPUs with more RAM allocated is that it can allow you to train models faster by either using a larger data batch size during training or allowing the capability to train multiple models at any one time. It is generally fine to also use CPU-only deep learning model training, but the model training time would just inevitably be much longer.

The GPU and CPU-based computing resources that are required during training are often considered overkill to be used during inference time when they are deployed. Different applications have different deployment computing requirements and the decision on what resource specification to allocate can be gauged by asking yourself the following three questions:

- How often are the inference requests made?

  - Many inference requests in a short period might signal the need to have more than one inference service up in multiple computing devices in parallel

- What is the average amount of samples that are requested for a prediction at any one time?

  - Device RAM requirements should match batch size expectations

- How fast do you need a reply?

  - GPUs are needed if it's seconds or a faster response time requirement

  - CPUs can do the job if you don't care about the response time

Resource planning is not restricted to just computing resource planning – it also expands to human resource planning. Assumptions for the number of deep learning engineers and data scientists working together in a team would ultimately affect the choices of software libraries and tools used in the model development process. The analogy of choosing these tools will be introduced in future sections.

The next step is to prepare your data.

Preparing data

Data is to machine learning models as is the fuel to your car, the electricity to your electronic devices, and the food for your body. A machine learning model works by trying to capture the relationships between the provided input and output data. Similar to how human brains work, a machine learning model will attempt to iterate through collected data examples and slowly build a memory of the patterns required to map the provided input data to the provided target output data. The data preparation stage consists of methods and processes required to prepare ready-to-use data to build a machine learning model that includes the following:

- Acquisition of raw input and targeted output data
- Exploratory data analysis of the acquired data
- Data pre-processing

We will discuss each of these topics in the following subsections.

Deep learning problem types

Deep learning can be broadly categorized into two problem types, namely **supervised learning** and **unsupervised learning**. Both of these problem types involve building a deep learning model that is capable of making informed predictions as outputs, given well-defined data inputs.

Supervised learning is a problem type where labels are involved that act as the source of truth to learn from. Labels can exist in many forms and can be broken down into two problem types, namely **classification** and **regression**. Classification is the process where a specific discrete class is predicted among other classes when given input data. Many more complex problems derive from the base classification problem types, such as **instance segmentation**, **multilabel classification**, and **object detection**. Regression, on the other hand, is the process where a continuous numerical value is predicted when given input data. Likewise, complex problem types can be derived from the base regression problem type, such as **multi-regression** and **image bounding box regression**.

Unsupervised learning, on the other hand, is a problem type where there aren't any labels involved and the goals can vary widely. Anomaly detection, clustering, and feature representation learning are the most common problem types that belong to the unsupervised learning category.

We will go through these two problem types separately for deep learning in *Chapter 8, Exploring Supervised Deep Learning*, and *Chapter 9, Exploring Unsupervised Deep Learning*.

Next, let's learn about the things you should consider when acquiring data.

Acquiring data

Acquiring data in the context of deep learning usually involves unstructured data, which includes image data, video data, text data, and audio data. Sometimes, data can be readily available and stored through some business processes in a database but very often, it has to be collected manually from the environment from scratch. Additionally, very often, labels for this data are not readily available and require manual annotation work. Along with the capability of deep learning algorithms to process and digest highly complex data comes the need to feed it more data compared to its machine learning counterparts. The requirement to perform data collection and data annotation in high volumes is the main reason why deep learning is considered to have a high barrier of entry today.

Don't rush into choosing an algorithm quickly in a machine learning project. Spend a quality amount of time formally defining the features that can be acquired to predict the target variable. Get help from domain experts during the process and brainstorm potential predictive features that relate to the target variable. In actual projects, it is common to spend a big portion of your time planning and acquiring the data while making sure the acquired data is fit for a machine learning model's consumption and subsequently spending the rest of the time in model building, model deployment, and model governance. A lot of research has been done into handling bad-quality data during the model development stage but most of these techniques aren't comprehensive and are limited in ways that they can cover up the inherent quality of the data. Displaying ignorance in quality assurance during the data acquisition stage and showing enthusiasm only in the data science portion of the workflow is a strong indicator that the project would be doomed to failure right from the inception stage.

Formulating a data acquisition strategy is a daunting task when you don't know what it means to have good-quality data. Let's go through a few pillars of data quality you should consider for your data in the context of actual business use cases and machine learning:

- **Representativeness**: How representative is the data concerning the real-world data population?
- **Consistency**: How consistent are the annotation methods? Does the same pattern match the same label or are there some inconsistencies?
- **Comprehensiveness**: Are all variations of a specific label covered in the collected dataset?
- **Uniqueness**: Does the data contain a lot of duplicated or similar data?
- **Fairness**: Is the collected data biased toward any specific labels or data groups?
- **Validity**: Does the data contain invalid fields? Do the data inputs match up with their labels? Is there missing data?

Let's look at each of these in detail.

Representativeness

Data should be collected in a way that it mimics what data you will receive during model deployment as much as possible. Very often in research-based deep learning projects, researchers collect their data in a closed environment with controlled environmental variables. One of the reasons researchers prefer collecting data from a controlled environment is that they can build stabler machine learning models and generally try to prove a point. Eventually, when the research paper is published, you see amazing results that were applied using handpicked data to impress. These models, which are built on controlled data, fail miserably when you apply them to random uncontrolled real-world examples. Don't get me wrong – it's great to have these controlled datasets available to contribute toward a stabler machine learning model at times, but having uncontrolled real-world examples as a main part of the training and evaluation datasets is key to achieving a generalizable model.

Sometimes, the acquired training data has an expiry date and does not stay representative forever. This scenario is called data drift and will be discussed in more detail in the *Managing risk* section closer to the end of this chapter. The representativeness metric for data quality should also be evaluated based on the future expectations of the data the model will receive during deployment.

Consistency

Data labels that are not consistently annotated make it harder for machine learning models to learn from them. This happens when the domain ideologies and annotation strategies differ among multiple labelers and are just not defined properly. For example, "Regular" and "Normal" mean the same thing, but to the machine, it's two completely different classes; so are "Normal" and "normal" with just a capitalization difference!

Practice formalizing a proper strategy for label annotation during the planning stage before carrying out the actual annotation process. Cleaning the data for simple consistency errors is possible post-data annotation, but some consistency errors can be hard to detect and complex to correct.

Comprehensiveness

Machine learning thrives in building a decisioning mechanism that is robust to multiple variations and views of any specific label. Being capable and accomplishing it are two different things. One of the prerequisites of decisioning robustness is that the data that's used for training and evaluation itself has to be comprehensive enough to provide coverage for all possible variations of each provided label. How can comprehensiveness be judged? Well, that depends on the complexity of the labels and how varied they can present themselves naturally when the model is deployed. More complex labels naturally require more samples and less complex labels require fewer samples.

A good point to start with, in the context of deep learning, is to have at least 100 samples for each label and experiment with building a model and deriving model insights to see if there are enough samples for the model to generalize on unseen variations of the label. When the model doesn't produce convincing results, that's when you need to cycle back to the data preparation stage again to acquire more data variations of any specific label. The machine learning life cycle is inherently a cyclical process where you will experiment, explore, and verify while transitioning between stages to obtain the answers you need to solve your problems, so don't be afraid to execute these different stages cyclically.

Uniqueness

While having complete and comprehensive data is beneficial to build a machine learning model that is robust to data variations, having duplicated versions of the same data variation in the acquired dataset risks creating a biased model. A biased model makes biased decisions that can be unethical and illegal and sometimes renders such decisions meaningless. Additionally, the amount of data acquired for any specific label is rendered meaningless when all of them are duplicated or very similar to each other.

Machine learning models are generally trained on a subset of the acquired data and then evaluated on other subsets of the data to verify the model's performance on unseen data. When the part of the dataset that is not unique gets placed in the evaluation partition of the acquired dataset by chance, the model risks reporting scores that are biased against the duplicated data inputs.

Fairness

Does the acquired dataset represent minority groups properly? Is the dataset biased toward the majority groups in the population? There can be many reasons why a machine learning model turns out to be biased, but one of the main causes is data representation bias. Making sure the data is represented fairly and equitably is an ethical responsibility of all machine learning practitioners. There are a lot of types of bias, so this topic will have its own section and will be introduced along with methods of mitigating it in *Chapter 13, Exploring Bias and Fairness*.

Validity

Are there outliers in the dataset? Is there missing data in the dataset? Did you accidentally add a blank audio or image file to the properly collected and annotated dataset? Is the annotated label for the data input considered a valid label? These are some of the questions you should ask when considering the validity of your dataset.

Invalid data is useless for machine learning models and some of these complicate the pre-processing required for them. The reasons for invalidity can range from simple human errors to complex domain knowledge mistakes. One of the methods to mitigate invalid data is to separate validated and unvalidated data. Include some form of automated or manual data validation process before a data sample gets included in the validated dataset category. Some of this validation logic can be derived from business processes or just common sense. For example, if we are taking age as input data, there are acceptable age ranges and there are age ranges that are just completely impossible, such as 1,000 years old. Having simple guardrails and verifying these values early when collecting them makes it possible to correct them then and there to get accurate and valid data. Otherwise, these data will likely be discarded when it comes to the model-building stage. Maintaining a structured framework to validate data ensures that the majority of the data stays relevant and usable by machine learning models and free from simple mistakes.

As for more complex invalidity, such as errors in the domain ideology, domain experts play a big part in making sure the data stays sane and logical. Always make sure you include domain experts when defining the data inputs and outputs in the discussion about how data should be collected and annotated for model development.

Making sense of data through exploratory data analysis (EDA)

After the acquisition of data, it is crucial to analyze the data to inspect its characteristics, patterns that exist, and the general quality of the data. Knowing the type of data you are dealing with allows you to plan a strategy for the subsequent model-building stage. Plot distribution graphs, calculate statistics, and perform univariate and multivariate analysis to understand the inherent relationships between the data that can help further ensure the validity of the data. The methods of analysis for different variable types are different and can require some form of domain knowledge beforehand. In the following subsections, we will be practically going through **exploratory data analysis (EDA)** for text-based data to get a sense of the benefits of carrying out an EDA task.

Practical text EDA

In this section, we will be manually exploring and analyzing a text-specific dataset using Python code, with the motive of building a deep learning model later in this book using the same dataset. The dataset we use will predict the categories of an item on an Indian e-commerce website based on its textual description. This use case will be useful to automatically group advertised items for user recommendation usage and can help increase purchasing volume on the e-commerce website:

1. Let's start by defining the libraries that we will use in a notebook. We will be using pandas for data manipulation and structuring, matplotlib and seaborn for plotting graphs, tqdm for visualizing iteration progress, and lingua for text language detection:

    ```
    import pandas as pd
    import matplotlib.pyplot as plt
    import seaborn as sns
    from tqdm import tqdm
    from lingua import Language, LanguageDetectorBuilder
    tqdm.pandas()
    ```

2. Next, let's load the text dataset using pandas:

    ```
    dataset = pd.read_csv('ecommerceDataset.csv')
    ```

3. pandas has some convenient functions to visualize and describe the loaded dataset; let's use them. Let's start by visualizing three rows of the raw data:

    ```
    dataset.head(3)
    ```

This will display the following figure in your notebook:

	category	description
0	Household	Paper Plane Design Framed Wall Hanging Motivational Office Decor Art Prints (8.7 X 8.7 inch) - Set of 4 Painting made up in synthetic frame with uv textured print which gives multi effects and attracts towards it. This is an special series of paintings which makes your wall very beautiful and gives a royal touch. This painting is ready to hang, you would be proud to possess this unique painting that is a niche apart. We use only the most modern and efficient printing technology on our prints, with only the and inks and precision epson, roland and hp printers. This innovative hd printing technique results in durable and spectacular looking prints of the highest that last a lifetime. We print solely with top-notch 100% inks, to achieve brilliant and true colours. Due to their high level of uv resistance, our prints retain their beautiful colours for many years. Add colour and style to your living space with this digitally printed painting. Some are for pleasure and some for eternal bliss.so bring home this elegant print that is lushed with rich colors that makes it nothing but sheer elegance to be to your friends and family.it would be treasured forever by whoever your lucky recipient is. Liven up your place with these intriguing paintings that are high definition hd graphic digital prints for home, office or any room.
1	Household	SAF 'Floral' Framed Painting (Wood, 30 inch x 10 inch, Special Effect UV Print Textured, SAO297) Painting made up in synthetic frame with UV textured print which gives multi effects and attracts towards it. This is an special series of paintings which makes your wall very beautiful and gives a royal touch (A perfect gift for your special ones).
2	Household	SAF 'UV Textured Modern Art Print Framed' Painting (Synthetic, 35 cm x 50 cm x 3 cm, Set of 3) Color:Multicolor | Size:35 cm x 50 cm x 3 cm Overview a beautiful painting involves the action or skill of using paint in the right manner; hence, the end product will be a picture that can speak a thousand words they say. Arts have been in trend for quite some time now. It can give different viewer different meanings style and design the SAF wood matte abstract painting with frame is quite abstract and mysteriously beautiful. The painting has a nice frame to it. You can gift this to a family or a friend. The painting has various forms of certain figures on it as seen in the image. You can add a good set of lights to the place where the painting is and the decor will give a different feel and look to the place. Quality and durability the painting has a matte finish and includes a good quality frame and will last for a long period. However, it does not include glass along with the frame. Specifications you can purchase SAF wood matte abstract painting with frame on amazon.in. It is the most customer-friendly platform with a wide range of products to choose from, and shopping is just a click away!

Figure 1.3 – Visualizing the text dataset samples

4.   Next, let's describe the dataset by visualizing the data column-based statistics:

```
dataset.describe()
```

This will display the following figure in your notebook:

	category	description
count	50425	50424
unique	4	27802
top	Household	Think & Grow Rich About the Author NAPOLEON HILL, born in Pound, Southwest Virginia in 1883, was a very successful American author in the area of the new thought movement—one of the earliest producers of the modern genre of personal-success literature. He is widely considered to be one of the great writers on success. The turning point in Hill's life occurred in the year 1908 when he interviewed the industrialist Andrew Carnegie—one of the most powerful men in the world at that time, as part of an assignment—an interview which ultimately led to the publication of Think and Grow Rich, one of his best-selling books of all time. the book examines the power of personal beliefs and the role they play in personal success. Hill, who had even served as the advisor to President Franklin D. Roosevelt from 1933-36, passed away at the age of 87.
freq	19313	30

Figure 1.4 – Showing the statistics of the dataset

5.   With these visualizations, it's obvious that the description of the dataset aligns with what exists in the dataset, where the **category** column contains four unique class categories paired with the text description data column named **description** with evidence showing that they are strings. One important insight from the describing function is that there are duplicates in the text description. We can remove duplicates by taking the first example row among all the duplicates, but we also have to make sure that the duplicates have the same category, so let's do that:

```
for i in tqdm(range(len(unique_description_information))):
    assert(
    len(
      dataset[
        dataset['description'] ==
        unique_description_information.keys()[i]
```

```
    ]['category'].unique()
  ) == 1
)
dataset.drop_duplicates(subset=['description'], inplace=True)
```

6. Let's check the data types of the columns:

```
dataset.dtypes
```

This will display the following figure in your notebook:

```
category       object
description    object
dtype: object
```

Figure 1.5 – Showing the data types of the dataset columns

7. When some samples aren't inherently a string data type, such as empty data or maybe numbers data, pandas automatically use an Object data type that categorizes the entire column as data types that are unknown to pandas. Let's check for empty values:

```
dataset.isnull().sum()
```

This gives us the following output:

```
category       0
description    1
dtype: int64
```

Figure 1.6 – Checking empty values

8. It looks like the description column has one empty value, as expected. This might be rooted in a mistake when acquiring the data or it might truly be empty. Either way, let's remove that row as we can't do anything to recover it and convert the columns into strings:

```
dataset.dropna(inplace=True)
for column in ['category', 'description']:
    dataset[column] = dataset[column].astype("string")
```

9. Earlier, we discovered four unique categories. Let's make sure we have a decent amount of samples for each category by visualizing its distribution:

```
sns.countplot(x="category", data=dataset)
```

This will result in the following figure:

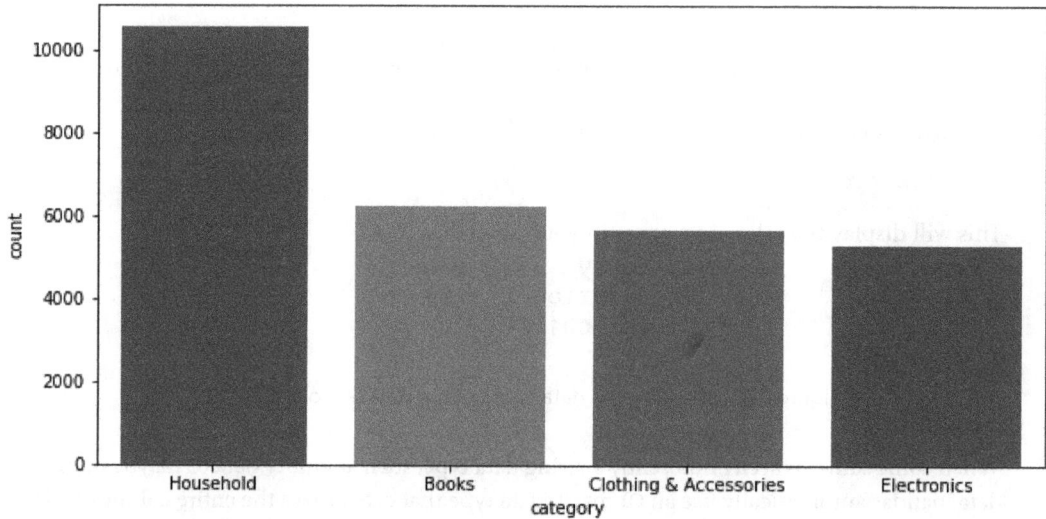

Figure 1.7 – A graph showing category distribution

Each category has a good amount of data samples and doesn't look like there are any anomaly categories.

10. The goal here is to predict the category of the selling item through the item's description on the Indian e-commerce website. From that context, we know that Indian citizens speak Hindi, so the dataset might not contain only English data. Let's try to estimate and verify the available languages in the dataset using an open sourced language detector tool called Lingua. Lingua uses both rule-based and machine learning model-based methods to detect more than 70 text languages that work great for short phrases, single words, and sentences. Because of that, Lingua has a better runtime and memory performance. Let's start by initializing the language detector instance from the lingua library:

```
detector = LanguageDetectorBuilder.from_all_languages(
        ).with_preloaded_language_models().build()
```

11. Now, we will randomly sample a small portion of the dataset to detect language as the detection algorithm takes time to complete. Using a 10% fraction of the data should allow us to adequately understand the data:

```
sampled_dataset = dataset.sample(frac=0.1, random_state=1234)
sampled_dataset['language'] = sampled_dataset[
    'description'
].progress_apply(lambda x: detector.detect_language_of(x))
```

12. Now, let's visualize the distribution of the language:

```
sampled_dataset['language'].value_counts().plot(kind='bar')
```

This will show the following graph plot:

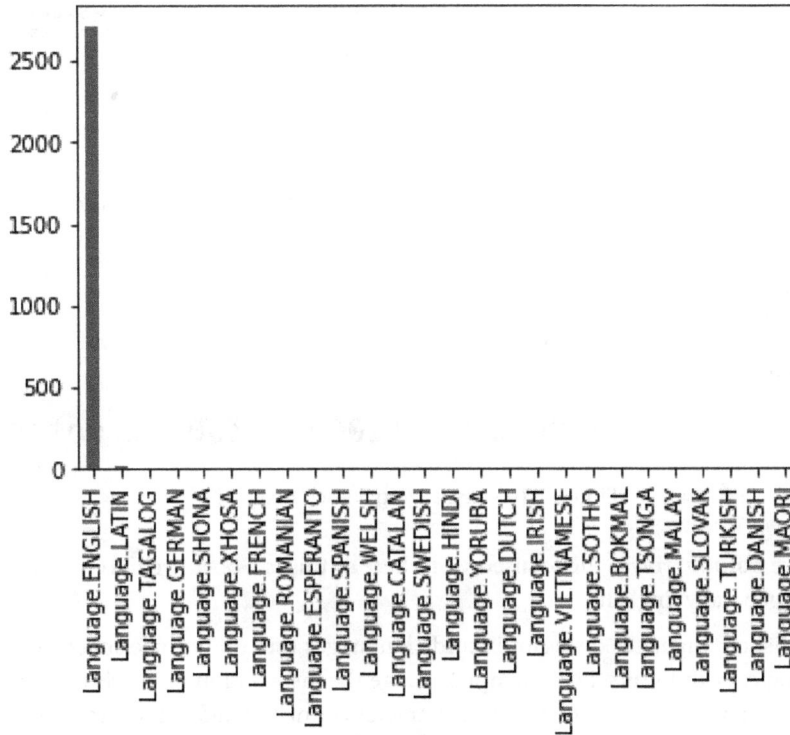

Figure 1.8 – Text language distribution

13. Interestingly, Lingua detected some anomalous samples that aren't English. The anomalous languages look like they might be mistakes made by Lingua. Hindi is also detected among them; this is more convincing than the other languages as the data is from an Indian e-commerce website. Let's check these samples out:

```
sampled_dataset[
    sampled_dataset['language'] == Language.HINDI
].description.iloc[0]
```

This will show the following text:

'Real Estate Evam Estate Planning रियल एस्टेट एवं एस्टेट प्लानिंग विषय पर बाजार में कई पुस्तकें उपलब्ध हैं, लेकिन लेखक ने इस पुस्तक के द्वारा बैंकिंग विषय का समावेश करते हुए इन दोनों क्षेत्रों के जटिल पहलुओं को सीधी-सरल भाषा में स्पष्टता से समझाने की कोशिश की है। पुस्तक में प्रॉपर्टी बाजार की जानकारी, विभिन्न होम लोन प्रोडक्ट की जानकारी एवं प्रॉपर्टी के खरीदने से लेकर बेचने तक के दौरान रखी जानेवाली सावधानियों को उदाहरण सहित समझाया गया है। इसके अलावा पुस्तक में कैपिटल गेन स्कीम, प्रधानमंत्री आवास योजना (अर्बन) एवं रियल एस्टेट रेगुलेशन ऐक्ट की जानकारी भी समाहित की गई है। विश्वास है कि पुस्तक प्रॉपर्टी में निवेश एवं एस्टेट प्लानिंग में रुचि रखनेवालों की जिज्ञासाओं, संशयों तथा प्रश्नों का निराकरण करेगी। About the Author Dr. Yogesh Sharma शिक्षा : बी.वी.एस.सी. एंड ए.एच., पोस्ट ग्रेजुएट डिप्लोमा इन बिजनेस एडमिनिस्ट्रेशन (फाइनेंस), एडवांस वैल्थ मैनेजमेंट, सी.ए.आई.आई.बी. भाग-II, सर्टिफाइड फाइनेंशियल प्लानर। मार्च 2007 से भारतीय स्टेट बैंक में अधिकारी के रूप में बैंकिंग करियर का प्रारंभ। लेखक प्रमाणित वित्तीय सलाहकार हैं। इन्हें बैंकिंग, रियल एस्टेट एवं फाइनेंशियल प्लानिंग में दस वर्षों का अनुभव है एवं इन विषयों पर प्रमुख पत्रिकाओं में लेख भी लिखते हैं। संप्रति : यूनियन बैंक ऑफ इंडिया में वरिष्ठ प्रबंधक (विपणन) के पद पर कार्यरत। अनुक्रम 1. रियल एस्टेट में निवेश–9 2. रियल एस्टेट इन्वेस्टमेंट ट्रस्ट (रीट)–21 3. नाबालिग बच्चों के लिए रियल एस्टेट में निवेश–24 4. या पत्नी के नाम निवेश करना चाहिए?–26 5. मकान किराया भत्ता (एच.आर.ए.)–28 6. रियल एस्टेट प्लानिंग में अविभक्त हिंदू परिवार (हिंदू अनडिवाइडेड फेमिली/एच.यू.एफ.) की भूमिका–31 7. होम लोन–35 (Continue).....'

Figure 1.9 – Visualizing Hindi text

14. It looks like there is a mix of Hindi and English here. How about another language, such as French?

```
sampled_dataset[
    sampled_dataset['language'] == Language.FRENCH
].description.iloc[0]
```

This will show the following text:

'Iris Lemon Grass Potpourri (100gm) 100 gm potpourri'

Figure 1.10 – Visualizing French text

15. It looks like **potpourri** was the focused word here as this is a borrowed French word, but the text is still generally English.

16. Since the list of languages does not include languages that do not use space as a separator between logical word units, let's attempt to gauge the distribution of words by using a space-based word separation. Word counts and character counts can affect the parameters of a deep learning neural network, so it will be useful to understand these values during EDA:

```
dataset['word_count'] = dataset['description'].apply(
    lambda x: len(x.split())
)
plt.figure(figsize=(15,4))
sns.histplot(data=dataset, x="word_count", bins=10)
```

This will show the following bar plot:

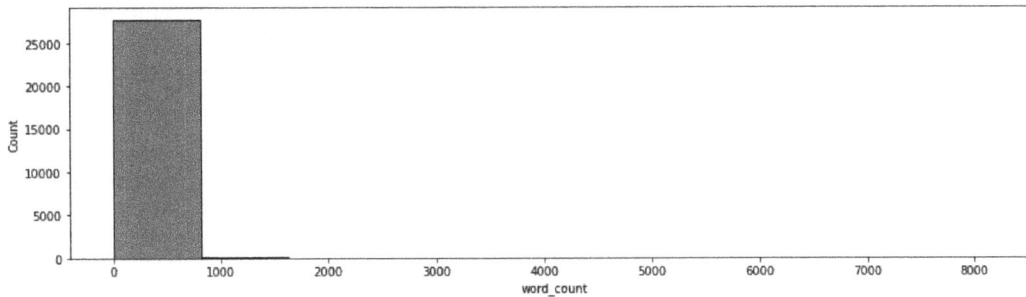

Figure 1.11 – Word count distribution

From the exploration and analysis of the text data, we can deduce a couple of reasons that will help set up the model type and structure we should use during the model development stage:

- The labels are decently sampled with 5,000-11,000 worth of samples per label, making them suitable for deep learning algorithms.

- The original data is not clean, has missing data, and duplicates but is fixable through manual processing. Using it as-is for model development would have the potential of creating a biased model.

- The dataset has more than one language but mostly English text; this will allow us to make appropriate model choices during the model development stage.

- An abundance of samples has fewer than 1,000 words, and some samples have 1,000-8,000 words. In some non-critical use cases, we can safely cap the number of words to around 1,000 words so that we can build a model with better memory and runtime performance.

The preceding practical example should provide a simple experience of performing EDA that will be sufficient to understand the benefit and importance of running an in-depth EDA before going into the model development stage. Similar to the practical text EDA, we prepared a practical EDA sample workflow for other datasets that includes audio, image, and video datasets in our Packt GitHub repository that you should explore to get your hands dirty.

A major concept to grasp in this section is the importance of EDA and the level of curiosity you should display to uncover the truth about your data. Some methods are generalizable to other similar datasets, but treating any specific EDA workflow as a silver bullet blinds you to the increasing research people are contributing to this field. Ask questions about your data whenever you suspect something of it and attempt to uncover the answers yourself by doing manual or automated inspections however possible. Be creative in obtaining these answers and stay hungry in learning new ways you can figure out key information on your data.

In this section, we have methodologically and practically gone through EDA processes for different types of data. Next, we will explore what it takes to prepare the data for actual model consumption.

Data pre-processing

Data pre-processing involves data cleaning, data structuring, and data transforming so that a deep learning model will be capable of using the processed data for model training, evaluation, and inferencing during deployment. The processed data should not only be prepared just for the machine learning model to accept but should generally be processed in a way that optimizes the learning potential and increases the metric performance of the machine learning model.

Data cleaning is a process that aims to increase the quality of the data acquired. An EDA process is a prerequisite to figuring out anything wrong with the dataset before some form of data cleaning can be done. Data cleaning and EDA are often executed iteratively until a satisfactory data quality level is achieved. Cleaning can be as simple as duplicate values removal, empty values removal, or removing values that don't make logical sense, either in terms of common sense or through business logic. These are concepts that we explained earlier, where the same risks and issues are applied.

Data structuring, on the other hand, is a process that orchestrates the data ingest and loading process from the stored data that is cleaned and verified of its quality. This process determines how data should be loaded from a source or multiple of them and fed into the deep learning model. Sounds simple enough, right? This could be very simple if this is a small, single CSV dataset where there wouldn't be any performance or memory issues. In reality, this could be very complex in cases where data might be partitioned and stored in different sources due to storage limitations. Here are some concrete factors you'd need to consider in this process:

- Do you have enough RAM in your computer to process your desired batch size to supply data for your model? Make sure you also take your model size into account so that you won't get memory overloads and **Out of Memory** (**OOM**) errors!

- Is your data from different sources? Make sure you have permission to access these data sources.

- Is the speed latency when accessing these sources acceptable? Consider moving this data to a better hardware resource that you can access with higher speeds, such as a **solid-state drive** (**SSD**) instead of a **hard disk drive** (**HDD**), and from a remote network-accessible source to a direct local hardware source.

- Do you even have enough local storage to store this data? Make sure you have enough storage to store this data, don't overload the storage and risk performance slowdowns or worse, computer breakdowns.

- Optimize the data loading and processing process so that it is fast. Store and cache outputs of data processes that are fixed so that you can save time that can be used to recompute these outputs.

- Make sure the data structuring process is deterministic, even when there are processes that need randomness. **Randomly deterministic** is when the randomness can be reproduced in a repeat of the cycle. Determinism helps make sure that the results that have been obtained can be reproduced and make sure model-building methods can be compared fairly and reliably.

- Log data so that you can debug the process when needed.

- Data partitioning methods. Make sure a proper cross-validation strategy is chosen that's suitable for your dataset. If a time-based feature is included, consider whether you should construct a time-based partitioning method where the training data consists of earlier time examples and the evaluation data is in the future. If not, a stratified partitioning method would be your best bet.

Different deep learning frameworks, such as **PyTorch** and **TensorFlow**, provide different **application programming interfaces** (**APIs**) to implement the data structuring process. Some frameworks provide simpler interfaces that allow for easy setup pipelines while some frameworks provide complex interfaces that allow for a higher level of flexibility. Fortunately, many high-level libraries attempt to simplify the complex interfaces while maintaining flexibility, such as `keras` on top of TensorFlow, `Catalyst` on top of PyTorch, `fast ai` on top of PyTorch, `pytorch lightning` on top of PyTorch, and `ignite` on top of PyTorch.

Finally, data transformation is a process that applies unique data variable-specific pre-processing to transform the raw cleaned data into more a representable, usable, and learnable format. An important factor to consider when attempting to execute the data structuring and transformation process is the type of deep learning model you intend to use. Any form of data transformation is often dependent on the deep learning architecture and dependent on the type of inputs it can accept. The most widely known and common deep learning model architectures are invented to tackle specific data types, such as convolutional neural networks for image data, transformer models for sequence-based data, and basic multilayer perceptrons for tabular data. However, deep learning models are considered to be flexible algorithms that can twist and bend to accept data of different forms and sizes, even in multimodal data conditions. Through collaboration with domain experts from the past few years, deep learning experts have been able to build creative forms of deep learning architectures that can handle multiple data modalities and even multiple unstructured data modalities that succeeded in learning cross-modality patterns. Here are two examples:

- *Robust Self-Supervised Audio-Visual Speech Recognition*, by Meta AI (formerly Facebook) (https://arxiv.org/pdf/2201.01763v2.pdf):

  - This tackled the problem of speech recognition in the presence of multiple speeches by building a deep learning transformer-based model that can take in both audio and visual data called AV-HuBERT

  - Visual data acted as supplementary data to help the deep learning model discern which speaker to focus on.

  - It achieved the latest state-of-the-art results on the LRS3-TED visual and audio lip reading dataset

- *Unifying Architectures, Tasks, and Modalities Through a Simple Sequence-to-Sequence Learning Framework*, by DAMO Academy and Alibaba Group (`https://arxiv.org/pdf/2202.03052v1.pdf`):

  - They built a model that took in text and image data and published a pre-trained model

  - At achieved state-of-the-art results on an image captioning task on the COCO captions dataset

With that being said, data transformations are mainly differentiated into two parts: **feature engineering** and **data scaling**. Deep learning is widely known for its feature engineering capabilities, which replace the need to manually craft custom features from raw data for learning. However, this doesn't mean that it always makes sense to not perform any feature engineering. Many successful deep learning models have utilized engineered forms of features as input.

Now that we know what data pre-processing entails, let's discuss and explore different data pre-processing techniques for unstructured data, both theoretically and practically.

Text data pre-processing

Text data can be in different languages and exist in different domains, ranging from description data to informational documents and natural human text comments. Some of the most common text data pre-processing methods that are used for deep learning are as follows:

- **Stemming**: A process that removes the suffix of words in an attempt to reduce words into their base form. This promotes the cross-usage of the same features for different forms of the same word.

- **Lemmatization**: A process that reduces a word into its base form that produces real English words. Lemmatization has many of the same benefits as stemming but is considered better due to the linguistically valid word reduction outputs it produces.

- **Text tokenization**, by **Byte Pair Encoding** (**BPE**): Tokenization is a process that splits text into different parts that will be encoded and used by the deep learning models. BPE is a sub-word-based text-splitting algorithm that allows common words to be outputted as a single token but rare words get split into multiple tokens. These split tokens can reuse representations from matching sub-words. This is to reduce the vocabulary that can exist at any one time, reduce the amount of out-of-vocabulary tokens, and allow token representations to be learned more efficiently.

One uncommon pre-processing method that will be useful to build more generalizable text deep learning models is text data augmentation. Text data augmentation can be done in a few ways:

- **Replacing verbs with their synonyms**: This can be done by using the set of synonym dictionaries from the NLTK library's WordNet English lexical database. The obtained augmented text will maintain the same meaning with verb synonym replacement.

- **Back translation**: This involves translating text into another language and back to the original language using translation services such as Google or using open sourced translation models. The obtained back-translated text will be in a slightly different form.

Audio data pre-processing

Audio data is essentially sequence-based data and, in some cases, multiple sequences exist. One of the most commonly used pre-processing methods for audio is raw audio data transformed into different forms of spectrograms using **Short-Time Fourier Transform** (**STFT**), which is a process that converts audio from the time domain into the frequency domain. A spectrogram audio conversion allows audio data to be broken down and represented in a range of frequencies instead of a single waveform representation that is a combination of the signals from all audio frequencies. These spectrograms are two-dimensional data and thus can be treated as an image and fed into convolutional neural networks. Data scaling methods such as log scaling and log-mel scaling are also commonly applied to these spectrograms to further emphasize frequency characteristics.

Image data pre-processing

Image data augmentation is a type of image-based feature engineering technique that is capable of increasing the comprehensiveness potential of the original data. A best practice for applying this technique is to structure the data pipeline to apply image augmentations randomly during the training process by batch instead of providing a fixed augmented set of data for the deep learning model. Choosing the type of image augmentation requires some understanding of the business requirements of the use case. Here are some examples where it doesn't make sense to apply certain augmentations:

- When the orientation of the image affects the validity of the target label, orientation modification types of augmentation such as rotation and image flipping wouldn't be suitable
- When the color of the image affects the validity of the target label, color modification types of augmentation such as grayscale, channel shuffle, hue saturation shift, and RGB shift aren't suitable

After excluding obvious augmentations that won't be suitable, a common but effective method to figure out the best set of augmentations list is iterative experiments and model comparisons.

Developing deep learning models

Let's start with a short recap of what deep learning is. Deep learning's core foundational building block is a neural network. A neural network is an algorithm that was made to simulate the human brain. Its building blocks are called neurons, which mimic the billions of neurons the human brain contains. Neurons, in the context of neural networks, are objects that store simple information called weights and biases. Think of these as the memory of the algorithm.

Deep learning architectures are essentially neural network architectures that have three or more neural network layers. Neural network layers can be categorized into three high-level groups – the input layer, the hidden layer, and the output layer. The input layer is the simplest layer group and whose functionality is to pass the input data to subsequent layers. This layer group does not contain biases and can be considered passive neurons, but the group still contains weights in its connections to neurons from subsequent layers. The hidden layer comprises neurons that contain biases and weights in their connections to neurons from subsequent layers. Finally, the output layer comprises neurons that relate to the number of classes and problem types and contains bias. A best practice when counting neural network layers is to exclude the input layer when doing so. So, a neural network with one input layer, one hidden layer, and one output layer is considered to be a two-layer neural network. The following figure shows a basic neural network, called a **multilayer perceptron (MLP)**, with a single input layer, a single hidden layer, and a single output layer:

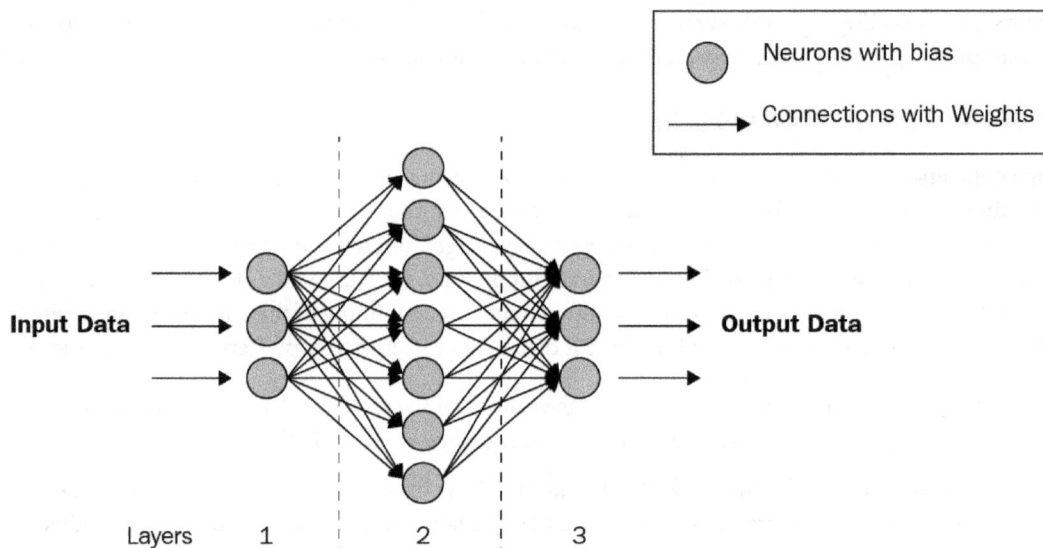

Figure 1.12 – A simple deep learning architecture, also called an MLP

Being a subset of the wider machine learning category, deep learning models are capable of learning patterns from the data through a loss function and an optimizer algorithm that optimizes the loss function. A loss function defines the error made by the model so that its memory (weights and biases) can be updated to perform better in the next iteration. An optimizer algorithm is an algorithm that decides the strategy to update the weights given the loss value.

With this short recap, let's dive into a summary of the common deep learning model families.

Deep learning model families

These layers can come in many forms as researchers have been able to invent new layer definitions to tackle new problem types and almost always comes with a non-linear activation function that allows the model to capture non-linear relationships between the data. Along with the variation of layers come many different deep learning architecture families that are meant for different problem types. A few of the most common and widely used deep learning models are as follows:

- **MLP** for tabular data types. This will be explored in *Chapter 2, Designing Deep Learning Architectures.*

- **Convolutional neural network** for image data types. This will be explored in *Chapter 3, Understanding Convolutional Neural Networks.*

- **Autoencoders** for anomaly detection, data compression, data denoising, and feature representation learning. This will be explored in *Chapter 5, Understanding Autoencoders.*

- **Gated recurrent unit (GRU)**, **Long Short-Term Memory (LSTM)**, and **Transformers** for sequence data types. These will be explored in *Chapter 4, Understanding Recurrent Neural Networks*, and *Chapter 6, Understanding Neural Network Transformers*, respectively.

These architectures will be the focus of *Chapters 2 to 6*, where we will discuss their methodology and go through some practical evaluation. Next, let's discover the problem types we can tackle in deep learning.

The model development strategy

Today, deep learning models are easy to invent and create due to the advent of deep learning frameworks such as PyTorch and TensorFlow, along with their high-level library wrappers. Which framework you should choose at this point is a matter of preference regarding their interfaces as both frameworks are matured with years of improvement work done. Only when there is a pressing need for a very custom function to tackle a unique problem type will you need to choose the framework that can execute what you need. Once you've chosen your deep learning framework, the deep model creation, training, and evaluation process is pretty much covered all around.

However, model management functions do not come out of the box from these frameworks. Model management is an area of technology that allows teams, businesses, and deep learning practitioners to reliably, quickly, and effectively build models, evaluate models, deliver model insights, deploy models to production, and govern models. Model management can sometimes be referred to as **machine learning operations** (**MLOps**). You might still be wondering why you'd need such functionalities, especially if you've been building some deep learning models off Kaggle, a platform that hosts data and machine learning problems as competitions. So, here are some factors that drive the need to utilize these functionalities:

- It is cumbersome to compare models manually:
  - Manually typing performance data in an Excel sheet to keep track of model performance is slow and unreliable
- Model artifacts are hard to keep track of:
  - A model has many artifacts, such as its trained weights, performance graphs, feature importance, and prediction explanations
  - It is also cumbersome to compare model artifacts
- Model versioning is needed to make sure model-building experiments are not repeated:
  - Overriding the top-performing model with the most reliable model insights is the last thing you want to experience
  - Versioning should depend on the data partitioning method, model settings, and software library versions
- It is not straightforward to deploy and govern models

Depending on the size of the team involved in the project and how often components need to be reused, different software and libraries would fit the bill. These software and libraries are split into paid and free (usually open sourced) categories. **Metaflow**, an open sourced software, is suitable for bigger data science teams where there are many chances of components needing to be reused across other projects and **MLFlow** (open sourced software) would be more suitable for small or single-person teams. Other notable model management tools are **Comet** (paid), **Weights & Biases** (paid), **Neptune** (paid), and **Algorithmia** (paid).

With that, we have provided a brief overview of deep learning model development methodology and strategy; we will dive deeper into model development topics in the next few chapters. But before that, let's continue with an overview of the topic of delivering model insights.

Delivering model insights

Model metric performance, when used exclusively for model comparisons and the model choosing process, is often not the most effective way to reliably obtain the true best model. When people care about the decisions that can potentially be made by the machine learning model, they typically require more information and insights to eventually put their trust in the ability of the model to make decisions. Ultimately, when models are not trusted, they don't get deployed. However, trust doesn't just depend on insights of the model. Building trust in a model involves ensuring accurate, reliable, and unbiased predictions that align with domain expertise and business objectives, while providing stakeholders with insights into the model's performance metrics, decision-making logic, and rationale behind its predictions. Addressing potential biases and demonstrating fairness are crucial for gaining confidence in the model's dependability. This ongoing trust-building process extends beyond initial deployment, as the model must consistently exhibit sound decision-making, justify predictions, and maintain unbiased performance. By fostering trust, the model becomes a valuable and reliable tool for real-world applications, leading to increased adoption and utilization across various domains and industries.

Deliver model insights that matter to the business. Other than delivering model insights with the obvious goal of ensuring model trust and eliminating trust issues, actual performance metrics are equally important. Make sure you translate model metrics into layman's business metrics whenever possible to effectively communicate the potential positive impact that the model can bring to the business. Success metrics, which are defined earlier in the planning phase, should be reported with actual values at this stage.

The process of inducing trust in a model doesn't stop after the model gets deployed. Similar to how humans are required to explain their decisions in life, machine learning models (if expected to replace humans to automate the decisioning process) are also required to do so. This process is called prediction explanations. In some cases, model decisions are expected to be used as a reference where there is a human in the loop that acts as a domain expert to verify decisions before the decisions are carried out. Prediction explanations are almost always a necessity in these conditions where the users of the model are interested in why the model made its predictions instead of using the predictions directly.

Model insights also allow you to improve a model's performance. Remember that the machine learning life cycle is naturally an iterative process. Some concrete examples of where this condition could happen are as follows:

- You realize that the model is biased against a particular group and either go back to the data acquisition stage to acquire more data from the less represented group or change to a modeling technique that is robust to bias
- You realize that the model performs badly in one class and go back to the model development stage to use a different deep learning model loss function that can focus on the badly performing class

Deep learning models are known to be a black box model. However, in reality, today, there have been many published research papers on deep learning explanation methods that have allowed deep learning to break the boundaries of a black box. We will dive into the different ways we can interpret and provide insights for deep learning models in *Part 2* of this book.

Now that we have more context of the processes involved in the deep learning life cycle, in the next section, we will discuss risks that can exist throughout the life cycle of the project that you need to worry about.

Managing risks

Deep learning systems are exposed to a multitude of risks starting early, from inception to system adoption. Usually, most people who are assigned to work on a deep learning project are responsible only for a specific stage of the machine learning life cycle, such as model development or data preparation. This practice can be detrimental when the work of one stage in the life cycle generates problems at a later stage, which often happens in cases where the team members involved have little sense of the bigger picture in play. Risks involved in a deep learning system generally involve interactions between stages in the machine learning life cycle and follow the concept of *Garbage in, garbage out*. Making sure everyone working on building the system has a sense of accountability for what eventually gets outputted from the entirety of the system instead of just the individual stages is one of the foundational keys to managing risks in a machine learning system.

But what are the risks? Let's start with something that can be handled way before anything tangible is made – something that happens when a use case is evaluated for worthiness.

Ethical and regulatory risks

Deep learning can be applied in practically any industry but some of the hardest industries to get deep learning adopted in are the highly regulated industries, such as the medical and financial industries. The regulations that are imposed in these industries ultimately determine what a deep learning system can or cannot do in the industry. Regulations are mostly introduced by governments and most commonly involve ethical and legal considerations. In these highly regulated industries, it is common to experience audits being conducted monthly, weekly, or even daily to make sure the companies are compliant with the regulations imposed. One of the main reasons certain industries are regulated more aggressively is that the repercussions of any actions in these industries bear a heavy cost to the well-being of the people or the country. Deep learning systems need to be built in a way that they will be compliant with these regulations to avoid the risk of facing decommission, the risk of not getting adopted at all, and, worst of all, the risk of getting a huge fine by regulatory officials.

At times, deep learning models can perform way better than their human counterparts but, on the other side, no deep learning model is perfect. Humans make mistakes, everybody knows that for a fact, but another reality we need to realize is that machines undoubtedly also make mistakes. The highest risk is when humans trust the machines so much that they give 100% trust in them to make decisions! So, how can we account for these mistakes and who will be responsible for them?

Let's see what we humans do in situations without machines. When the stakes are high, important decisions always go through a hierarchy of approvals before a decision can be final. These hierarchies of approvals signal the need to make important decisions accountable and reliable. The more approvals we get for a decision, the more we can say that we are confident in making that important decision. Some examples of important decisions that commonly require a hierarchy of approvals include the decision to hire an employee in a company, the decision on insurance premiums to charge, or the decision of whether to invest money into a certain business. With that context in mind, deep learning systems need to have similar workflows for obtaining approvals when a deep learning model makes a certain predictive decision in high-stakes use cases. These workflows can include any form of insight, and an explanation of the predictions can be used to make it easier for domain experts to judge the validity of the decisions. Adding a human touch makes the deep learning system many times more ethical and trustable enough to be part of the high-stake decision workflow.

Let's take a medical industry use case – for instance, a use case to predict lung disease through X-ray scans. From an ethical standpoint, it is not right that a deep learning model can have the complete power to strip a person's hope of life by predicting extremely harmful diseases such as end-stage lung cancer. If the predicted extreme disease is misdiagnosed, patients may have grieved unnecessarily or spent an unnecessary amount of money on expensive tests to verify the claims. Having an approval workflow in the deep learning system allows doctors to use these results as an assistive method would solve the ethical concerns of using an automated decisioning system.

Business context mismatch

Reiterating the previous point in the *Defining success* section, aligning the desired deep learning input data and target label choice to the business context and how the target predictions are consumed makes deep learning systems adoptable. The risk involved here is when the business value is either not properly defined or not matched properly by the deep learning system. Even when the target is appropriate for the business context, how these predictions are consumed holds the key to deriving value. Failing to match the business context in any way simply risks the rejection of the systems.

Understanding the business context involves understanding who the targeted user groups are and who they are not. A key step in this process is documenting and building user group personas. User personas hold information about their workflows, history, statistics, and concerns that should provide the context needed to build a system that aligns with the specific needs of potential users. Don't be afraid to conduct market research and make proper validation of the targeted user needs in the process of building these user personas. After all, it takes a lot of effort to build a deep learning system and it would not be great to waste time building a system that nobody wants to use.

Data collection and annotation risks

Your machine learning model will only be as good as the quality of your data. Deep learning requires a substantially larger quantity of data compared to machine learning. Additionally, very often for new deep learning use cases, data is not available off the shelf and must be manually collected and meticulously annotated. Making sure that the data is collected and annotated in a way that ensures quality is upheld is a very hard job.

The strategies and methods that are used to collect and annotate data for deep learning vary across use cases. Sometimes, the data collection and annotation process can happen at the same time. This can happen either when there are natural labels or when a label is prespecified before data is collected:

- **Natural labels** are labels that usually come naturally after some time passes. For example, this can be the gender of a baby through an ultrasound image; another use case is the price of a house as a label with the range of images of the property as the input data.

- **Prespecified labels** are labels where the input data is collected with predetermined labels. For example, this can be the gender label of some speech data that was collected in a controlled environment just for building a machine learning model or the age of a person in a survey before taking a facial photo shot to build a machine learning model.

These two types of labels are relatively safe from risks since the labels are likely to be accurate.

A final method of data collection and data annotation that poses a substantial amount of risk is when the labels are post-annotated after the data is collected. **Post annotation** requires some form of domain knowledge about the characteristics of the desired label and does not always result in a 100% truth value due to human evaluation errors. Unconscious labeling errors happen when the labeler simply makes an error by accident. Conscious labeling errors, on the other hand, happen when the labeler actively decides to make a wrong label decision with intent. This intent can be from a strong or loosely rooted belief tied to a certain pattern in the data, or just a plain error made purposely for some reason. The risk being presented here is the potential of the labels being mistakenly annotated. Fortunately, there are a few strategies that can be executed to eliminate these risks; these will be introduced in the following paragraphs.

Unconscious labeling errors are hard to circumvent as humans are susceptible to a certain amount of error. However, different people generally have varying levels of focus and selective choice of the labelers could be a viable option for some. If the labelers are hired with financial remuneration just to annotate the labels for your data, another viable option is to periodically provide data that is already labeled in secret in between providing data that needs labeling for the labelers to label the data. This strategy allows you to evaluate the performance of individual labelers and provide compensation according to the accuracy of their annotation work (yes, we can use metrics that are used to evaluate machine learning models too). As a positive side effect, labelers also would be incentivized to perform better in their annotation work.

Conscious labeling errors are the most dangerous risk here due to their potential to affect the quality of the entire dataset. When a wrong consensus on the pattern of data associated with any target label is used for the entire labeling process, the legitimacy of the labels won't be discovered until the later stages of the machine learning life cycle. Machine learning models are only equipped with techniques to learn the patterns required to map the input data to the provided target labels and will likely be able to perform very well, even in conditions where the labels are incorrect. So long as there is a pattern that the labeler enforced during the labeling process, no matter right or wrong, machine learning models will do their best to learn the patterns needed to perform the exact input-to-output mapping. So, how do we mitigate this risk? Making sure ideologies are defined properly with the help of domain experts plays an important part in the mitigation process, but this strategy by itself holds varying levels of effectiveness, depending on the number of domain experts involved, the expertise level of the domain experts themselves, the number of labelers, and the varying degrees of anomalies or quality of the collected data.

Even domain experts can be wrong at times. Let's take doctors as an example – how many times have you heard about a doctor giving wrong prescriptions and wrong medical assessments? How about the times when you misheard someone's speech and had to guess what the other person just said? In the last example, the domain expert is you, and the domain is language speech comprehension. Additionally, when the number of labelers exceeds one person and forms a labeling team, one of the most prominent risks that can happen is a mismatch between different preferred domain ideologies to label the data. Sometimes, this happens because of the different variations of a certain label that can exist in a digital format or the existence of confusing patterns that can deter the analytical capabilities of the labeler. Sometimes, this can also happen due to the inherent bias the labeler or the domain expert has toward a certain label or input data. Subsequently, bias in the data creates bias in the model and creates ethical issues that demote trust in the decisions of a machine learning model. When there is a lack of trust in a machine learning model, the project will fail to be adopted and will lose its business value. The topic of bias, fairness, and trust will be discussed more extensively in *Chapter 13*, *Exploring Bias and Fairness*, which will elaborate on its origins and ways to mitigate it.

Data with wrong labels that still have correct labels are called noisy labels. Fortunately, there are methods in deep learning that can help you work your way around noisy labels, such as weakly supervised learning. However, remember that it's always a good idea to fix the issue at the source, which is when the data is collected and annotated, rather than after. Let's dive into another strategy we can use to make the labeling process less risky. The data labeling process for deep learning projects usually involves using a software tool that allows the specific desired labels to be annotated. Software tools make annotation work faster and easier, and using a labeling collaboration software tool can make annotation work more error-proof. A good collaboration-based labeling tool will allow labelers to align their findings of the data with each other in some way, which will promote a common ideology when labeling. For instance, automatically alerting the entire labeler team when an easily misinterpreted pattern is identified can help prevent more data from being mislabeled and make sure all the previously related data gets rereviewed.

As a final point here, not as a risk, unlabeled data presents huge untapped potential for machine learning. Although it lacks specific labels to achieve a particular goal, there are inherent relationships between the data that can be learned. In *Chapter 9*, *Exploring Unsupervised Deep Learning*, we will explore how to use unsupervised learning to leverage this data as a foundation for subsequent supervised learning, which forms the workflow more widely known as semi-supervised learning.

Next, we will go through the next risk related to security in the consumption of data for machine learning models.

Data security risk

Security, in the context of machine learning, relates to preventing the unauthorized usage of data, protecting the privacy of data, and preventing events or attacks that are unwanted relating to the usage of the data. The risk here is when data security is compromised and can result in failure to comply with regulatory standards, the downfall of the model's performance, or the corruption and destruction of business processes. In this section, we will go through four types of data security risk, namely sensitive data handling, data licensing, software licensing, and adversarial attacks.

Sensitive data handling

Some data is more sensitive than others and can be linked to regulatory risks. Sensitive data gives rise to data privacy regulations that govern the usage of personal data in different jurisdictions. Specifics of regulations vary but they generally revolve around lawfully protecting the rights of the usage of personal data and requiring consent for any actions taken on these types of data along with terms required when using such data. Examples of such regulations are the **General Data Protection Regulation (GDPR)**, which covers the European Union, the **Personal Data Protection Act (PDPA)**, which covers Singapore, the **California Consumer Privacy Act (CCPA)**, which covers only the state of California in the United States, and the **Consumer Data Protection Act (CDPA)**, which covers the state of Virginia in the United States. This means that you can't just collect data that is categorized as personal, annotate it, build a model, and deploy it without adhering to these regulations as doing so would be considered a crime in some of these jurisdictions. Other than requesting consent, one of the common methods that's used to mitigate this risk is to anonymize data so that the data can't be identified by any single person. However, anonymization has to be done reliably so that the key general information is maintained while reliably removing any possibility of reconstructing the identity of a person. Making sure sensitive and personal data is handled properly goes a long way in building trust in the decisions of a machine learning model. Always practice extreme caution in handling sensitive and personal data to ensure the longevity of your machine learning project.

Data and software licensing

Particularly for deep learning projects, a lot of data is required to build a good quality model. The availability of publicly available datasets helps shorten the time and complexity of a project by partially removing the cost and time needed to collect and label data. However, most datasets, like software, have licenses associated with them. These licenses govern how the data associated with the license can be used. The most important criterion of data license about machine learning models for business problems is whether the data allows for commercial usage or not. As most business use cases are considered to be commercial use cases due to profits derived from them, datasets with a license that prevents commercial usage cannot be used. Examples of data licenses with terms that prevent commercial usage are all derivatives of the **Creative Commons Attribution-NonCommercial (CC-BY-NC)** license. Similarly, open sourced software also poses a risk to deep learning projects. Always make sure you triple-check the licenses before using any publicly available data or software for your project. Using data or code that has terms that prevent commercial usage in your commercial business project puts your project at risk of being fined or sued for license infringement.

Adversarial attacks

When an application of machine learning is widely known, it exposes itself to targeted attacks meant to maliciously derail and manipulate the decisions of a model. This brings us to a type of attack that can affect a deployed deep learning model, called an **adversarial attack**. Adversarial attacks are a type of attack that involves manipulating the data input in a certain way to affect the predictions from the machine learning model. The most common adversarial attacks are caused by adversarial data examples that are modified from the actual input data in a way that they appear to maintain their legitimacy as input data but are capable of skewing the model's prediction. The level of risk of such an attack varies across different use cases and depends on how much a user can interact with the system that eventually passes the input data to the machine learning model. One of the most widely known adversarial examples for a deep learning model is an optimized image that looks like random color noise and, when used to perturb pixel values of another image by overlaying its own pixel values, generates an adversarial example. The original image that's obtained after the perturbation looks as though it's untouched visually to a human but is capable of producing erroneous misclassification. The following figure shows an example of this perturbation as part of a tutorial from *Chapter 14, Analyzing Adversarial Performance*. This figure depicts the result of a neural network called ResNet50, trained to classify facial identity, that, when given a facial image, correctly predicted the right identity. However, when added together with another noise image array that was strategically and automatically generated just with access to the predicted class probabilities, the model mistakenly predicts the identity of the facial image, even when the combined image looks visually the same as the original image:

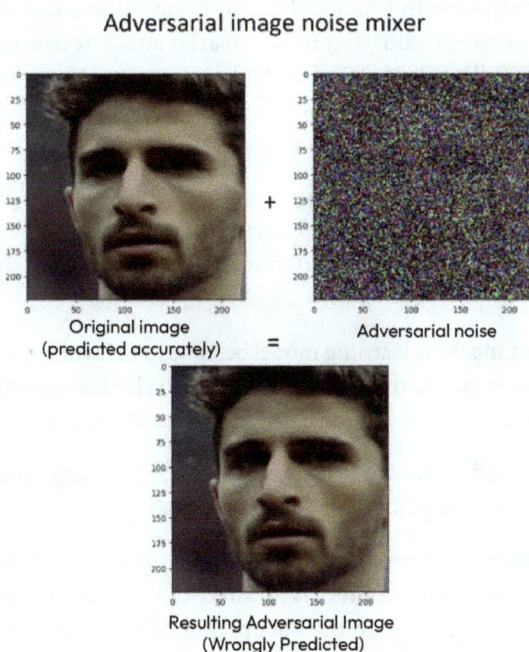

Figure 1.13 – Example of a potential image-based adversarial attack

Theoretically, adversarial attacks can be made using any type of data and are not limited to images. As the stakes of machine learning use cases grow higher, it's more likely that attackers are willing to invest in a research team that produces novel adversarial examples.

Some of the adversarial examples and the methods that generate them rely on access to the deep learning model itself so that a targeted adversarial example can be made. This means that it is not remotely possible for a potential attacker to be able to succeed in confusing a model created by someone else unless they have access to the model. However, many businesses utilize publicly available pre-trained models as-is and apply transfer learning methods to reduce the amount of work needed to satisfy a business use case. Any publicly available pre-trained models will also be available to the attackers, allowing them to build targeted adversarial examples for individual models. Examples of such pre-trained models include all the publicly available ImageNet pre-trained convolutional neural network models and weights.

So, how do we attempt to mitigate this risk?

One of the methods we can use to mitigate the risk of an adversarial attack is to train deep learning models with examples from the known set of methods from public research. By training with the known adversarial examples, the model will learn how to ignore the adversarial information through the learning process and be unfazed by such examples during validation and inference time. Evaluating different variations of the adversarial examples also helps set expectations on when the model will fail.

In this section, we discussed an in-depth take on the security issues when dealing with and consuming data for machine learning purposes. In *Chapter 14*, *Analyzing Adversarial Performance*, we will go through a more in-depth practical evaluation of adversarial attack technologies for different data modalities and how to mitigate them in the context of deep learning. Next, we will dive into another category of risk that is at the core of the model development process.

Overfitting and underfitting

During the model development process, which is the process of training, validating, and testing a machine learning model, one of the most foundational risks to handle is overfitting a model and underfitting a model.

Overfitting is an event where a machine learning model becomes so biased toward the provided training dataset examples and learned patterns that it can only exclusively distinguish examples that belong in the training dataset while failing to distinguish any examples in the validation and testing dataset.

Underfitting, on the other hand, is an event where a machine learning model fails to capture any patterns of the provided training, validation, and testing dataset.

Learning a generalizable pattern to output mapping capability is the key to building a valuable and usable machine learning model. However, there is no silver bullet to achieving a nicely fitted model, and very often, it requires iterating between the data preparation, model development, and deliver model insights stages.

Here are some tips to prevent overfitting and ensure generalization in the context of deep learning:

- Augment your data in expectation of the eventual deployment conditions
- Collect enough data to cover every single variation possible of your targeted label
- Collaborate with domain experts and understand key indicators and patterns
- Use cross-validation methods to ensure the model gets evaluated fairly on unseen data
- Use simpler neural network architectures whenever possible

Here are some tips to prevent underfitting:

- Evaluate a variety of different models
- Collaborate with domain experts and understand key indicators and patterns
- Make sure the data is clean and has as low an error as possible
- Make sure there is enough data
- Start with a small set of data inputs when building your model and build your way up to more complex data to ensure models can fit appropriately on your data

In this section, we have discussed issues while building a model. Next, we will be discussing a type of risk that will affect the built model after it has ben trained.

Model consistency risk

One of the major traits of the machine learning process is that it is cyclical. Models get retrained constantly in hopes of finding better settings. These settings can be a different data partitioning method, a different data transformation method, a different model, or the same model with different model settings. Often, the models that are built need to be compared against each other fairly and equitably. Model consistency is the one feature that ensures that a fair comparison can be made between different model versions and performance metrics. Every single process before a model is obtained is required to be consistent so that when anybody tries to execute the same processes with the same settings, the same model should be obtainable and reproducible. We should reiterate that even when some processes require randomness when building the model, it needs to be randomly deterministic. This is needed so that the only difference in the setup is the targeted settings and nothing else.

Model consistency doesn't stop with just the reproducibility of a model – it extends to the consistency of the predictions of a model. The predictions should be the same when predicted using a different batch size setting, and the same data input should always produce the same predictions. Inconsistencies in model predictions are a major red flag that signals that anything the model produces is not representative of what you will get during deployment and any derived model insights would be misinformation.

To combat model consistency risk, always make sure your code produces consistent results by seeding the random number generator whenever possible. Always validate model consistency either manually or automatically in the model development stage after you build the model.

Model degradation risk

When you have built a model, verified it, and demonstrated its impact on the business, you then take it into the model deployment stage and position it for use. One mistake is to think that this is the end of the deep learning project and that you can take your hands off and just let the model do its job. Sadly, most machine learning models degrade, depending on the level of generalization your model achieved during the model development stage on the data available in the wild. A common scenario that happens when a model gets deployed is that, initially, the model's performance and the characteristics of the data received during deployment stay the same, and change over time. Time has the potential to change the conditions of the environment and anything in the world. Machines grow old, seasons change, and people change, and expecting that conditions and variables surrounding the machine learning model will change can allow you to make sure models stay relevant and impactful to the business.

How a model can degrade can be categorized into three categories, namely **data drift**, **concept drift**, and **model drift**. Conceptually, drift can be associated with a boat slowly drifting away from its ideal position. In the case of machine learning projects, instead of a boat drifting away, it's the data or model drifting away from its original perceived behavior or pattern. Let's briefly go through these types of drift.

Data drift is a form of degradation that is associated with the input data the model needs to provide a prediction value. When a deployed model experiences data drift, it means that the received that's data during deployment doesn't belong to the inherent distribution of the data that was trained and validated by the machine learning model. If there were ways to validate the model on the new data supplied during deployment, data drift would potentially cause a shift in the original expected metric performance obtained during model validation. An example of data drift in the context of a deep learning model can be a use case requiring the prediction of human actions outdoors. In this use case, data drift would be that the original data that was collected consisted of humans in summer clothing during the summer season, and due to winter, the people are wearing winter clothing instead.

Concept drift is a form of degradation that is associated with the change in how the input data interacts with the target output data. In the planning stage, domain experts and machine learning practitioners collaborate to define input variables that can affect the targeted output data. This defined input and output setup will subsequently be used for building a machine learning model. Sometimes, however, not all of the context that can affect the targeted output data is included as input variables due to either the availability of that data or just the lack of domain knowledge. This introduces a dependence on the conditions of the missing context associated with the data collected for machine learning. When the conditions of the missing context drift away from the base values that exist in the training and validation data, concept drift occurs, rendering the original concept irrelevant or shifted. In simpler terms, this means that the same input doesn't map to the same output from the training data anymore. In the context of deep learning, we can take a sentiment classification use case that is based on textual data. Let's say that a comment or speech can be a negative, neutral, or positive sentiment based on both the jurisdiction and the text itself. This means that in some jurisdictions, people have different thresholds on what is considered negative, neutral, or positive and grade things differently. Training the model with only textual data wouldn't allow itself to generalize across jurisdictions accurately and thus would face concept drift any time it gets deployed in another jurisdiction.

Lastly, model drift is a form of degradation associated with operational metrics and easily measurable metrics. Some of the factors of degradation that can be categorized into model drift are model latency, model throughput, and model error rates. Model drift metrics are generally easy to measure and track compared to the other two types of drift.

One of the best workflows for mitigating these risks is by tracking the metrics under all of these three types of drift and having a clear path for a new machine learning model to be built, which is a process I call **drift reset**. Now that we've covered a brief overview of model degradation, in *Chapter 16, Governing Deep Learning Models*, and *Chapter 17, Managing Drift Effectively in a Dynamic Environment*, we will go more in-depth on these risks and discuss practically how to mitigate this risk in the context of deep learning.

Summary

In this chapter, you learned what it takes to complete a deep learning project successfully in a repeatable and consistent manner at a somewhat mid to high-level view. The topics we discussed in this chapter were structured to be more comprehensive at the earlier stages of the deep learning life, which covers the planning and data preparation stages.

In the following chapters, we will explore the mid to final stages of the life cycle more comprehensively. This involves everything after the data preparation stage, which includes model development, model insights, model deployment, and, finally, model governance.

In the next chapter, we will start to explore common and widely used deep learning architectures more extensively.

Further reading

- Bowen Shi, Wei-Ning Hsu, Abdelrahman Mohamed. *Robust Self-Supervised Audio-Visual Speech Recognition*, 2022. Toyota Technological Institute at Chicago, Meta AI: `https://arxiv.org/pdf/2201.01763v2.pdf`.

- Peng Wang, An Yang, Rui Men, Junyang Lin, Shuai Bai, Zhikang Li, Jianxin Ma, Chang Zhou, Jingren Zhou, Hongxia Yang. *Unifying Architectures, Tasks, and Modalities Through a Simple Sequence-to-Sequence Learning Framework*. DAMO Academy, Alibaba Group: `https://arxiv.org/pdf/2202.03052v1.pdf`.

2

Designing Deep Learning Architectures

In the previous chapter, we went through the entire deep learning life cycle and understood what it means to make a deep learning project successful from end to end. With that knowledge, we are now ready to dive further into the technicalities of deep learning models. In this chapter, we will dive into common deep learning architectures used in the industry and understand the reasons behind each architecture's design. For intermediate and advanced readers, this will be a brief recap to ensure alignment in the definitions of terms. For beginner readers, architectures will be presented in a way that is easy to digest so that you can get up to speed on the useful neural architectures in the world of deep learning.

Grasping the methodologies behind a wide variety of architectures allows you to innovate custom architectures specific to your use case and, most importantly, gain the skill to choose an appropriate foundational architecture based on the data input or problem type.

In this chapter, the focus will be on the **Multilayer Perceptron** (**MLP**) network architecture. The comprehensive coverage of MLPs, along with some key concepts in general related to neural network implementations, such as gradients, activation functions, and regularization methods, will set the stage for exploring other, more complex architecture types in later chapters. Specifically, the following topics will be covered in this chapter:

- Exploring the foundations of neural networks using an MLP
- Understanding neural network gradients
- Understanding gradient descent
- Implementing an MLP from scratch
- Implementing an MLP using deep learning frameworks
- Designing an MLP

Technical requirements

This chapter includes some practical implementations in the **Python** programming language. To complete it, you will need to have a computer with the following libraries installed:

- pandas
- Matplotlib
- Seaborn
- Scikit-learn
- NumPy
- Keras
- PyTorch

The code files are available on GitHub: https://github.com/PacktPublishing/The-Deep-Learning-Architect-Handbook/tree/main/CHAPTER_2.

Exploring the foundations of neural networks using an MLP

A deep learning architecture is created when at least three perceptron layers are used, excluding the input layer. A perceptron is a single-layer network consisting of neuron units. Neuron units hold a bias variable and act as nodes for vertices to be connected. These neurons will interact with other neurons in a separate layer with weights applied to the connections/vertices between neurons. A perceptron is also known as a **fully connected layer** or **dense layer**, and MLPs are also known as **feedforward neural networks** or **fully connected neural networks**.

Let's refer back to the MLP figure from the previous chapter to get a better idea.

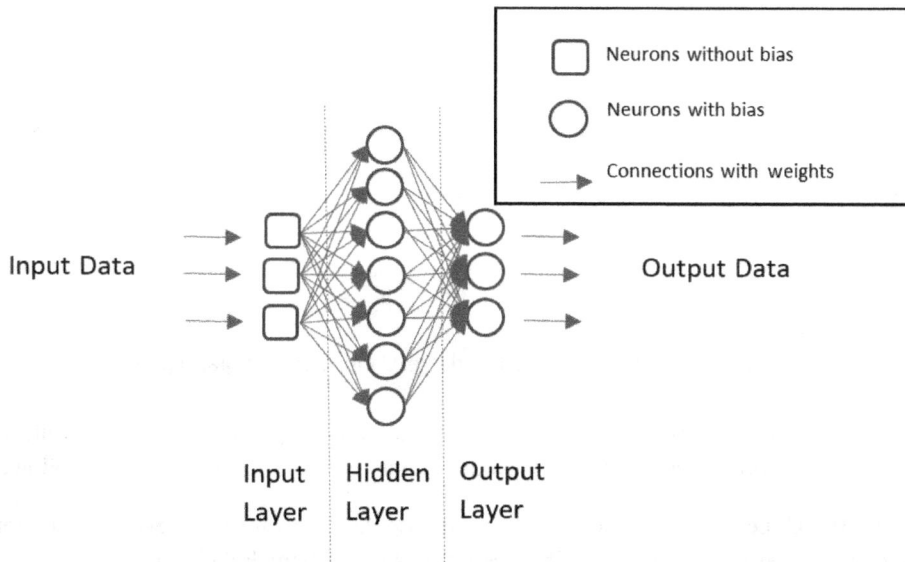

Figure 2.1 – Simple deep learning architecture, also called an MLP

The figure shows how three data column inputs get passed into the input layer, then subsequently get propagated to the hidden layer, and finally, through the output layer. Although not depicted in the figure, an additional activation function is applied at the hidden and output layer outputs. The activation function at the hidden layers adds non-linearity to the model and allows the neural network to capture non-linear relationships between the input data and output data. The activation function used at the output layer depends on the problem type and will be discussed in more detail in *Chapter 8, Exploring Supervised Deep Learning.*

Before we dive into the relevant hidden activation methods, we need to first be aware of the vanishing gradient problem. The vanishing gradient problem is a challenge that arises when gradients of the loss function with respect to the model's parameters become very small during backpropagation. This can lead to slow learning and poor convergence, as the weights update minimally or not at all. The vanishing gradient problem is particularly prominent when using activation functions that squash input values into a narrow range. To address this issue, the **Rectified Linear Unit (ReLU)** activation function has been widely adopted due to its ability to mitigate the vanishing gradient problem to a certain extent. ReLU maps negative values into zeros and maintains positive values, as depicted in *Figure 2.2.*

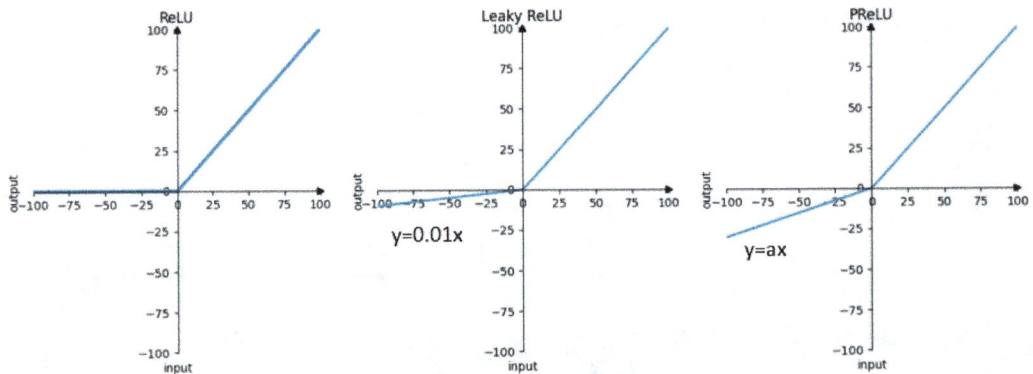

Figure 2.2 – ReLU, Leaky ReLU, and PReLU input/output graph plot

Apart from ReLU, there are other useful hidden layer activation functions that can help alleviate the vanishing gradient problem while offering various benefits. Some of these include the following:

- **Leaky ReLU**: Leaky ReLU is a variation of the ReLU function that allows a small, non-zero gradient for negative input values. This helps mitigate the "dying ReLU" problem, where neurons become inactive and stop learning if their input values are consistently negative. Leaky ReLU introduces a small slope for negative inputs, ensuring that gradients do not vanish entirely.

- **Parametric ReLU (PReLU)**: PReLU is another variation of the ReLU function, where the negative slope is learned during the training process, allowing the model to adapt its behavior. This flexibility can lead to better performance but at the cost of increased complexity and the risk of overfitting.

Additionally, we will be exploring more hidden activation functions as we dive into different prominent architectures in this book. Each of these activation functions has its strengths and weaknesses, and the choice of activation function depends on the specific problem being addressed and the architecture being employed. Understanding, experimenting with, and assessing these activation functions is crucial for selecting the most suitable one for a given task within a neural network's hidden layers. Furthermore, the recommended method to assess any model-building-related experiments will be explored in *Chapter 8*, *Exploring Supervised Deep Learning*.

Moving on, the process of propagating values from one layer to another is called a forward pass or forward propagation, where the formula can be generally defined as follows:

$$a = g\left(\sum_{n=0}^{neurons} wx + b\right)$$

Here, a represents the outputs of the neural network layer (called an **activation**), g represents the non-linear activation function, w represents the weights between neuron connections, x represents the input data or activation, and b represents the bias of the neuron. Different types of neural network layers consume and output data in different ways but generally still use this formula as a foundation.

Understanding neural network gradients

The goal of machine learning for an MLP is to find the weights and biases that will effectively map the inputs to the desired outputs. The weights and biases generally get initialized randomly. In the training process, with a provided dataset, they get updated iteratively and objectively in batches to minimize the loss function, which uses gradients computed with a method called **backward propagation**, also known as **backpropagation**. A batch is a subset of the dataset used for training or evaluation, allowing the neural network to process the data in smaller groups rather than the entire dataset at once. The loss function is also known as the error function or the cost function.

Backpropagation is a technique to find out how sensitive a change of weights and bias of every neuron is to the overall loss by using the partial derivative of the loss with respect to the weights and biases. Partial derivatives from calculus are a measure of the rate of change of a function with respect to a variable, which uses a technique called **differentiation**, and is effectively applied in neural networks. A convenient method called the **chain rule** allows you to obtain the derivative of neural networks by calculating the partial derivatives of each function, a forward pass in the case of neural networks, separately. To be clear, derivatives can be called the sensitivity of change, the gradients, and the rate of change. The idea is that when we know which model parameter affects the error the most, we can update its weights proportionally according to its magnitude and direction. Let's take a simple case of a two-layer MLP with one neuron in each layer as an example to get an idea of this, as depicted in *Figure 2.3*.

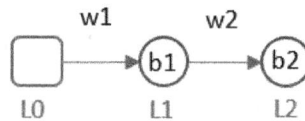

Figure 2.3 – A diagram of a two-layer MLP

For clarity, **w** indicates the weight of neuron connections, **b** indicates the bias of the neurons, and **L** indicates layers. Different problem types require a different loss function, but for explanation purposes, let's assume this is an MLP for a regression problem, where we will use the mean squared error as a loss function to compute the loss component from the final layer activation and the numerical target value. The loss function can then be defined as the following:

$$L = \frac{1}{n} \sum_{i=1}^{n} (a2 - y)^2$$

Here, n is the total size of neurons. To obtain the rate of change of the loss function with respect to the output layer weight **w2**, which is $\frac{\delta L}{\delta w2}$, let's define the formula based on the chain rule. Consider the following:

$$z2 = w2 \cdot a1 + b2$$

So, if $a2 = g(z2)$, where g is a ReLU function, the gradients with respect to the output layer weight $w2$ will be defined as follows:

$$\frac{\delta L}{\delta w2} = \frac{\delta L}{\delta a2} \cdot \frac{\delta a2}{\delta z2} \cdot \frac{\delta z2}{dw2}$$

The rate of change of the loss function with respect to $w2$ can be computed by multiplying the three independent change components: namely, the change of the loss function with respect to the second-layer outputs, the change of the activation outputs with respect to a wrapped $z2$ that is a forward pass without the activation, and the change of the wrapped $z2$ with respect to $w2$. Let's define these components next. Now consider the following:

$$e = a2 - y$$

Based on the chain rule, the first change component will be defined as the following:

$$\frac{\delta L}{\delta a2} = \frac{\delta L}{\delta e} \cdot \frac{\delta e}{\delta a2}$$

$$\frac{\delta L}{\delta e} = 2e$$

$$\frac{\delta e}{\delta a2} = 1$$

Putting this in the simplified component representation will result in the following equation:

$$\frac{\delta L}{\delta a2} = \frac{2}{n}(a2 - y)$$

For the second change component, it can be defined with the following formula:

$$\frac{\delta a2}{\delta z2} = g'(z2) = 1$$

There is no activation function applied at the output layer in this case.

For the third and last change component, it can be defined with the following formula:

$$\frac{\delta z2}{\delta w2} = a1$$

Finally, placing the simplified representation of the three components into the formula to obtain the gradients of the output layer weights $w2$ will result in the following equation:

$$\frac{\delta L}{\delta w2} = \frac{2}{n}(a2 - y) \cdot a1$$

All you need to do now is to plug in the actual values to obtain the layer 2 weight gradients. The same formula structure can be adapted similarly to the hidden layer's weight $w1$ as follows:

$$a1 = g(z1)$$

$$z1 = w1 \cdot a0 + b1$$

$$\frac{\delta L}{\delta w1} = \frac{\delta L}{\delta a1} \cdot \frac{\delta a1}{\delta z1} \cdot \frac{\delta z1}{\delta w1}$$

After expanding $\frac{\delta L}{\delta a1}$ using the chain rule, the formula can then be the following:

$$\frac{\delta L}{\delta w1} = \frac{\delta L}{\delta a2} \cdot \frac{\delta a2}{\delta z2} \cdot \frac{\delta z2}{\delta a1} \cdot \frac{\delta a1}{\delta z1} \cdot \frac{\delta z1}{\delta w1}$$

The additional individual components can be defined as follows:

$$\frac{\delta z2}{\delta a1} = w2$$

$$\frac{\delta a1}{\delta z1} = g'(z1) = 0 \; if \; a2 < 0, 1 \; if \; a2 > 0$$

$$\frac{\delta z1}{\delta w1} = a0$$

$a0$ here is the input data. Now, let's define the gradient of the hidden layer weights $w1$ with the representation where we can plug in actual values to compute it, as follows:

$$\frac{\delta L}{\delta w1} = \frac{2}{n}(a2 - y) \cdot w2 \cdot g'(z1) \cdot a0$$

The same process can be repeated for the bias term to obtain its gradients. Only the partial derivative of z with respect to the weights needs to be replaced with a partial derivative of z with respect to the biases, as shown here:

$$\frac{\delta z2}{\delta b2} = \frac{\delta z1}{\delta b1} = 1$$

$$\frac{\delta L}{\delta b2} = \frac{\delta L}{\delta a2} \cdot \frac{\delta a2}{\delta z2} \cdot \frac{\delta z2}{\delta b2}$$

$$\frac{\delta L}{\delta b2} = 2(a2 - y)$$

$$\frac{\delta L}{\delta b1} = \frac{\delta L}{\delta a2} \cdot \frac{\delta a2}{\delta z2} \cdot \frac{\delta z2}{\delta a1} \cdot \frac{\delta a1}{\delta z1} \cdot \frac{\delta z1}{\delta b1}$$

Now, let's define the gradient of the first bias term with the representation where we can plug in actual values to compute it, as follows:

$$\frac{\delta L}{\delta b1} = \frac{2}{n} \cdot w2 \cdot (a2 - y) \cdot g'(z1)$$

The previously defined formulae were meant to be specific to the example neural network for layers with one neuron. In practical usage, these layers usually contain more than one neuron in each of the layers. To compute the loss and derivatives for layers with more than one neuron, and for more than one data sample, you simply need to obtain an average of all the values.

Once the gradients or derivatives are obtained, different strategies can be used to update the weights. The algorithm used to optimize the weights and biases of the neural network is called the optimizer. There are many optimizer options today and each has its own pros and cons. As gradients are used to optimize weights and biases, this process of optimization is called **gradient descent**.

Understanding gradient descent

A good way to think about loss for a deep learning model is that it exists in a three-dimensional loss landscape that has many different hills and valleys, with valleys being more optimal, as shown in *Figure 2.4*.

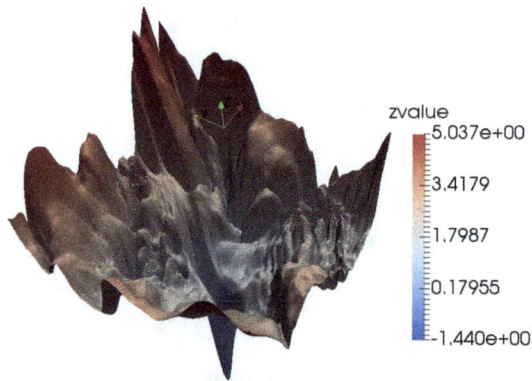

Figure 2.4 – An example loss landscape

In reality, however, we can only approximate these loss landscapes as the parameter values of the neural networks can exist in an infinite number of ways. The most common way practitioners use to monitor the behavior of loss during each epoch of training and validation is to simply plot a two-dimensional line graph with the x axis being the epochs executed and the y axis being the loss performance. An epoch is a single iteration through the entire dataset during the training process of a neural network. The loss landscape in *Figure 2.4* is an approximation of the loss landscape in three dimensions of a neural network. To visualize the three-dimensional loss landscape in *Figure 2.4*, we can use two randomly initialized parameters and one fully trained parameter from the same neuron positions within the neural network. The loss can be calculated by performing a weighted summation of these three parameters. The weight of the fully trained parameter remains constant, while the weights of the two randomly initialized parameters are adjusted. This process allows us to approximate the 3D loss landscape shown in *Figure 2.4*. In this figure, x axis and y axis are the weights of the two randomly initialized parameters of the same neural network and the z axis is the loss value. The goal of gradient descent is to attempt to find the *global* deepest valleys and not be stuck in *local* valleys or *local* minima. The gradients computed provide the suggested directions needed to nudge and update the weights and biases iteratively. One thing to note is that gradients provide the direction to increase the loss function in the fastest way, so for the descent, the parameters are subtracted from the gradients. Let's go through a simple form of gradient descent that controls how the weights and biases should be updated:

$$w = w - \alpha \cdot \frac{\delta L}{\delta w}$$
$$b = b - \alpha \cdot \frac{\delta L}{\delta b}$$

Here, α refers to the learning rate, which controls how aggressive you want the deep learning model to be. A learning rate is a hyperparameter that controls the speed at which a neural network learns and updates its weights and biases during the optimization process. The higher the learning rate, the bigger the steps taken by the deep learning model in the loss landscape. By iteratively applying this parameter update step, the neural network will slowly move downhill so that the learned set of parameters can allow the network to effectively map the input to the desired target values. The gradients are obtained for all the data samples and averaged together to obtain a single update direction for the weight and biases update.

Datasets can sometimes be too big and cause a slow learning process from basic gradient descent due to the need to compute the gradients from every sample before an update can be done to the neural network parameters. **Stochastic gradient descent (SGD)** was created to tackle this problem. The idea is simply to learn from the dataset in batches and iteratively learn the entire dataset with a different data batch partition instead of waiting for the gradients to be obtained from the entire dataset before updating the parameters of the network. This way, the learning process can be efficient even for a big-sized dataset with the added benefit of seeing initial results quickly.

There are a lot more variations of gradient descent that offer different advantages and are suited for specific situations. Here, we will list gradient descent algorithms that, on average across a variety of datasets, work well and are relevant:

- **Momentum**: A variation of SGD, Momentum incorporates a "momentum" term that helps the optimizer navigate through the loss landscape more effectively. This momentum term is a moving average of the gradients, which helps the optimizer overcome local minima and converge faster. The momentum term also adds some inertia to the optimizer, causing it to take bigger steps in directions that have consistent gradients, which can speed up convergence.

- **Root Mean Square Propagation (RMSProp)**: RMSProp is an adaptive learning rate optimization algorithm that adjusts the learning rate for each parameter individually. By dividing the learning rate by an exponentially decaying average of squared gradients, RMSProp helps to prevent the oscillations observed in the convergence of SGD. This results in a more stable and faster convergence toward the optimal solution.

- **Adaptive Moment Estimation (Adam)**: Adam is another popular optimization algorithm that combines the advantages of both Momentum and RMSProp. It maintains separate adaptive learning rates for each parameter, as well as incorporating a momentum term. This combination allows Adam to converge quickly and find more accurate solutions, making it a popular choice for many deep learning tasks.

While there are many gradient descent algorithms available, choosing the right one depends on the specific problem and dataset at hand. In general, Adam is often recommended as a good starting point due to its adaptive nature and combination of Momentum and RMSProp features. To determine the best fit for your specific deep learning task, it is essential to experiment with different algorithms and their hyperparameters and validate their performance. Next, we will code up an MLP using Python.

Implementing an MLP from scratch

Today, the process to create a neural network and its layers along with the backpropagation process has been encapsulated in deep learning frameworks. The differentiation process has been automated, where there is no actual need to define the derivative formulas manually. Removing the abstraction layer provided by the deep learning libraries will help to solidify your understanding of neural network internals. So, let's create this neural network manually and explicitly with the logic to forward pass and backward pass instead of using the deep learning libraries:

1. We'll start by importing numpy and the methods from the scikit-learn library to load sample datasets and perform data partitioning:

    ```python
    import numpy as np
    from sklearn import datasets
    from sklearn.model_selection import train_test_split
    ```

2. Next, we define ReLU, the method that makes an MLP non-linear:

    ```python
    def ReLU(x):
        return np.maximum(x, 0)
    ```

3. Now, let's define partially the class needed to initialize an MLP model with a single hidden layer and an output layer that can perform a forward pass. The layers are represented by weights, where w1 is the weight of the hidden layer, and w2 is the weight of the output layer. Additionally, b1 is the bias for the connection between the input layer and the hidden layer and b2 is the bias for the connection between the hidden layer and the output layer:

    ```python
    class MLP(object):
      def __init__(
        self, input_layer_size, hidden_layer_size, output_layer_
    size, seed=1234
    ):
        rng = np.random.RandomState(seed)
        self.w1 = rng.normal(
          size=(input_layer_size, hidden_layer_size)
        )
        self.b1 = np.zeros(hidden_layer_size)
        self.w2 = rng.normal(
          size=(hidden_layer_size, output_layer_size)
        )
        self.b2 = np.zeros(output_layer_size)
        self.output_layer_size = output_layer_size
        self.hidden_layer_size = hidden_layer_size
      def forward_pass(self, x):
        z1 = np.dot(x, self.w1) + self.b1
        a1 = ReLU(z1)
        z2 = np.dot(a1, self.w2)  + self.b2
        a2 = z2
        return z1, a1, z2, a2
    ```

4. To allow the MLP to learn, we will now implement the backward pass method to generate the average gradients for the biases and weights:

```
def ReLU_gradient(x):
    return np.where(x > 0, 1, 0)
def backward_pass(self, a0, z1, a1, z2, a2, y):
    number_of_samples = len(a2)
    average_gradient_w2 = (
      np.dot(a1.T, (a2 - y)) *
      (2 / (number_of_samples * self.output_layer_size))
    )
    average_gradient_b2 = (
      np.mean((a2 - y), axis=0) * (2 / self.output_layer_size)
    )
    average_gradient_w1 = np.dot(
      a0.T, np.dot((a2 - y), self.w2.T) * ReLU_gradient(z1)
    ) * 2 / (number_of_samples * self.output_layer_size)
    average_gradient_b1 = np.mean(
      np.dot((a2 - y), self.w2.T) * ReLU_gradient(z1), axis=0
    ) *  2 / self.output_layer_size
    return (
      average_gradient_w2, average_gradient_b2, average_
gradient_w1, average_gradient_b1
    )
```

Notice that the derivative of the ReLU function is $f'(x) = 1$ if $x > 0$ and $f'(x) = 0$ if $x <= 0$.

5. For the last class method, we will implement the gradient descent step utilizing the average gradients from the backward pass, which is the process that allows the bias and weights to be updated:

```
def gradient_descent_step(
    self, learning_rate, average_gradient_w2, average_gradient_
b2, average_gradient_w1, average_gradient_b1
  ):
    self.w2 = self.w2 - learning_rate * average_gradient_w2
    self.b2 = self.b2 - learning_rate * average_gradient_b2
    self.w1 = self.w1 - learning_rate * average_gradient_w1
    self.b1 = self.b1 - learning_rate * average_gradient_b1
```

6. Now that we have made a proper class for the MLP manually, let's set up a dataset and attempt to learn from it. The structure of MLP only allows for tabular structured dataset types that are both one-dimensional and numerical, so we will be using a dataset called `diabetes`, which contains 10 numerical features, that is, age, sex, body mass index, average blood pressure, and 6 blood serum measurements, as inputs along with a quantitative measure of diabetes disease

progression as the target data. The data is conveniently saved in the scikit-learn library, so let's first load the input DataFrame:

```
diabetes_data = datasets.load_diabetes(as_frame=True)
diabetes_df = diabetes_data['data']
```

7. Now, we will convert the DataFrame into NumPy array values so that it is ready to be used by a neural network:

```
X = diabetes_df.values
```

8. The final step for loading the data is to load the target data from the diabetes data and make sure it has an additional outer dimension as the PyTorch model outputs its predictions in this way:

```
target = np.expand_dims(diabetes_data['target'], 1)
```

9. Next, let's partition the dataset into 80% for training and 20% for validation:

```
X_train, X_val, y_train, y_val = train_test_split(
    X, target, test_size=0.20, random_state=42
)
```

10. Now that the data is prepared with training and evaluation partitions, let's initialize an MLP model from our defined class with a single 20-neuron hidden layer and a 1-neuron output layer:

```
mlp_model = MLP(
    input_layer_size=len(diabetes_df.columns),
    hidden_layer_size=20,
    output_layer_size=target.shape[1]
)
```

11. With the data and model ready, it is time to train the model we built from scratch. Since the dataset is small enough, with 442 samples, there isn't a runtime issue using gradient descent, so we will be using the full gradient descent here for 100 epochs. One epoch means a full round of training going through the entire training dataset once:

```
iterations = 100
training_error_per_epoch = []
validation_error_per_epoch = []
for i in range(iterations):
  z1, a1, z2, a2 = mlp_model.forward_pass(X_train)
  (
    average_gradient_w2,
    average_gradient_b2,
    average_gradient_w1,
    average_gradient_b1
  ) = mlp_model.backward_pass(X_train, z1, a1, z2, a2, y_train)
```

```
mlp_model.gradient_descent_step(
  learning_rate=0.1,
  average_gradient_w2=average_gradient_w2,
  average_gradient_b2=average_gradient_b2,
  average_gradient_w1=average_gradient_w1,
  average_gradient_b1=average_gradient_b1,
)
_, _, _, a2_val = mlp_model.forward_pass(X_val)
training_error_per_epoch.append(mean_squared_error(y_train,
a2)
validation_error_per_epoch.append(
  mean_squared_error(y_val, a2_val)
)
```

12. Let's plot the collected mean squared error for both the validation and training partitions using `matplotlib`:

```
plt.figure(figsize=(10, 6))
plt.plot(training_error_per_epoch)
plt.plot(validation_error_per_epoch,   linestyle = 'dotted')
plt.show()
```

This will produce the following plot:

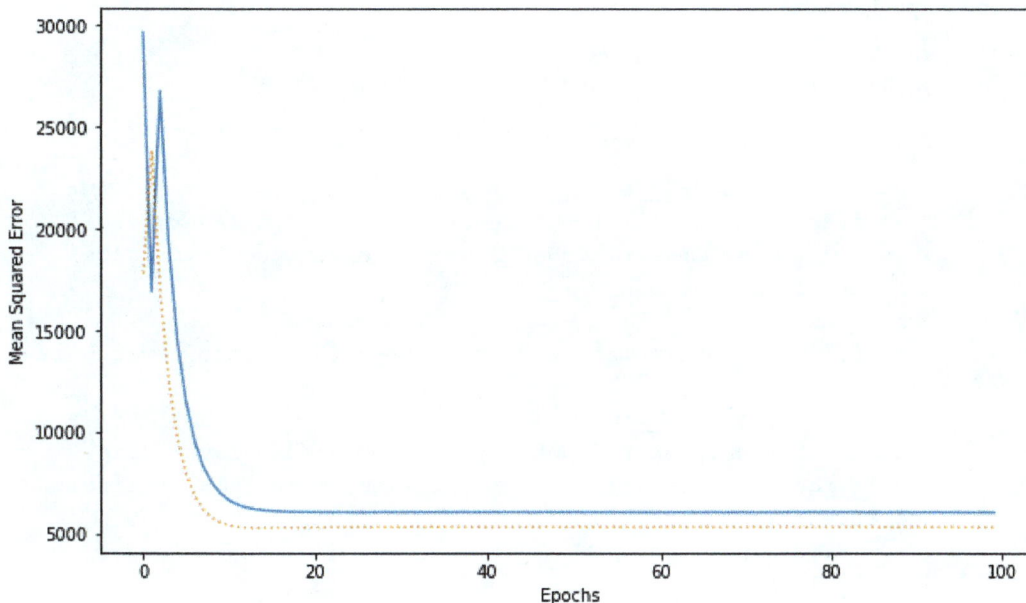

Figure 2.5 – Training and validation partition mean squared error versus epochs plots

With that, you've implemented an MLP and trained it from scratch without depending on deep learning frameworks! But is our implementation correct and sound? Let's verify this in the next topic.

Implementing MLP using deep learning frameworks

Deep learning frameworks are made to ease and expedite the development of deep learning models. They provide a plethora of commonly used neural network layers, optimizers, and tools generally used to build neural network models, along with very easily extensible interfaces to implement new methods. Backpropagation itself is abstracted away from the users of the frameworks as the gradients are computed automatically in the background when needed. Most importantly, they allow the usage of the GPU for efficient model training and prediction.

In this section, we will build the same MLP model as in the previous section, using a deep learning framework called PyTorch, and verify that both implementations produce the same results:

1. We'll start by importing the necessary libraries:

```
Import torch
import torch.nn as nn
import torch.nn.functional as F
```

2. Next, let's define the MLP class with the two fully connected layers along with the forward propagation method with arguments that allow us to set the input layer size, hidden layer size, and output layer size:

```
Class MLPPytorch(nn.Module):
  def __init__(
    self, input_layer_size, hidden_layer_size, output_layer_size
  ):
    super(Net, self).__init__()
    self.fc1 = nn.Linear(input_layer_size, hidden_layer_size)
    self.fc2 = nn.Linear(hidden_layer_size, output_layer_size)
class MLPPytorch(nn.Module):
  def __init__(
    self, input_layer_size, hidden_layer_size, output_layer_size
  ):
    super(Net, self).__init__()
    self.fc1 = nn.Linear(input_layer_size, hidden_layer_size)
    self.fc2 = nn.Linear(hidden_layer_size, output_layer_size)
  def forward(self, x):
    x = F.relu(self.fc1(x))
    x = self.fc2(x)
    return x
```

3. You will notice that there isn't a backward propagation function implemented, which reduces the amount of effort needed to define a neural network model. When you inherit from the Pytorch module class, the backward propagation functionality will already be provided out of the box with your defined Pytorch layers. Finally, let's initialize the MLP using a hidden layer size of 10 along with input and output sizes according to the diabetes dataset:

```
Net = MLPPytorch(
    input_layer_size=len(diabetes_df.columns),
    hidden_layer_size=10,
    output_layer_size=y_train.shape[1],
)
```

4. Now, let's check the forward and backward propagation functionality with our numpy variant. First, let's initialize the Pytorch MLP and copy the weights from the numpy-based MLP model:

```
with torch.no_grad():
        net.fc1.weight.copy_(
      torch.from_numpy(mlp_model.w1.T)
    )
    net.fc1.bias.copy_(
      torch.from_numpy(mlp_model.b1)
    )
    net.fc2.weight.copy_(
      torch.from_numpy(mlp_model.w2.T)
    )
    net.fc2.bias.copy_(
      torch.from_numpy(mlp_model.b2)
    )
```

5. Now, let's prepare the dataset into Tensor objects suitable for PyTorch model consumption:

```
torch_input = torch.from_numpy(X_train)
torch_target = torch.from_numpy(y_train)
```

6. To obtain the same gradients, we have to use the same MSE loss and apply backward propagation:

```
criterion = nn.MSELoss()
output = net(torch_input.float())
loss = criterion(output, torch_target.float())
loss.backward()
```

7. Now, let's verify the gradients for the two implementations:

```
np.testing.assert_almost_equal(output.detach().numpy(), a2,
decimal=3)
np.testing.assert_almost_equal(net.fc2.weight.grad.numpy(),
average_gradient_w2.T, decimal=3)
np.testing.assert_almost_equal(net.fc2.bias.grad.numpy(),
average_gradient_b2, decimal=3)
np.testing.assert_almost_equal(net.fc1.weight.grad.numpy(),
average_gradient_w1.T, decimal=3)
np.testing.assert_almost_equal(net.fc1.bias.grad.numpy(),
average_gradient_b1, decimal=3)
```

This solidifies your foundational neural network knowledge, along with knowledge of the MLP architecture, and prepares you for more advanced concepts in the realm of deep learning. Before we move on to look at a more advanced neural network, in the next section, we will explore the topic of regularization, and then finally, explore how to design an MLP with a practical use case.

Regularization

Regularization in deep learning has evolved into a state where it now means any addition or modification to the neural network, data, or training process that is used to increase the generalization of the build model to external data. All performant neural networks today have some form of regularization embedded into the architecture. Some of these regularization methods introduce some extra beneficial side effects, such as the speedup of training or the performance on the training dataset. But ultimately, the regularizer's main goal is to improve generalization, which in other words is to improve performance metrics and reduce errors on external data. As a quick recap, the following list shows a summary of some of the more common regularization methods:

- **Dropout layer**: During training, randomly remove information from all neural nodes according to a specified probability level by replacing neural node outputs with zeros, effectively nullifying information. This reduces over-reliance on any single node/information and increases the probability of generalization.

- **L1/L2 regularization**: These methods add a penalty term to the loss function, which discourages the model from assigning high weights to the features. L1 regularization, also known as Lasso, uses the absolute value of the weights, while L2 regularization, also known as Ridge, uses the squared value of the weights. By controlling the magnitude of the weights, these methods help to prevent overfitting and improve generalization. Typically, this is applied to the input features.

- **Batch normalization layer**: This method is standardizing the data in both training and inferencing on the external data stage by scaling the data to have a mean of zero and a standard deviation of one. This is done by removing the computed mean and dividing it by the computed standard deviation. The mean and standard deviation are computed and iteratively updated by mini-batch (based on the models determining the training batch size) during training.

During inference, the final learned running mean and standard deviation calculated during training are applied. This has the side effect of improving the training time, training stability, and generalization. Note that each element has its own mean and standard deviation. Research has shown that batch normalization smooths out the loss landscape, making it way easier to reach an optimum value.

- **Group normalization layer**: Instead of having an individual mean and standard deviation for each element across the batch size, group normalization standardizes the data by groups per sample, where each group has one mean and one standard deviation. The number of groups can be configured. Batch normalization degrades the performance when the number of samples in a batch is small due to hardware limitations. This layer is used over batch normalization when the amount of data per batch is a very small number as the mean and standardization updates do not depend on the batch. In large batches, however, batch normalization still triumphs.

- **Weight standardization**: This applies the same standardization process to the weights of the neural networks. The weights of the neural network might grow to very large numbers after training, which would create large output values. The idea is that if we use the batch normalization layer, the output values will be standardized anyway, so why don't we take a step back and apply the same process to the weights themselves, making sure the values are standardized in some form before becoming an output value? Some simple benchmarks have demonstrated that it works well when combined with a group normalization layer in low batch sizes, achieving a better performance than batch normalization with a high batch size setting.

- **Stochastic depth**: Instead of conceptually making a neural network with narrower layers during the training stage with dropout, stochastic depth reduces the depth of the network during training. This regularizing method leverages the concept of skip connections from ResNets, which will be introduced later, where outputs from earlier layers are additionally forwarded to the later layers. During training, the layers in between the skip connections are completely bypassed to simulate a shallower network randomly. This regularizer has the effect of a faster training time along with an increased generalization performance.

- **Label smoothing**: This is used for classification problems, including binary class, multiclass, or multilabel class problems. It introduces a relaxation for the actual one-hot-encoded labels to learn from. It modifies the actual label of 1 in the one-hot encoder vector by $1 - \varepsilon$, where ε is a reasonably small value. Additionally, 0 is replaced with $\frac{\varepsilon}{k-1}$, where k can be set using the number of classes. One example input and output would be [0, 0, 0, 1] and [0.0001, 0.0001, 0.0001, 0.9999], respectively. The idea is that we shouldn't train the model to be overconfident in its result, which will signal that it is overfitted to the training data and won't be able to generalize to external data. This method encourages representations of the last layer outputs to be closer to each other for samples in the same class and encourages the same outputs to be equally distant among samples from different classes. Additionally, this helps to mitigate overconfidence in samples that have inaccurate labels.

- **Data augmentation**: When the raw data does not adequately represent all the variations of the data of any label, data augmentation helps to computationally add variations in the data to be used for training. This effectively increases generalization simply due to the model being able to learn from more complete variations of the data. This can be applied to any data modality and will be introduced in more detail in *Chapter 8, Exploring Supervised Learning*.

Regularization is an important component in any neural network architecture and will be seen in all of the architectures that will be introduced in the chapter. When choosing regularization techniques for a specific problem, you should first consider the nature of your dataset and the problem you are trying to solve. For instance, if you have a small batch size, group normalization or weight standardization might be more suitable than batch normalization. If your dataset has limited variations, data augmentation can be used to improve generalization. To choose between these techniques, start with a simple regularization method such as dropout or L1/L2 regularization, and evaluate its performance. Then, you can experiment with other techniques, either individually or in combination, and compare their impact on the model's performance. It's essential to monitor the training and validation metrics to ensure that the chosen regularization methods are not causing overfitting or underfitting. Ultimately, the choice of regularization technique depends on a combination of experimentation and validation, domain knowledge, and understanding of the specific problem and dataset at hand.

Next, let's dive into the design of an MLP.

Designing an MLP

Tabular data is not where neural networks shine most, and more often than not, boosted decision trees outperform MLPs in terms of metric performance. However, sometimes, in some datasets, neural networks can outperform boosted trees. Make sure to benchmark MLPs with other non-neural network models when dealing with tabular data.

MLPs are the simplest form of neural networks and can be modified at a high level in two dimensions similar to all neural networks, which are the width of the network and the depth of the network. A common strategy when building standard MLP architectures from scratch is to start small with a shallow depth and narrow width and gradually increase both dimensions once a small baseline is obtained. Usually, for MLPs on tabular data, the performance benefits of increasing the depth of the neural network stagnate at around the fourth layer. ReLU is a standard activation layer that is proven to allow stable gradients and optimal learning of any task. However, if you have time to achieve practical value, consider replacing the activation layer with more advanced activation layers. At this point, the space of activation layer research is just too nuanced and results are mostly not standardized, with mixed responses on different datasets, so there is no guarantee of better performance when you use any advanced activation layers.

One adaptation of MLPs is to use a type of neural network called **denoising autoencoders** to generate denoised features that can be used as input to MLPs. This advancement will be described in more detail later in *Chapter 5, Understanding Autoencoders*. Training methods go hand in hand with the architecture when trying to achieve good performance. The methods are mostly generic and don't depend on any architecture specifically, so they will be covered in *Chapter 8, Exploring Supervised Deep Learning*, and *Chapter 9, Exploring Unsupervised Deep Learning*, separately.

Next, let's summarize what we've learned from this chapter.

Summary

MLPs are the foundational piece of architecture in deep learning that transcends just processing tabular data and is more than an old architecture that got superseded. MLPs are very commonly utilized as a sub-component in many advanced neural network architectures today to either provide more automatic feature engineering, reduce the dimensionality of large features, or shape the features into the desired shapes for target predictions. Look out for MLPs or, more importantly, the fully connected layer, in the next few architectures that are going to be introduced in the next few chapters!

The automatic gradient computation provided by deep learning frameworks simplifies the implementation of backpropagation and allows us to focus on designing new neural networks. It is essential to ensure that the mathematical functions used in these networks are differentiable, although this is often taken care of when adopting successful research findings. And that's the beauty of open source research coupled with powerful deep learning frameworks!

Regularization is a crucial aspect of neural network design, and while we have discussed it in detail in this chapter, upcoming chapters will showcase its application in different architectures without delving into further explanations.

In the next chapter, we will dive into a different kind of neural network, called the convolutional neural network, which is particularly suited for image-related tasks and has a wide range of applications.

3

Understanding Convolutional Neural Networks

An MLP is structured to accept one-dimensional data and cannot directly work with two-dimensional data or higher-dimensional data without preprocessing. One-dimensional data is also called tabular data, which commonly includes categorical data, numerical data, and maybe text data. Two-dimensional data, or data with higher dimensions, is some form of image data. Image data can be in two-dimensional format when it is a grayscale formatted image, in three-dimensional format when it has RGB layers that closely represent what humans see, or in more than three dimensions with hyperspectral images. Usually, to make MLP work for images, you would have to flatten the data and effectively represent the same data in a one-dimensional format. Flattening the data might work well in some cases, but throwing away the spatial characteristics that define that image removes the potential of capturing that relationship to your target. Additionally, flattening the data doesn't scale properly to large images. An important characteristic of image data is that the target to be identified can be present in any spatial position of the image. Simple MLPs are highly dependent on the position of the data input and won't be able to adapt to the ever-changing positions and orientation of the target in an image.

This is where the **convolutional neural network (CNN)** layer shines and is often the go-to method for experts dealing with image data using machine learning. To date, top convolutional architectures always exceed the performance of MLP architectures for image datasets. In this chapter, we will cover the following topics while focusing on CNNs:

- Understanding the convolutional neural network layer
- Understanding the pooling layer
- Building a CNN architecture
- Designing a CNN architecture for practical usage
- Exploring the CNN architecture families

Technical requirements

This chapter includes some practical implementations in the **Python** programming language. To complete it, you will need to have a computer with the following libraries installed:

- `pandas`
- `matplotlib`
- `seaborn`
- `scikit-learn`
- `numpy`
- `keras`
- `pytorch`

The code files for this chapter are available on GitHub: `https://github.com/PacktPublishing/The-Deep-Learning-Architect-Handbook/tree/main/CHAPTER_2`.

Understanding the convolutional neural network layer

Now, let's focus on the foundations of the convolutional layer, starting with *Figure 3.1*, which shows the operational process of a convolutional filter. A filter is a small matrix of weights that's used to extract features or patterns from an input array. A convolutional filter is a type of filter that slides over an image, performing convolution operations to extract features by calculating dot products:

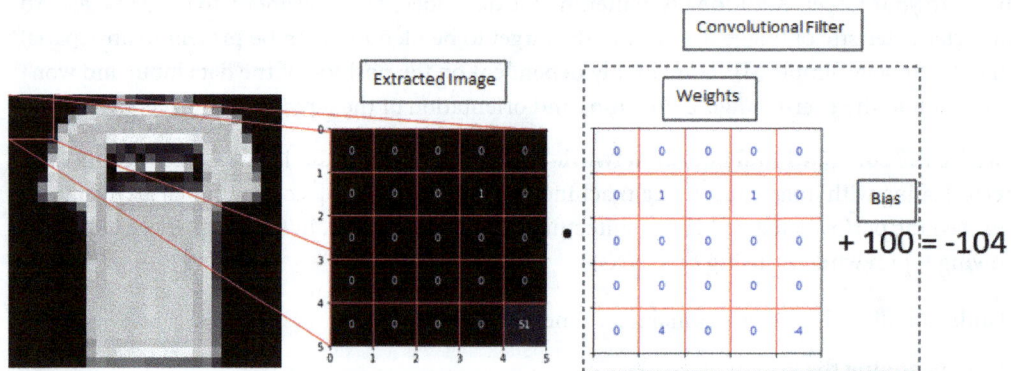

Figure 3.1 – Operation of a convolutional filter on an image of a t-shirt from the Fashion MNIST dataset

Convolutional layers are made out of multiple convolutional filters of the same sizes. Convolutional filters are the main pattern detectors in a CNN, where each filter will learn to identify multidimensional patterns that exist in an image. The patterns can range from low-level patterns such as lines and edges to mid-level patterns such as circles or squares and finally to high-level patterns such as specific t-shirt types or shoe types depending on the deepness level of the convolutional layer in a CNN. In *Chapter 12, Interpreting Neural Networks*, we will explore the patterns that are learned and evaluate the patterns qualitatively and objectively.

Convolutional filters can be in multiple dimensions but for plain images, two-dimensional convolution filters are more commonly used. Kernels are analogous to filters in terms of a CNN layer. These two-dimensional filters have the same number of weights as their size and apply a dot product on a part of the input image data with the same size to obtain a single value; subsequently, these are added with a bias term. By operating on the same operation using the filter in a sliding window manner systematically, from top to bottom and left to right with a defined stride, the convolutional filter will complete a forward pass and obtain a two-dimensional output with smaller dimensions. Here, stride means the number of pixel steps taken when sliding the filter systematically from left to right and top to bottom where the minimum has to be at least 1 pixel. This operation builds on the fact that the target can be present in any spatial position in an image and uses the same pattern identification method across the entire image.

Figure 3.1 shows a two-dimensional convolutional filter with the size of 5x5 pixels, where there are 25 weight components and one bias component that could be learned, and the input image of a t-shirt, which has a size of 28x28 pixels. The filter size is something that can be configured to other values that typically range from 1 to 7 and is commonly a square but can be set to be irregular rectangular shapes. The typical filter size values might seem too small of a **receptive field** to identify high-level features that are capable of predicting a shirt in this picture, but when multiple filters are applied one after another in a non-cyclical manner, the filters at the end of the operation have the receptive field of a larger part of the image. Receptive field refers to the region of the input space that a convolutional filter can "see" or respond to. It determines the spatial extent of the input that influences a particular output unit. To capture low-level patterns, the filter must have a small receptive field, and to capture high-level patterns, the filter must have a large receptive field. *Figure 3.2* depicts this concept for three filters applied one after another:

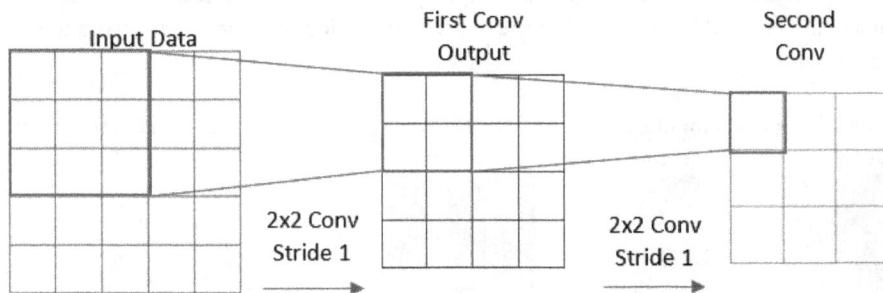

Figure 3.2 – The second convolutional filter has a receptive field size of 3x3 of the input data, even when it has only a 2x2 convolutional filter size

Now, the sliding window from left to right and top to bottom is just a good way to visualize the process. However, in reality, this process can be done in parallel in one go to take advantage of GPU parallel processing. *Figure 3.3* shows all the window positions of a convolutional filter with a 4x4-pixel dimension with a stride of 4 applied on the same 28x28 t-shirt image. This would result in a 7x7 data output size:

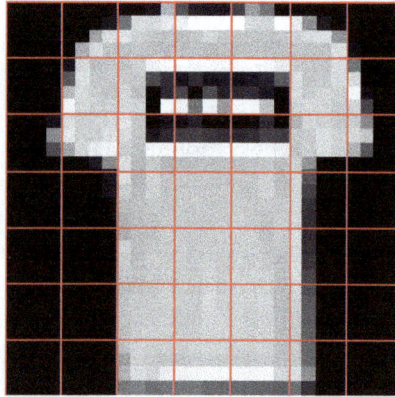

Figure 3.3 – All window positions using a 4x4 convolutional filter on a
28x28 image with a stride of 4 pixels in both dimensions

Consider a CNN layer with 16 filters with a size of 5x5 pixels with a stride of 1 pixel. Running a forward pass of the t-shirt image with this layer configuration will result in a total of 16x26x26 (depth x width x height) pixels of output. In the case where the image is an RGB image with three channels colored red, green, and blue, since the convolutional filters are two-dimensional, the same filter would be applied similarly to each of the three channels using a standard convolutional layer. The three-dimensional outputs from the filters, applied separately on the three channels, will then be added up. The number of convolutional filters will be the output data channel size and will serve as the input channel size for the subsequent convolutional layers. *Figure 3.4* depicts this channel-wise addition process for a 3x3 output from a convolutional filter. Note that the bias is only added once per filter across the channels and not by channel:

Figure 3.4 – Aggregation of a multichannel 3x3 output from a convolution filter

In the preceding figure, **Conv** is short for *convolutional layer* and will be a convention that will be used in the rest of this chapter to simplify figures. Since it is beneficial to be aware of the size of your neural network to ensure you have the computational resources to hold and process the network, let's also compute the number of parameters that will be held by this convolutional layer. The number of parameters can be computed as follows:

number of input channels x number of filters x ((width of filter x height of filter) + 1)

By putting the respective numbers into the equation, you will obtain 416 parameters. If these weights are stored as a **floating-point 32 (fp32)** format, in bytes, this would mean 416x32 bits/8=1,664 bytes. With an output data size of 16x26x26, the data size in spatial dimension is decreasing at a very slow rate and since the end goal will be to reduce these values to the size of the target data, we need something to reduce the size of the data. This is where another layer, called the **pooling layer**, comes into play.

Understanding the pooling layer

With just a forward pass from a CNN layer of an image, the size of the two-dimensional output data is likely reduced but is still a substantial size. To reduce the size of the data further, a layer type called a pooling layer is used to aggregate and consolidate the values strategically while still maintaining useful information. Think of this operation as an image-resizing method while maintaining as much information as possible. This layer has no parameters for learning and is mainly added to simply and meaningfully reduce the output data. The pooling layer works by applying a similar sliding window filter process with similar configurations as the convolutional layers but instead of applying a dot product and adding a bias, a type of aggregation is done. The aggregation function can be either maximum aggregation, minimum aggregation, or average aggregation. The layers that apply these aggregations are called max pooling, min pooling, and average pooling, respectively.

Consider an average pooling layer with a filter size of 4 and a stride of 2 applied after the first CNN layer. A forward pass of the 16x26x26 output will result in a data size of 16x12x12. That reduces the size considerably!

Another type of pooling layer applies the aggregation function globally. This means that the entire two-dimensional width and height component of the data will be aggregated into a single value. This variation of the pooling layer is commonly known as the global pooling layer. This layer is applied to completely break down the data into a one-dimensional structure so that it can be compatible with one-dimensional targets. This layer is directly available in the `keras` library but only available in `pytorch` indirectly through setting the pooling filter size to the same as the size of the input feature map.

Building a CNN architecture

CNN architectures are commonly made by stacking multiple conceptual logical blocks of layers one after another. These logical blocks are all structured the same way, with the same type of layer and layer connections, but they can be different in terms of their parameter configurations, such as the size of the filters, the stride, the type of padding used, and the amount of padding used. The simplest logical convolutional block is a convolutional layer, pooling layer, and activation function, in that order. **Padding** is a term that's used to refer to any extra pixels that are added around the input image to preserve its spatial dimensions after convolution. Logical blocks are a way for you to describe and reference the architecture simply and efficiently. They also allow you to build CNN architectures in a depth-wise scalable way without the need to create and set the settings of each layer one by one. Depth is the same as deepness and refers to the number of neural network layers.

The parameters can be designed depending on whether the goal is to gradually scale down the feature maps or to scale up the feature maps. For the case of one-dimensional targets, the goal is likely to slowly scale down the features into one-dimensional features so that the features can be passed into fully connected layers. These fully connected layers can then further map the (still large) dimensions to the dimensions that are suitable for the targets. You can see a simple design of such an architecture in *Figure 3.5*:

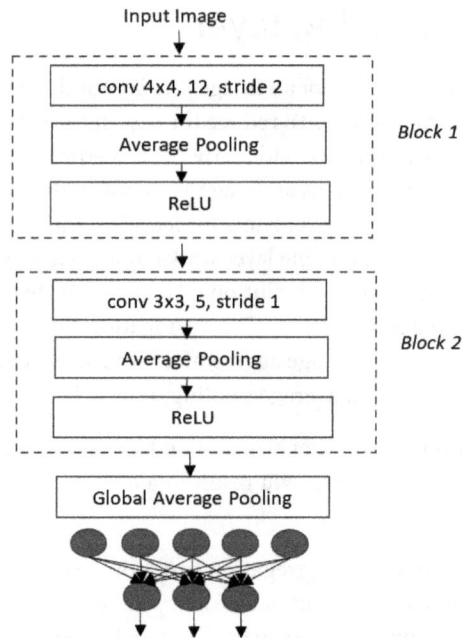

Figure 3.5 – A simple CNN architecture from scratch while following the logical block analogy

In the code for `pytorch`, this example would look like this:

```python
class ConvArch(nn.Module):
  def __init__(self):
      super(ConvArch, self).__init__()
      self.conv1 = nn.Conv2d(
          in_channels=1, out_channels=12, kernel_size=4,
```

```
            stride=2, padding=0
    )
    self.conv2 = nn.Conv2d(
            in_channels=12, out_channels=5, kernel_size=3,
            stride=1, padding=0
    )
    self.fc2 = nn.Linear(5, 3)
def forward(self, x):
    x = F.relu(
            F.avg_pool2d(
                    self.conv1(x), kernel_size=3, stride=2
            )
    )
    x = F.relu(
            F.avg_pool2d(
                    self.conv2(x), kernel_size=3, stride=1
            )
    )
    x = F.avg_pool2d(x, kernel_size=x.size()[2:])
    x = self.fc2(x.reshape((x.size()[:2])))
    return x
```

Again, the backpropagation of a CNN architecture will be handled automatically by the deep learning libraries.

The architecture we built was based on the basic classification problem type, which justifies the need to have a fully connected network at the end of the network; this is commonly known as the head. The set of logical blocks with convolutional layers that were used is called the backbone of the network. The head of the network can be switched out with other structures, depending on the problem type, but the backbone can be completely adapted into most architectures for any problem type, such as object detection, image generation, image captioning, or image recognition by representation learning. Some of these problem types will be discussed in *Chapter 8, Exploring Supervised Deep Learning*.

Now that we have successfully made a simple CNN manually, we have grounded and synced our theories toward the core algorithm of convolutional networks and will be ready to have more advanced CNN backbone architecture designs at our fingertips. But before that, this begs the question, how do we design a CNN for our use case?

Designing a CNN architecture for practical usage

For real-world use cases, CNNs should not be designed similarly to how an MLP is designed. Practical usage means that the goal is not to research a new innovative architecture for an unexplored problem type. Many advancements have been made today based on CNNs. Advancements usually come in one of two flavors:

- It sets a new baseline that completely redesigned the way CNN architectures are made
- It's built on top of existing baseline CNN architectures while complementing and improving the performance of the baseline architecture

The key difference between the ideal design approach of a CNN compared to an MLP is that the structures of published CNN architectures should be used instead of designing the architecture from scratch. The structures of CNN architectures define the type of layers and the way the different types of layers connect; they are usually implemented using logical blocks. Additionally, the uniqueness of a certain structure defines the family of CNN architecture. A new CNN research advancement usually comes in different size configurations so that architectures of suitable sizes can be chosen based on either the compute resource limitations, runtime requirements, or the dataset and problem complexity. Similar to the design of MLP, if resources and runtime are not a limitation, after choosing a CNN architecture structure, start with a reasonably small-sized CNN based on the complexity of the problem and the dataset's size. You can gradually test bigger CNNs to see if the performance improves when you increase the size, or vice versa if the performance decreases.

Different CNN architecture structures or architecture families are usually meant to capture different inherent architectural issues of the network. Some architectural families are designed in a way that they leverage better hardware resources, without which it wouldn't have been even possible to execute the architecture. A good practice to achieve good performance is to diversify the type of architectural structures you are using in the initial stage. Pick size variants of architectural structures with similar sizes in terms of the floating-point operation per second and run your experiments to obtain performance scores, ideally with a small size to maximize the efficiency of the exploration. Instead of the number of parameters of the model, it is more relevant to consider the floating-point operation per second as an indicator of the complexity of the model; parameter count doesn't consider the actual runtime of the model, which could benefit from parallelization. Once you've obtained these numbers, pick the top model families and try bigger size variants to benchmark with to find the best model variant for your use case.

Most research improvements on CNNs are based on a simple baseline architecture. This means that all the other individual improvements that are made on the same baseline architecture are not benchmarked together. Testing these improvements together can often be complementary but sometimes, it can be detrimental to the metric performance of the model. Iteratively benchmarking the different configurations will likely be the most systematic and grounded way to obtain a satisfactory performance improvement on your model.

How do researchers grade their improvements? To design a CNN architecture for practical usage, knowing how to evaluate your architecture will help you slowly work toward an acceptable metric performance. In *Chapter 10, Exploring Model Evaluation Methods*, we will discuss the strategies for evaluation more extensively. One of the main evaluation methods for improvements that's used for

CNN improvements is the top-1 predicted accuracy performance on a massive publicly available image dataset called `ImageNet`, which consists of millions of images with many classes. `ImageNet` is considered to be a highly complex problem-type use case, where each class has an infinite amount of variations possible in the wild, from indoors to outdoors and from real to synthetic data.

So, how do we consider whether improvements are valuable? Improvements are made to either improve performance in the top-1 accuracy, based on the `ImageNet` dataset, to improve the efficiency of running a forward pass of the model, or to specifically improve the training time of the network.

Ranking architectures by their top-1 accuracy metric performance improvements on `ImageNet` alone, however, is a biased evaluation as, more often than not, the absolute ranking of models differs when applying the same architectures on a separate image dataset. Using it as a starting point to choose ready-made architectures is wise, but make sure you evaluate a few of the top-performing `ImageNet` architectures to figure out which works the best. Furthermore, while `ImageNet` was curated with manual effort, involving querying search engines and passing candidate images through a validation step on Amazon Mechanical Turk, it still contains some label noise that can obfuscate the meaning behind the metric's performance.

As for the improvement direction of increasing the efficiency of running a forward pass of the model, this is usually done either by reducing the number of parameters of the architecture, breaking down a compute-intensive logical component into multiple components that reduces the amount of operations, or switching out layers for layers that have higher parallelism potential. The improvements that are made in this direction either maintain or improve the metric performance score on the validation dataset of `ImageNet`. This line of improvement is a main focus for the family of CNN architectures that have been built to run in low-resource edge devices. We will dive into this later.

An efficient way to explore different CNN architecture families in your use case to achieve better metric performance is to pick model families that have a publicly available implementation complete with pre-trained weights trained on `ImageNet`. Initializing your architecture using pre-trained weights presents a bunch of benefits that include faster training, less overfitting, and increased generalization, even when the dataset that was used to pre-train the weights is part of a different problem subset. This process is called transfer learning and we will learn about it in more detail in *Chapter 8, Exploring Supervised Deep Learning*, and *Chapter 9, Exploring Unsupervised Deep Learning*.

Exploring the CNN architecture families

Now, instead of going through the history of CNN through the years, let's look at a list of the different handpicked model architecture families. These architecture families are selectively chosen to be sufficiently different and diverse from each other. One thing to note is that neural networks are advancing at an astounding pace. With that in mind, the architecture families that will be introduced here are ensured to be relevant today. Additionally, only the most important information you need to know about the architecture family will be presented, simplifying the many pages of research papers in concise but sufficient detail.

Another thing to note before diving into this topic is that the metric performance on a dataset will often be the main comparison method among different architectures, so be aware that the metric performance of a model is achieved by the collective contribution of the training method and the architecture. The training method includes details not specifically related to the architecture of the model, such as the loss used, data augmentation strategy, and data resolution. These topics will be covered in *Chapter 8, Exploring Supervised Deep Learning*. The architecture families we'll introduce here are **ResNet**, **DenseNet**, **MobileNet**, **EfficientNet**, **ShuffleNet**, and **MicroNet**.

Understanding the ResNet model family

The ResNet architecture, from 2015, was made based on the pretext that deep networks are hard to train and aimed to change that. The vanishing gradient is a widely known problem that neural networks face, where the information from the data diminishes the deeper the network is. Plain deep CNN architectures, however, were identified to *not* suffer from vanishing gradients, with proof from verifying gradient information. The key cause of vanishing gradients is when we use too many activation functions that squish the data into very small value ranges. An example of this is the sigmoid function, which maps data from 0 to 1. So, when in doubt, use ReLU in your architecture!

The *Res* part of ResNet is named after the term **residuals**. The idea was that learning from residuals is much easier than learning from unmodified feature maps. Residuals are achieved by adding skip connections from the earlier layers to later layers. In layman's terms, that means adding (not concatenating) feature maps from earlier parts of the network to feature maps of later parts of the network. The result of this addition creates the residuals that will be learned by subsequent convolutional layers, which again apply more skip connections and create more residuals. Residuals can be easily applied to any CNN architecture with different configurations and can be considered an improvement on top of older baseline architectures. However, the authors also presented multiple variations of an architecture that utilized skip connections with a different number of convolutional layers, including ResNet-18, ResNet-34, ResNet-50, ResNet-101, and ResNet-152. The ResNet architecture family serves as a boilerplate for easy usage of residual networks and eventually led to it becoming the most popular baseline for research on new advancements. The actual architecture designs are not presented here formally as memorizing those designs won't have any impact on your grasp of CNN knowledge. Instead, *Figure 3.6* shows the residual computation of a single logical block:

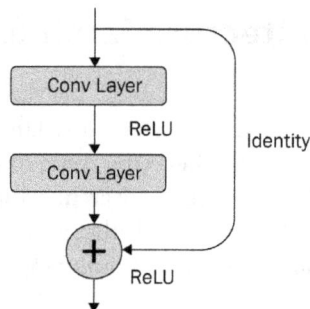

Figure 3.6 – Example of the actual residual connection method in the ResNet model family

To summarize the entire baseline ResNet size variants, the following table shows a summary configuration of all the different size variants:

layer name	output size	18-layer	34-layer	50-layer	101-layer	152-layer
conv 1	112 × 112	7 × 7, 64, stride 2				
		3 × 3 max pool, stride 2				
conv 2 x	56 × 56	$\begin{bmatrix} 3 \times 3,\ 64 \\ 3 \times 3,\ 64 \end{bmatrix} \times 2$	$\begin{bmatrix} 3 \times 3,\ 64 \\ 3 \times 3,\ 64 \end{bmatrix} \times 3$	$\begin{bmatrix} 1 \times 1,\ 64 \\ 3 \times 3,\ 64 \\ 1 \times 1,\ 256 \end{bmatrix} \times 3$	$\begin{bmatrix} 1 \times 1,\ 64 \\ 3 \times 3,\ 64 \\ 1 \times 1,\ 256 \end{bmatrix} \times 3$	$\begin{bmatrix} 1 \times 1,\ 64 \\ 3 \times 3,\ 64 \\ 1 \times 1,\ 256 \end{bmatrix} \times 3$
conv 3 x	28 × 28	$\begin{bmatrix} 3 \times 3,\ 128 \\ 3 \times 3,\ 128 \end{bmatrix} \times 2$	$\begin{bmatrix} 3 \times 3,\ 128 \\ 3 \times 3,\ 128 \end{bmatrix} \times 4$	$\begin{bmatrix} 1 \times 1,\ 128 \\ 3 \times 3,\ 128 \\ 1 \times 1,\ 512 \end{bmatrix} \times 4$	$\begin{bmatrix} 1 \times 1,\ 128 \\ 3 \times 3,\ 128 \\ 1 \times 1,\ 512 \end{bmatrix} \times 4$	$\begin{bmatrix} 1 \times 1,\ 128 \\ 3 \times 3,\ 128 \\ 1 \times 1,\ 512 \end{bmatrix} \times 4$
conv 4 x	14 × 14	$\begin{bmatrix} 3 \times 3,\ 256 \\ 3 \times 3,\ 256 \end{bmatrix} \times 2$	$\begin{bmatrix} 3 \times 3,\ 256 \\ 3 \times 3,\ 256 \end{bmatrix} \times 6$	$\begin{bmatrix} 1 \times 1,\ 256 \\ 3 \times 3,\ 256 \\ 1 \times 1,\ 1024 \end{bmatrix} \times 6$	$\begin{bmatrix} 1 \times 1,\ 256 \\ 3 \times 3,\ 256 \\ 1 \times 1,\ 1024 \end{bmatrix} \times 23$	$\begin{bmatrix} 1 \times 1,\ 256 \\ 3 \times 3,\ 256 \\ 1 \times 1,\ 1024 \end{bmatrix} \times 36$
conv 5 x	7 × 7	$\begin{bmatrix} 3 \times 3,\ 512 \\ 3 \times 3,\ 512 \end{bmatrix} \times 2$	$\begin{bmatrix} 3 \times 3,\ 512 \\ 3 \times 3,\ 512 \end{bmatrix} \times 3$	$\begin{bmatrix} 1 \times 1,\ 512 \\ 3 \times 3,\ 512 \\ 1 \times 1,\ 2048 \end{bmatrix} \times 3$	$\begin{bmatrix} 1 \times 1,\ 512 \\ 3 \times 3,\ 512 \\ 1 \times 1,\ 2048 \end{bmatrix} \times 3$	$\begin{bmatrix} 1 \times 1,\ 512 \\ 3 \times 3,\ 512 \\ 1 \times 1,\ 2048 \end{bmatrix} \times 3$
	1 × 1	average pool, 1000-d fc, softmax				
FLOPs		1.8×10^9	3.6×10^9	3.8×10^9	7.6×10^9	11.3×10^9

Figure 3.7 – Base ResNet different size variants

The brackets denoted in the table are meant to signify the set of serial convolutional layers that are defined as the logical blocks. ResNet has also gone the extra mile to group multiple base logical blocks to a higher-level category they call "layer name." The layer name has the same number of groups across different size variants. Logical blocks and higher-level grouping methods are just a way for you to describe and reference the complicated architecture simply and efficiently. The first two numbers, which with multiplication in between in the logical block, define the convolution filter size; the number that comes after it with a comma defines the number of filters.

Skip connections were found to smoothen the loss landscape of any neural network, making the learning process much easier and stabler. This allows for the neural network to converge to the optimum more easily. *Figure 3.8* shows how the loss landscape for a neural network without skip connections has an uneven terrain with many hills and valleys on the left. It becomes a smooth terrain with an obvious single valley when skip connections are added using a variant of ResNet:

Figure 3.8 – Loss landscape with no skip connections on the left and with skip connections on the right

One notable piece of information you can derive from the table is that, starting from ResNet-50 onwards, the architecture utilizes 1x1 convolutional filters. As this filter operates exclusively with a single-dimensional filter weight across the channel dimension, the operation is equivalent to applying a fully connected layer in a windowing fashion in the channel dimension. Since the fully connected layer is a network itself, along with the convolutional network, this operation is often called a **network in a network**. Next, let's explore the different improvements with the ResNet architecture as a base.

Improving ResNets

As mentioned previously, ResNet is considered a model family and contains many different varieties, not just by size but with different architectural improvements. Any CNN architecture that is based on the ResNet belongs to this model family. Here, we have mentioned some of the few notable architectural advancements and provided descriptions of their main advancements based on ResNet; they are ordered by the year they were made:

- **ResNeXt** (2016): This converted serial convolutional layers in the logical blocks of ResNets into parallel branches of convolutional layers and merged the parallel layers at the end of the logical block. It increases the metric performance on ImageNet compared to ResNet variants, even when the number of parameters is the same and also in general.

- **Squeeze-and-Excitation Networks (SE-ResNet)** (2017): Since two-dimensional convolutions don't consider the inter-channel relationships and only consider the local and spatial (width and height) information of the feature maps, a method to leverage the channel relationships was made that includes a squeeze block and an excitation block. This is a method that can be repeatedly applied to many parts of an existing CNN architecture. *Figure 3.9* shows the structure of this method, where a global average pooling is applied in the width and height dimensions and, subsequently, two fully connected networks are applied to scale down and scale up the features back to the same size to allow channel information to be combined:

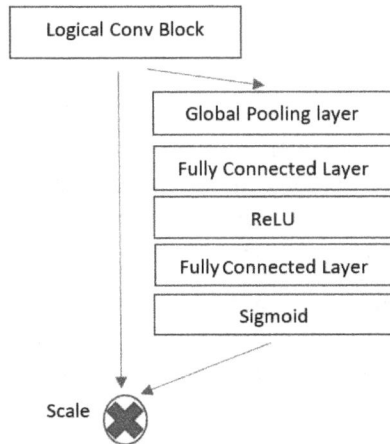

Figure 3.9 – Squeeze and excitation structure

The scaling part of the structure is where the values are multiplied together. When combined with ResNet, the architecture is called SE-ResNet.

- **ResNet-D** (2019): This advancement made simple architecture parameter tweaks to improve metric performance while maintaining the number of parameters, albeit at the slightly justifiable increase of the FLOP specification. As some path of ResNet utilizes 1x1 convolutions with a stride of 2, ¾ of the information is discarded and one of the tweaks was to the stride sizes to ensure no information gets removed explicitly. They also reduced computational load by changing a 7x7 convolution to three separate 3x3 convolutions serially.

- **ResNet-RS** (2021): This advancement combines ResNet-D and the squeeze and excitation network and uses an image size that depends on the size of the network. However, it grows slower than EfficientNets (these will be introduced later).

Understanding the DenseNet architecture family

DenseNet is an architecture family that was introduced in the early months of 2018. The architecture is based on the idea of skip connections, similar to the ResNet architecture family but on steroids, which means it uses a lot of skip connections. The skip connections in this family of architecture differ in that they use **concatenation** instead of residual connections by **summation**. Summation allows earlier information to be directly encoded into the outputs of future layers without the need to modify the number of neurons, albeit at the slight informational disadvantage of needing the future layers to learn to decode this information. Concatenation adds to the size of the architecture as you need to create extra neurons to account for the extra information, allowing the model to work on the raw data. Both provide similar advantages to using skip connections. Logical blocks are created called **dense blocks**, where each subsequent layer in the block has access to all the outputs of the layers before it in the block by feature map concatenation. In these blocks, **zero-padding** is used to ensure that the spatial dimensions of the outputs of each layer are maintained so that feature maps can be concatenated. This setup promotes a lot of feature reuse between layers in the same block and allows the number of model parameters to stay the same while increasing the model's learning capacity. The number of filters for each subsequent layer in a block for all blocks is fixed at a constant number called the **growth rate**, as there needs to be a structured way to add layers to not exponentially increase the number of channels. Taking a constant of 32 filters, the input of the second layer in the block will be a feature map with 32 channels, the input of the third layer in the block will be 64 and concatenated, and so on.

To create a full network architecture, multiple dense blocks were stacked one after another with a separate convolutional layer and pooling layer in between the dense blocks to gradually reduce the spatial dimensions of the feature map. *Figure 3.10* shows the network structure of the four different DenseNet model architectures under the DenseNet model family:

Layers	Output Size	DenseNet-121	DenseNet-169	DenseNet-201	DenseNet-264
Convolution	112 x 112	7 x 7 conv, stride 2			
Pooling	56 x 56	3 x 3 max pool, stride 2			
Dense Block (1)	56 x 56	[1 x 1 conv / 3 x 3 conv] x 6	[1 x 1 conv / 3 x 3 conv] x 6	[1 x 1 conv / 3 x 3 conv] x 6	[1 x 1 conv / 3 x 3 conv] x 6
Transition Layer (1)	56 x 56	1 x 1 conv			
	28 x 28	2 x 2 average pool, stride 2			
Dense Block (2)	28 x 28	[1 x 1 conv / 3 x 3 conv] x 12	[1 x 1 conv / 3 x 3 conv] x 12	[1 x 1 conv / 3 x 3 conv] x 12	[1 x 1 conv / 3 x 3 conv] x 12
Transition Layer (2)	28 x 28	1 x 1 conv			
	14 x 14	2 x 2 average pool, stride 2			
Dense Block (3)	14 x 14	[1 x 1 conv / 3 x 3 conv] x 24	[1 x 1 conv / 3 x 3 conv] x 24	[1 x 1 conv / 3 x 3 conv] x 24	[1 x 1 conv / 3 x 3 conv] x 24
Transition Layer (3)	14 x 14	1 x 1 conv			
	7 x 7	2 x 2 average pool, stride 2			
Dense Block (4)	7 x 7	[1 x 1 conv / 3 x 3 conv] x 16	[1 x 1 conv / 3 x 3 conv] x 32	[1 x 1 conv / 3 x 3 conv] x 32	[1 x 1 conv / 3 x 3 conv] x 48
Classification Layer	1 x 1	7 x 7 global average pool			
		1000D fully-connected, softmax			

Figure 3.10 – The DenseNet model family where "conv" corresponds to sequential layers of batch normalization, ReLU, and the convolutional layer

This allows DenseNet to improve upon its predecessor network architectures in terms of top-1 ImageNet accuracy.

Understanding the EfficientNet architecture family

Made in 2020, EfficientNet created a family of architectures by using an automated **neural architecture search (NAS)** method to create a small base efficient architecture and utilize an easy-to-use compound scaling method to scale the depth and width of the architecture, as well as the resolution of the image. The neural architecture search is searched in a way that it balances FLOPS and accuracy. The NAS method that's used is from another research called **MnasNet** and will be introduced properly in *Chapter 7, Deep Neural Architecture Search*.

The compound scaling method is simple and can be extended to any other network, though ResNet-RS demonstrates that scaling resolution slower provides more value. The scaling method for depth, width, and resolution is defined in the following equation, where the result will be multiplied by the original base architecture parameters to scale up the architecture:

$$depth = \alpha \, \varphi \, , width = \beta \, \varphi, resolution = \gamma \, \varphi$$

Here, φ is the coefficient that can be scaled to different values based on requirements and the other variables are constants that should be set to optimize something such as the FLOPS increase rate when we change the coefficient. For EfficientNet, the constants are constrained to satisfy the following condition:

$$\alpha \cdot \beta 2 \cdot \gamma 2 \approx 2$$

This will constrain the increase of the coefficient to approximately increase the FLOPS by 2 φ. EfficientNet sets this to $\alpha = 1.2, \beta = 1.1, \gamma = 1.15$. This compound scaling strategy allowed seven EfficientNets to be made, named B0 to B7. *Figure 3.11* shows the structure of the EfficientNet-B0:

State i	Operator $\hat{F}_i$	Resolution $\hat{H}_i \times \hat{W}_i$	#Channels $\hat{C}_i$	#Layers $\hat{L}_i$
1	Conv 3×3	224 ×224	32	1
2	MBConv1, k3×3	112 × 112	16	1
3	MBConv6, k3×3	112 × 112	24	2
4	MBConv6, k5×5	56 × 56	40	2
5	MBConv6, k3×3	28 × 28	80	3
6	MBConv6, k5×5	14 × 14	112	3
7	MBConv6, k5×5	14 × 14	192	4
8	MBConv6, k3×3	7 × 7	320	1
9	Conv1×1 & Pooling & FC	7 × 7	1280	1

Figure 3.11 – EfficientNet-B0 architecture structure

Note that **MBConv** is also known as an inverted residual block and will be properly introduced in another network later in the *Understanding MobileNetV2* section.

EfficientNetV2, made in 2021, identified that large image resolution slows down training time, where depthwise convolutions are slow in early layers, and scaling up depth, width, and resolution at the same time is not optimal. EfficientNetV2 also uses NAS to find a base architecture but with the addition of a modification of an MBConv block that has more parameters and operations with the reason that it can be faster sometimes, depending on the input and output data shape, where they are in the entire architecture, and how the data is transferred to the computing processer. This will be explained in more detail when we formally introduce MBConv later. EfficientNetV2 also utilizes the original compound scaling method but adds a few improvements by setting a maximum image resolution to 480 and adding extra layers to the last few stages of the architecture when it wants to increase the network capacity to higher proportions. A new training method was added to reduce training time; we'll discuss this in more detail in the next chapter. These improvements resulted in four different EfficientNetV2 models called **EfficientNetV2-S**, **EfficientNetV2-M**, **EfficientNetV2-L**, and **EfficientNetV2-XL**. These exceeded the top-1 accuracy compared to the original EfficientNet at similar FLOP values:

State	Operator	Stride	#Channels	#Layers
0	Conv 3×3	2	24	1
1	Fused-MBConv1, k3×3	1	24	2
2	Fused-MBConv4, k3×3	2	48	4
3	Fused-MBConv4, k3×3	2	64	4
4	MBConv4, k3×3, SE0.25	2	128	6
5	MBConv6, k3×3, SE0.25	1	160	9
6	MBConv6, k3×3, SE0.25	2	256	15
7	Conv1×1 & Pooling & FC	-	1280	1

Figure 3.12 – EfficientNetV2-S architecture structure

EfficientNetV2-S served as the base architecture, similar to how EfficientNetB0 was the base, where the architecture structure is shown in *Figure 3.12*.

Understanding small and fast CNN architecture families for small-scale edge devices

One of the clear groups of architectures that holds a place in the world of CNN architectures is the group that is built not for scalability purposes or beating the ImageNet benchmark, but to be used for small devices. Small devices are usually called **edge devices** as they are small and compact enough to be mobile or to be physically deployed with actual data processing capabilities where the data originated. Our mobile smartphones are examples of mobile devices that are capable of producing images. Examples of edge devices that *aren't* mobile include CCTV video cameras and doorbell cameras. Some of the key benefits of deploying models to the edge are listed here:

- **Reduced communication latency**: Images and real-time video feeds are large compared to simple numerical or categorical data. By using computation at the edge, less data needs to be transferred to a centralized server, thus reducing the time needed to transfer data. Sometimes, the need for a transfer can be eliminated when computation is done at the edge, thus significantly simplifying a system.

- **Reduced bandwidth requirements**: When the images are processed where they are produced, only simple data formats such as numerical or categorical will need to be returned, thus removing the need to have expensive equipment for high bandwidth requirements.

- **Increased redundancy**: A centralized server also means a single point of failure. Distributing processing to the individual edge devices makes sure the failure of any one device won't affect the entire system.

Small CNN architectures meant for edge devices typically engineer the entire structure of a model without a single notable baseline, so there isn't a proper model family to categorize each of these architectures. For ease of referencing architectures meant for this purpose, it is recommended to think of the following architectures as under *architecture for the edge*. There are other techniques that can be applied architecture-wide to further optimize the efficiency of the model that we will explore in *Chapter 15, Deploying Deep Learning Models in Production*, but for now let's explore two of these architectures, namely **SqueezeNet** and **MobileNet**.

Understanding SqueezeNet

SqueezeNet was developed in 2016 to build small and fast CNNs with the benefits described in the previous section but with an emphasis on deploying to hardware with limited memory. The three strategies that can be employed are as follows:

- For a convolutional layer with 3x3 filters, replace the filters partially with 1x1 filters. This means that the 1x1 filter and 3x3 filters co-exist theoretically in a single layer applied to the same input. However, the implementation details will differ as parallel branch paths for the 1x1 filters and 3x3 filters are applied on the same input. This is called the expand layer.

- Decrease the data input channels that are passed into the parallel 1x1 and 3x3 filters by using a small number of 1x1 filters. This is called the squeeze layer.

- Downsample the feature maps late in the architecture so that the majority of the convolutional layers have access to large feature maps based on findings and improve metric performance. This is done by using pooling layers with strides of 2 pixels less frequently in earlier stages and more frequently in later stages and larger intervals. Pooling layers are not used after every convolutional layer.

A logical block called `Fire` was made that provides the ease of creating multiple blocks to create architectures. This block enabled the configuration of the number of 1x1-sized filters in the squeeze layer, 1x1 filters in the expand layer, and 3x3 filters in the expand layer. This is depicted in *Figure 3.13*. Note that padding is applied to the outputs of the 3x3 convolutional layer to ensure it can be concatenated with the outputs from the 1x1 convolutional layer in the expand layer:

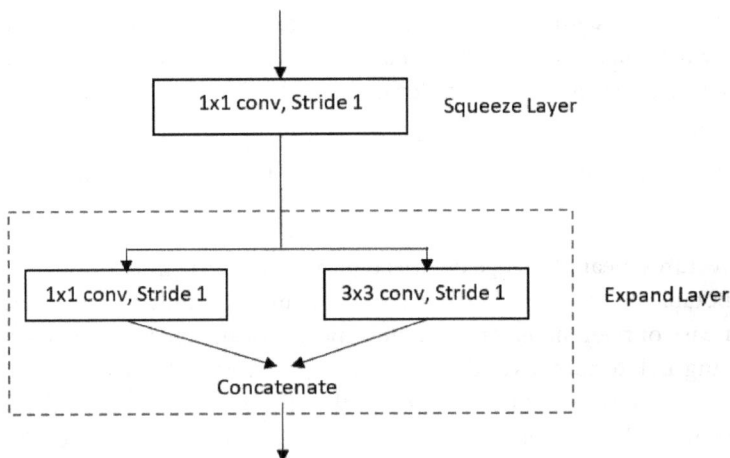

Figure 3.13 – The Fire module/logical block

Eight `Fire` blocks were used to build an architecture called SqueezeNet that maintained the performance of `ImageNet` from the historical architecture known as **AlexNet** (not introduced in this book as it is not practically useful nowadays) while being 50x smaller.

Understanding MobileNet

The first MobileNet version, called MobileNetV1, was introduced in 2017 for mobile and embedded devices by focusing on optimizing latency issues and getting small-sized networks as a side effect instead of vice versa.

The depthwise separable convolutional layer was used extensively in the entire architecture except in the first layer in the first iteration of MobileNet, which consisted of 28 layers. This layer is built on factorizing the standard convolutional layer into two layers called the depthwise convolution and the pointwise convolution. This new two-layer setup is based on the idea that using a standard convolutional filter is expensive to compute. The depthwise convolutional layer uses one unique filter for one unique input channel, where each input channel has only one filter. Having one filter per channel is a unique case of **grouped convolutions**, which will be introduced more extensively later. The output from the depthwise convolutional layer is subsequently passed into a standard 1x1 convolution (that they call pointwise convolution) that combines the information between channels. Depthwise separable convolutional layers as a whole reduce the computation runtime by eight to nine times compared to the standard convolutional layer operation we introduced earlier while only reducing the accuracy on `ImageNet` by 1%. This breakdown also reduces the number of parameters by around 5 times. This process is essentially a kind of factorization. The depthwise convolutional logical block is depicted in the following figure:

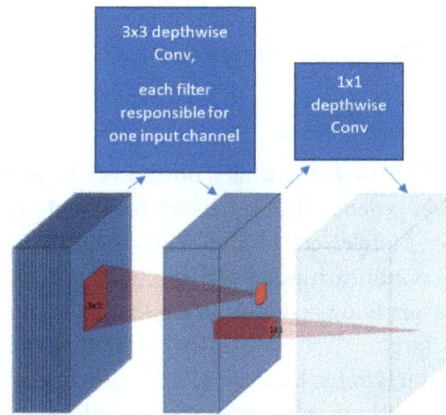

Figure 3.14 – Depthwise separable convolutional layer as a logical block

MobileNet also made two parameters that can be configured to reduce the model's size and computational requirements while trading off some metric performance. The first is a width multiplier that can configure the input and output channels across the entire architecture. The second is a resolution multiplier between 0 and 1 that applies to the original 224x224 image size to reduce the input and output feature map sizes when passing the image into the network. These parameters can also be adapted to other architectures if a faster runtime is needed. *Figure 3.15* shows the MobileNetV1 architecture's structure and layer configurations:

MobileNet Body Architecture

Type / Stride	Filter Shape	Input Size
Conv / s2	3 × 3 × 3 × 32	224 × 224 × 3
Conv dw / s1	3 × 3 × 32 dw	112 × 112 × 32
Conv s1	1 × 1 × 32 × 64	112 × 112 × 32
Conv dw / s2	3 × 3 × 64 dw	112 × 112 × 64
Conv / s1	1 × 1 × 64 × 128	56 × 56 × 64
Conv dw / s1	3 × 3 × 128 dw	56 × 56 × 128
Conv s1	1 × 1 × 128 × 128	56 × 56 × 128
Conv dw / s2	3 × 3 × 128 dw	56 × 56 × 128
Conv s1	1 × 1 × 128 × 256	28 × 28 × 128
Conv dw / s1	3 × 3 × 256 dw	28 × 28 × 256
Conv s1	1 × 1 × 256 × 256	28 × 28 × 256
Conv dw / s2	3 × 3 × 256 dw	28 × 28 × 256
Conv s1	1 × 1 × 256 × 512	14 × 14 × 256
Conv dw / s1	3 × 3 × 512 dw	14 × 14 × 512
Conv s1	1 × 1 × 512 × 512	14 × 14 × 512
Conv dw / s2	3 × 3 × 512 dw	14 × 14 × 512
Conv / s1	1 × 1 × 512 × 1024	7 × 7 × 512
Conv dw / s2	3 × 3 × 1024 dw	7 × 7 × 1024
Conv / s1	1 × 1 × 1024 × 1024	7 × 7 × 1024
Avg Pool / s1	Pool 7 × 7	7 × 7 × 1024
FC/ s1	1024 × 1000	1 × 1 × 1024
Softmax / s1	Classifier	1 × 1 × 1000

Figure 3.15 – MobileNetV1 architecture

MobileNet is considered to be a unique model family with two improvements that are based on the first version. They are named MobileNetv2 and MobileNetv3-small and both were introduced in 2019.

Understanding MobileNetV2

Before we go through MobileNetV2, let's define what a **bottleneck layer** is, a core idea that's utilized in advancement. A bottleneck layer is generally a layer with fewer output feature maps compared to the layers before and after the layer. MobileNetV2 is built upon the idea that bottlenecks are where the information of interest will exist; nonlinearities destroy too much information in bottleneck layers, so a linear layer is applied, finally applying residuals using shortcut connections on the bottleneck layers. MobileNetV2 builds upon the base depthwise separable building block, adds a linear bottleneck layer without ReLU, and adds residuals to the bottleneck layers. This building block is depicted in the following figure. This is called the **bottleneck inverted residual block**:

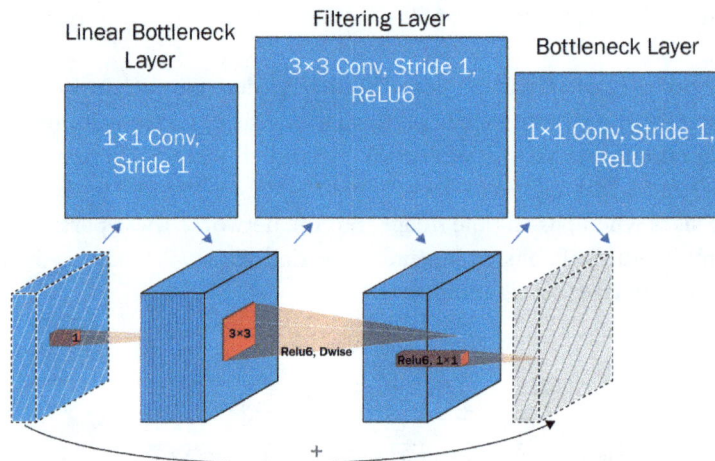

Figure 3.16 – The bottleneck inverted residual block for MobileNetV2, also called MBConv

As for the entire network architecture-wise, all depthwise separable layers from the first MobileNet were replaced with the new block except for the first convolution layer with a 3x3 kernel size and 32 filters, as shown in *Figure 3.16*. One small extra detail is that they used the **ReLU6** activation function, which is robust to low-precision computation. The MobileNetV2 architecture used the depicted logical block to create many repeated layer blocks with different settings. This architecture allowed for an improvement in the performance curve on ImageNet compared to MobileNetV2 at around 5% to 10% at the same FLOPs. Remember that EfficientNetV2 uses this block and also another version of this block that fuses back the linear bottleneck layer with the filtering layer together. The purpose of having two layers instead of one was to reduce the number of operations needed but again, for edge devices, the actual latency might differ due to the bottleneck in memory access cost. Sometimes, using a fused version might result in a faster runtime with the benefits of having more parameters to learn more information from.

Understanding MobileNetV3-small

For **MobileNetV3-small**, a few changes were made to MobileNetV2:

- It used a more advanced non-linearity called **hard-swish**, which improved the top-1 accuracy of ImageNet.

- Expensive computation was reduced even further in the initial and last few layers.

- For the first layer, as shown in *Figure 3.15*, 32 filters was reduced to 16 and hard-swish nonlinearity was used to get 2 milliseconds runtime and 10 million FLOPS savings.

- For the last few layers, the last 1x1 bottleneck convolution layer is moved to after the final average pooling layer, and the previous bottleneck (1x1) and filtering (3x3) layer are also removed.

- It used a modified version of network architecture search that is platform-aware called **Mnasnet** and a post-search layer reduction for latency reduction called **NetAdapt** to automatically find an optimized architecture based on the building blocks from the MobileNetV1, MobileNetV2, and squeeze and excitation networks while considering latency and accuracy performance. NetAdapt and MnasNet will be introduced in *Chapter 7, Deep Neural Architecture Search*.

MobileNetV3-small ended up achieving a higher top-1 ImageNet accuracy with the same parameters and FLOPS compared to MobileNetV2.

Understanding the ShuffleNet architecture family

ShuffleNet has two versions, ShuffleNetV1 and ShuffleNetV2, which we will discuss separately.

ShuffleNetV1, from 2017, reuses a known variant of convolution called **grouped convolutions**, where each convolutional filter is responsible only for a subset of input data channels. MobileNet uses a special variant of this by using one filter for each channel. Grouped convolutions save computation costs by operating only on a small subset of input channel features. However, when stacked together one after another, the information between channels doesn't interact – ultimately causing accuracy degradation. ShuffleNetV1 uses channel shuffling as an operation in between stacking multiple grouped convolutional layers to manually shift information without sacrificing FLOPs. This allows for an efficient and small network.

ShuffleNetV2, from 2018, builds upon ShuffleNetV1 and focuses on the practical runtime efficiency of the architecture in real life while considering factors such as memory access cost, data **input-output (I/O)**, and the degree of network parallelism. The following four design strategies were used to craft the new architecture:

- Equal channel width from input to output to minimize memory access cost.

- Decrease group convolution to minimize memory access cost.

- Reduce network fragmentation to increase parallelism. One example is the number of convolutional and pooling operations in a single building block.

- Reduce element-wise operations such as ReLU as they have heavy memory access costs:

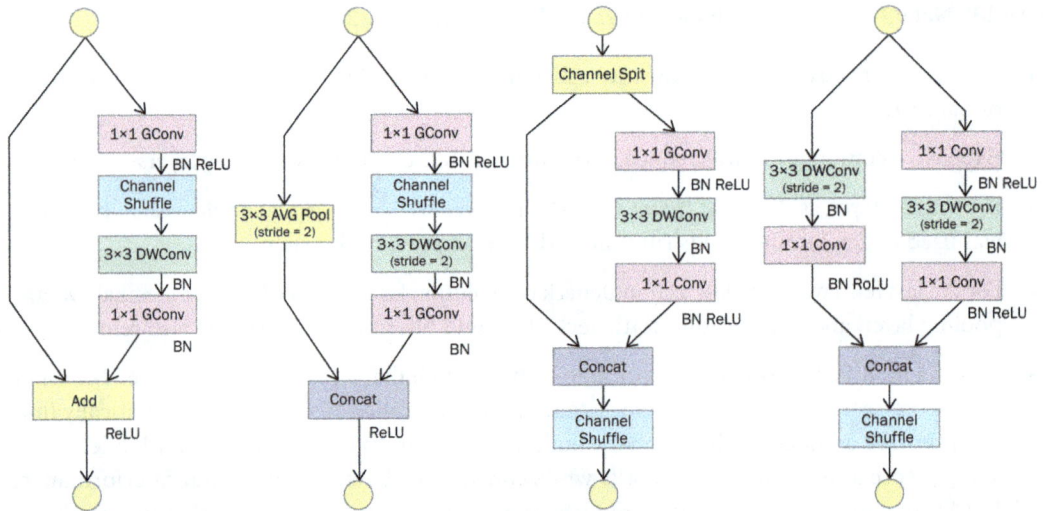

Figure 3.17 – Two building blocks of ShuffleNetv1 on the left and
two building blocks of ShuffleNetv2 on the right

In *Figure 3.17*, the first two structures on the left show the two building blocks of ShuffleNetV1, while the last two structures on the right show the two building blocks of ShuffleNetV2.

Understanding MicroNet, the current state-of-the-art architecture for the edge

Created in 2021, MicroNet is the current state of the art in terms of latency and achievable top-1 `ImageNet` accuracy performance, for a very low FLOP range of 4 million to 21 million. The novelty of MicroNet is two-fold:

- It introduced factorized versions of the bottleneck/pointwise convolution layer and depthwise convolutional layers from MobileNet, called **micro-factorized convolutions**, in a way that the number of connections/paths for input data to output data is reduced. This is achieved by using multiple grouped convolutions and some dilated convolutions. Dilated convolutions are simply convolutions with fixed spacing in the kernels. Take these techniques as a form of sparse computation and only compute what's needed most efficiently to ensure minimal input-to-output path redundancy.

- It introduced a new activation function called **dynamic shift-max** that leverages the output of the grouped convolutions in a way that it applies a higher order of non-linearity (two times) and strengthens connections between groups at the same time. This is implemented by using the grouped outputs of the blocks of squeeze and excitation (produce a single value per channel) as a weighting mechanism to obtain the maximum of a group-based weighted addition. Take this as an improvement on top of channel shuffling in ShuffleNet. *Figure 3.18* shows the operation structure of dynamic shift-max for a single example feature map of 12 channels from the output of four groups using the grouped convolution operation:

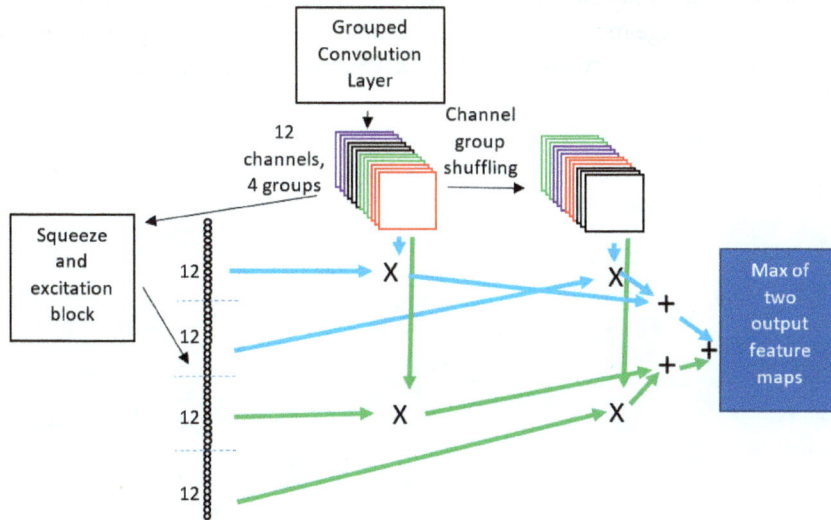

Figure 3.18 – Dynamic shift-max general operation flow

MicroNet utilizes concepts from ShuffleNet (the channel shuffling mechanism), ResNet (skip connections), SENet (squeeze and excitation networks), and MobileNet (create factorized versions out of the already factorized convolutions) on top of its novelties to create networks that are highly efficient by focusing on the concept of sparsity and improvements in efficient information flow. The specifics of this network can be overwhelming and, frankly, hard to comprehend, so the information presented here does not contain all the details:

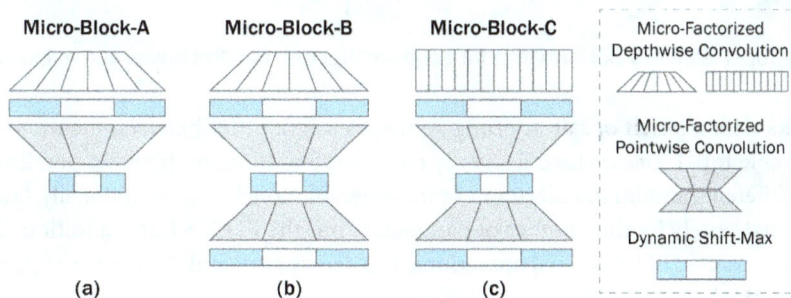

Figure 3.19 – Diagram of three logical blocks, called micro-blocks, that
are used to build different size variants of MicroNets

However, *Figure 3.19* shows how logical blocks are, again, in the most advanced network today to build networks with different sizes based on the same ideas.

Summarizing CNN architectures for the edge

To summarize the architectures for the edge, you now have the intuitive knowledge that was used by experts in the field to build highly effective CNN architectures capable of running at an amazingly tiny footprint compared to large models such as GPT-3 today. The following figure shows the overall top-1 ImageNet accuracy performance versus FLOPS graph of multiple different architecture families for edge computation:

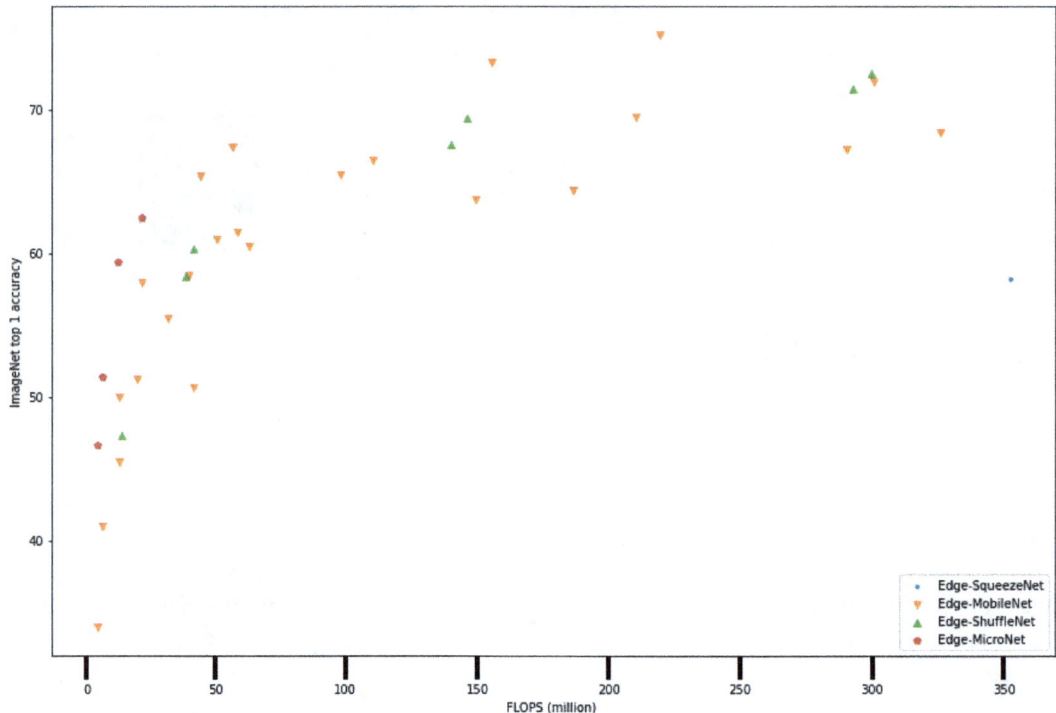

Figure 3.20 – Top-1 accuracy performance for edge architecture families below 400 million FLOPS

Take these results with a pinch of salt; training strategies might differ between the models and can affect the achievable top-1 ImageNet accuracy considerably, along with giving possibly differing results between different random initializations of the different model runs. Additionally, latency is not directly represented purely by the number of parameters nor the FLOPS but is additionally affected by the memory access cost of individual operations, I/O access cost, and the degree of parallelism of the different operations.

To recap, *Figure 3.21* shows the overall performance plots based on the FLOPS of all the CNN model families that were introduced in this chapter:

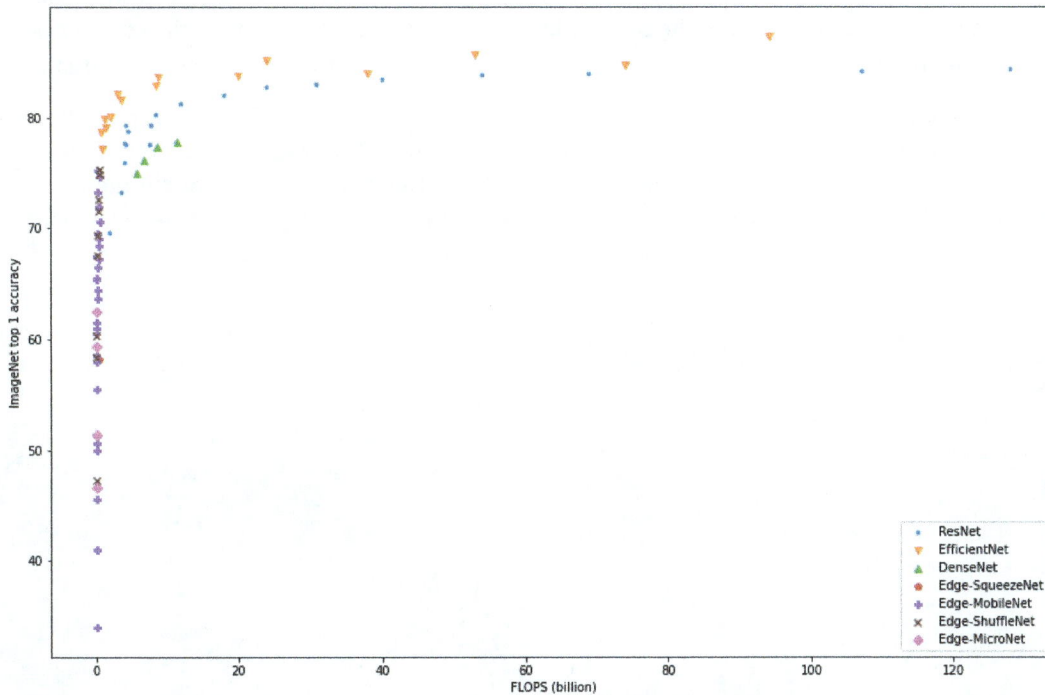

Figure 3.21 – Overall CNN model family performance in terms of ImageNet top-1 accuracy based on FLOPS

Again, note that we should take the results presented here with a pinch of salt as the training techniques that were performed against the `ImageNet` dataset are not exactly standardized across different benchmarks. Variation in the training technique can result in widely different results and will be covered more extensively in *Chapter 8*, *Exploring Supervised Deep Learning*, and *Chapter 9*, *Exploring Unsupervised Deep Learning*. Another important thing to note is that even though `ImageNet` is considered to be a large enough image dataset to be considered as a benchmark, maintain a level of skepticism toward the results as the data itself has been proven to have noisy labels with systematic errors in some cases. A corrected form of `ImageNet` has been published called `ImageNet Real` but not all models are benchmarked or pre-trained against it. Train it on your dataset to be 100% sure which architecture, when pre-trained on certain datasets, performs better! Additionally, the FLOPS indicator does not fully represent the actual latency of the model, which can vary widely based on how the code is structured, how the model is distributed through multiple devices, how many GPUs or CPUs are available, and how parallel the model architecture is.

Summary

CNNs are the go-to model for capturing patterns in image data. The handpicked architectures that were introduced in this chapter are the core backbones that can be subsequently utilized as a base for solving more custom downstream tasks such as image object detection and image generation.

The CNNs that were covered here will be used practically in later chapters as a basis to help you learn other deep learning-based knowledge. Take your time and look into how different architectures are implemented in a deep learning library offline in this book's GitHub repository; we won't be presenting the actual implementation code here. Now that we have covered CNNs in intermediate to low-level detail, in the next chapter, we'll shift gears and look at recurrent neural networks.

4
Understanding Recurrent Neural Networks

A **recurrent neural network (RNN)** is a neural network that is made to process sequential data while being aware of the sequence of the data. Sequential data can involve time series based data and data that has a sequence but does not have a time component, such as text data. The applications of such a neural network are built upon the nature of the data itself. For time-series data, this can be either for nowcasting (predictions made for the current time with both past and present data) or forecasting targets. For text data, applications such as speech recognition and machine translation can utilize these neural networks.

Research in recurrent neural networks has slowed in the past few years with the advent of neural networks that can capture sequential data while removing recursive connections completely and achieving better performance, such as transformers. However, RNNs are still used extensively in the real world today to serve as a good baseline or just an alternative model for faster computations due to their lower number of computations and low memory requirements with reasonable metric performance.

The two most prominent RNN layers are **Long Short-Term Memory (LSTM)** and **Gated Recurrent Units (GRU)**. We will not be going through the vanilla and the original recurrent neural networks in this book, but we will show LSTM and GRU as a refresher. The main operations of LSTM and GRU provide a mechanism to keep only relevant memory and ignore data that is not useful, which is a key inductive bias crafted for time series or sequential data.

In this chapter, we will dive deeper into these two RNN networks more extensively. Specifically, we will cover the following topics:

- Understanding LSTM
- Understanding GRU
- Understanding advancements over the standard GRU and LSTM layers

Technical requirements

This chapter is short and sweet but still covers some practical implementations in the **Python** programming language to realize the RNN architecture. To complete it, you will need to have a computer with the `Pytorch` library installed.

You can find the code files for this chapter on GitHub at `https://github.com/PacktPublishing/The-Deep-Learning-Architect-Handbook/tree/main/CHAPTER_4`.

Understanding LSTM

LSTM was invented in 1997 but remains a widely adopted neural network. LSTM uses the `tanh` activation function as it provides nonlinearities while providing second derivatives that can be preserved for a longer sequence. The `tanh` function helps to prevent exploding and vanishing gradients. An LSTM layer uses a sequence of LSTM cells sequentially connected. Let's take an in-depth look at what the LSTM cell looks like in *Figure 4.1*.

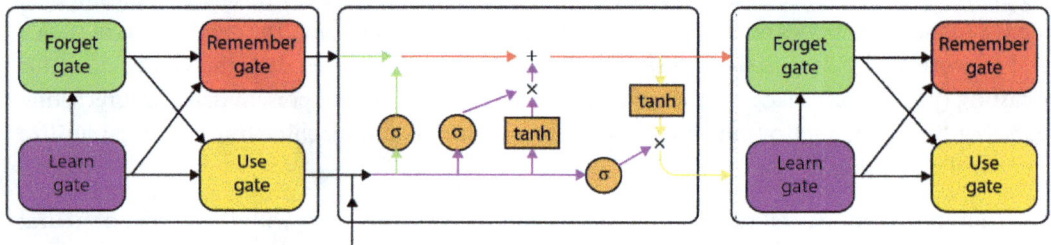

Figure 4.1 – A visual deep dive into an LSTM cell among a sequence of LSTM cells that forms an LSTM layer

The first LSTM cell on the left depicts the high-level structure of an LSTM cell and the second LSTM cell on the left depicts the medium-level operations, connections, and structure of an LSTM cell, while the third cell on the right is just another LSTM cell to emphasize that LSTM layers are made of multiple LSTM cells sequentially connected to each other. Think of an LSTM cell as containing four gating mechanisms that provide a way to **forget**, **learn**, **remember**, and **use** the sequential data. One notable thing you might wonder about is why the sigmoid is shown as a separate process in three paths from the input and past the hidden state to the forget gate, the remember gate, and the use gate. The structure shown in *Figure 4.1* is a famous depiction of the LSTM cell but does not contain information about the weights of the connections. This is due to the fact that the inputs go through a weighted addition process to combine the previous cell's hidden state and the current sequence data with different sets of weights for each of the four connections. *Figure 4.2* shows the final low-level structure of a single LSTM cell with weights considered:

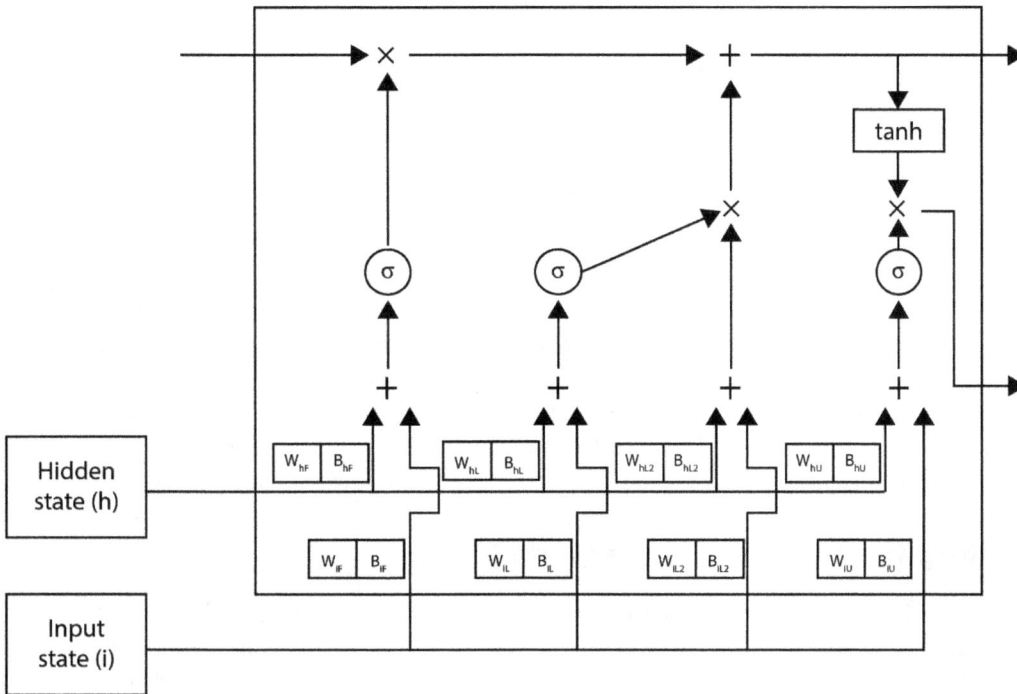

Figure 4.2 – Low-level structure of the LSTM cell

In *Figure 4.2*, *W* and *B* represent the weights and biases, respectively. The two small letters represent the data type and gate mechanisms, respectively. The data type is split into two – the hidden state represented by *h* and the input data represented by *i*. The gate mechanisms involved are the forget mechanism represented by *F*, the learning mechanism represented by *L*, where there are two weights and biases associated with learning, and the use mechanism represented by *U*. To properly perceive how many parameters an LSTM cell has we still have to decode the dimensions of the hidden state and input state weight vectors. For ease of reference, let's take the input vector size as *n* and the hidden size as *m*.

The hidden state weights have a dimension of:

$$nm$$

While the input state weights have a dimension of:

$$n^2$$

Bias, on the other hand, is the size of the input vector size. For Tensorflow and Keras with Tensorflow, bias is only added once per mechanism. For PyTorch, the bias is added for each hidden state and input state weight. For PyTorch, the number of parameters for bias can be defined as:

$$2n$$

As there are four mechanisms, as depicted in *Figure 4.2*, this means that the number of parameters for an LSTM in the PyTorch implementation can then be computed according to the following formula:

$$number\ of\ parameters = 4(nm + n^2 + 2n)$$

Now that we understand where the actual parameters live in the cell, let's dive into each of these gating mechanisms.

Decoding the forget mechanism of LSTMs

The forget mechanism is accomplished by using a sigmoid activation function multiplied against the previous cell state. The name of the gating mechanism implies that it determines the information to be removed based on a combination of the current input sequence and the previous cell output. A way to think of it is, on a scale of 0 to 1, how relevant is the information from the past? The sigmoid mechanism forces the scale to be between 0 and 1. Values closer to 0 forget more of the previous cell state (long-term memory) and values closer to 1 forget less of the previous cell state memory.

Decoding the learn mechanism of LSTMs

The learn mechanism employs a combination of sigmoid activation of previous cell output and tanh activation of previous cell output added on the outputs of the forget gate and multiplied by the outputs of the use gate. This mechanism is also known as the input gate. This mechanism allows information learning from the current input sequence. The information learned then gets passed into the remembering mechanism. Additionally, the information learned will also get passed into the mechanism that allows information usage for the next LSTM cell. Both of these mechanisms will be introduced sequentially next.

Decoding the remember mechanism of LSTMs

The remember mechanism is achieved by simply adding up information from what's left of the forget process and what has been learned, which is the output of the learning gate. The output of this gate will then be considered the current cell state of the LSTM cell. The cell state contains what is known as the long-term memory of the LSTM sequence. Take this mechanism simply as an operation that allows the network to selectively choose which part of the input to maintain and remember.

Decoding the "information-using" mechanism of LSTMs

The information-using mechanism is achieved by applying nonlinearities using the tanh activation function on the current cell state and again using the current input sequence and previous cell output as a weighting mechanism to determine how much relevant information from the past and present should be used. The output of applying the use gate gives us the hidden state that will also be used as the previous cell output for the next LSTM cell.

Building a full LSTM network

Usually, to create a full LSTM network, multiple LSTM layers are concatenated to each other using the sequence of hidden states from multiple LSTM cells as the subsequent sequence data to apply to the next LSTM layer. After a couple of LSTM layers, the hidden state sequence of the previous layer will usually then be passed into a fully connected layer to form the basis of supervised learning based simple LSTM architecture. *Figure 4.3* shows a visual structure of how this is done:

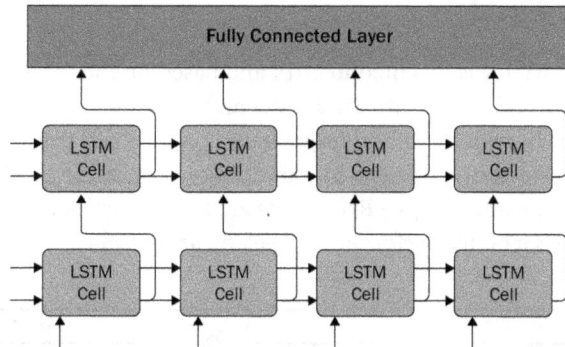

Figure 4.3 – A simple LSTM network with two LSTM layers fed into a fully connected layer

Based on the network depicted in *Figure 4.3*, the implementation in PyTorch will look like the following:

1. Let's first import the PyTorch library's handy nn module:

    ```
    import torch.nn as nn
    ```

2. Now, we will define the network architecture based on *Figure 4.3*, using the sequential API this time instead of the class method:

    ```
    RNN = nn.Sequential(
        nn.LSTM(
            input_size=10, hidden_size=20,
            num_layers=2, dropout=0,
        ),
        nn.Linear(in_features=10, out_features=10),
        nn.Softmax(),
    )
    ```

3. The input size, hidden sizes, and layer number of the LSTM, along with the output feature size of the linear layer, can be configured according to your input dataset and desire. Note that each timestep or sequential step of the input data can have a size greater than one. This allows us to easily map original features into more representative feature embeddings and leverage their descriptive power. Additionally, the dropout regularizer can be added easily by setting

the `dropout` parameter to a value between 0 and 1, which will introduce the dropout layer at each layer except the last layer at the specified probability. The RNN defined in `pytorch` in *step 2* can now be trained as usual, like any PyTorch models defined in a class.

As usual, PyTorch has made building RNNs so much simpler and faster. Next, we will step into the next type of RNN, called **Gated Recurrent Units**.

Understanding GRU

Gated recurrent units (GRU) was invented in 2014 and based on the ideas implemented in LSTM. GRU was made to simplify LSTM and provide a faster and more efficient way of achieving the same goals as LSTM to adaptively remember and forget based on past and present data. In terms of the learning capacity and metric performance achievable, there isn't a clear silver-bullet winner among the two and often in the industry, the two RNN units are benchmarked against each other to figure out which method provides a better performance level. *Figure 4.4* shows the structure of GRU.

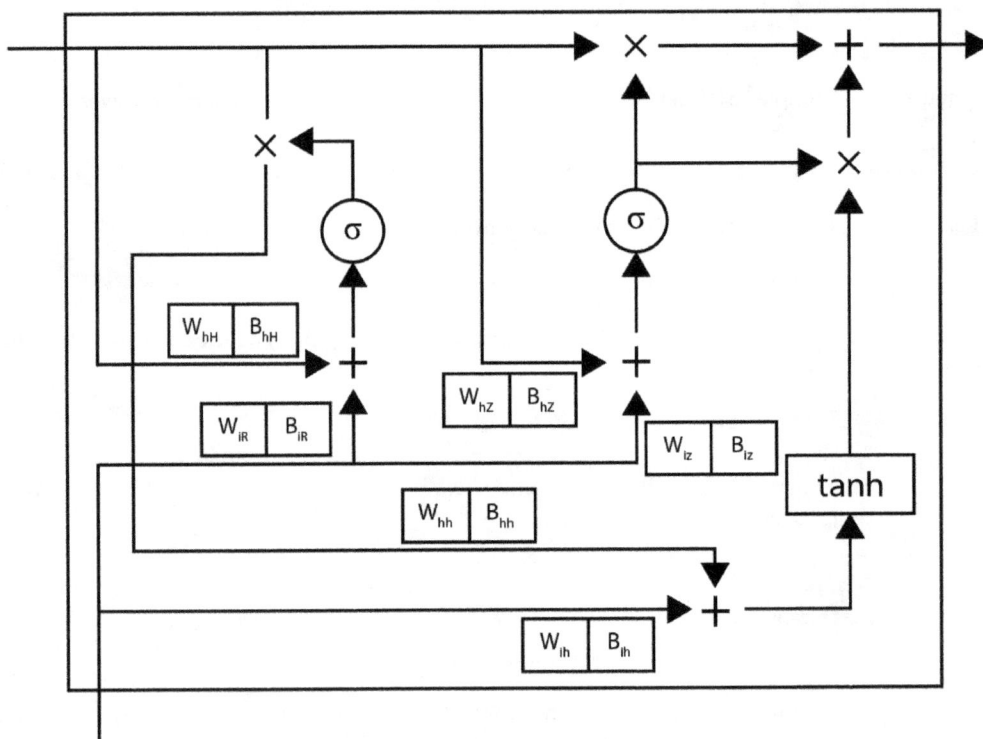

Figure 4.4 – A low-level depiction of GRU

Figure 4.4 adopts the same weights and bias notations as the LSTM depicted in *Figure 4.2*. There are three different names here for the final small letter notation. R being the reset gate, z representing the update gate, and h representing weights used to obtain the next hidden states. This means a GRU cell has fewer parameters than an LSTM cell, with three sets of weights and biases instead of four. This allows GRU networks to be slightly faster than LSTM networks.

Although depicted as a single cell, the same theory that required multiple LSTM cells to be sequentially connected also applies to GRU: a GRU network layer will have multiple GRU cells connected sequentially together. GRU contains only two mechanisms, called the **reset gate** and the **update gate**, and only has one input from the previous GRU cell, and one output to the next GRU cell. The one input-output itself is obviously more efficient than LSTM, as we require fewer operations to be carried out. Now, let's dive into these two mechanisms.

Decoding the reset gate of GRU

The reset gate of GRU serves as a mechanism to forget the long-term information, also called the hidden state, of the previous cell. The goal of this mechanism is similar to the forget gate in the LSTM cell. Similarly, this will exist on a scale of 0 to 1, based on the current input sequence and the previous cell state, which decides how much we should reduce and remove the previously gained long-term information.

However, the reset gates of GRU are functionally different from the forget gate of LSTM. While the forget gate of LSTM decides what information to forget from the long-term memory, the reset gate of GRU decides how much of the previous hidden state to forget.

Decoding the update gate of GRU

The update gate of GRU controls the amount of information from the long-term memory to be transferred to the currently maintained memory. This is similar to the remember gate in LSTMs and helps the network to remember long-term information. Each weight associated with the previous cell's hidden unit will learn to capture both short-term dependencies and long-term dependencies. The short-term dependencies usually have reset gates output values that are closer to 0 to forget former information more frequently, and vice versa with weights and hidden state positions that learn long-term dependencies.

In terms of the difference from the LSTM remember gate, while the remember gate of LSTM decides what information to remember from the current input and previous hidden state, the update gate of GRU decides how much of the previous hidden state to remember.

GRU is a simple RNN to consider with more efficient operations compared to LSTMs. Now that we have decoded both LSTMs and GRU, instead of repeating another GRU-only full network similar to LSTM, let's discover improvements that can be made using these two methods as a base.

Understanding advancements over the standard GRU and LSTM layers

GRU and LSTM are the most widely used RNN methods today, but one might wonder how to push the boundaries achievable by a standard GRU or a standard LSTM. One good start to building this intuition is to understand that both of the layer types are capable of accepting sequential data, and to build a network you need multiple RNN layers. This means that it is entirely possible to combine GRU and LSTM layers in the same network. This, however, is not credible enough to be considered an advancement as a fully LSTM network or a fully GRU network can exceed the performance of a combined LSTM and GRU network at any time. Let's dive into another simple improvement you can make on top of these standard RNN layers, called **bidirectional RNN**.

Decoding bidirectional RNN

Both GRU and LSTM rely on the sequential nature of the data. This order of the sequence can be forward in increasing time steps and also can be backward in decreasing time steps. Which direction to use usually comes down to an act of trial and error, and more often than not, the natural direction to use will be the forward time order.

In the year 1997, an improvement was made called bidirectional RNNs, which combine both forward-ordered RNNs with backward-ordered RNNs in an effort to maximize the input data that can be processed by an RNN model. The original idea was to estimate the value at a current timestep using both future information and historical information, given that both future and historical information was available with two of the RNNs taking in different sets of data. This naturally allowed for the capacity to achieve better prediction performance on such data setups. Today, this idea has extended to be a general layer applied to the same sequential data estimating and is also proven to provide prediction performance improvements. *Figure 4.5* shows an example of bidirectional RNNs using GRU where the hidden states from the forward and backward GRU are concatenated:

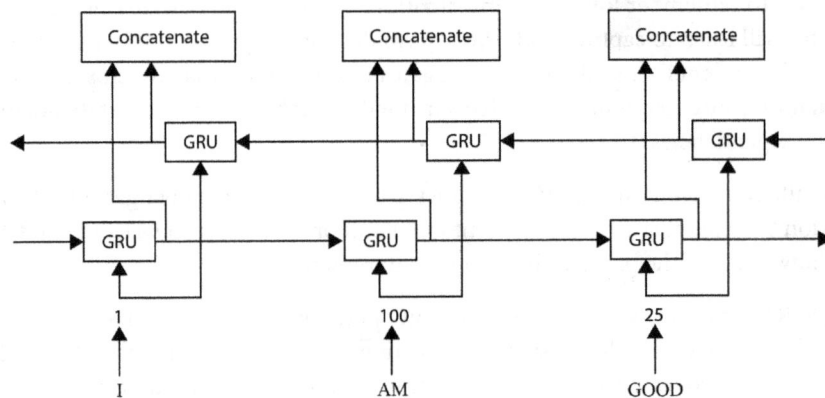

Figure 4.5 – Bidirectional GRU

The concatenated hidden states can then be passed into fully connected layers for standard supervised learning objectives. An example implementation in PyTorch of a bidirectional GRU is shown:

```
RNN = nn.Sequential(
    nn.GRU(
        input_size=10, hidden_size=20,
        num_layers=2, bidirectional=True
    ),
    nn.Linear(in_features=10, out_features=10),
    nn.Softmax(),
)
```

Next, let's discover an improvement that is made based on LSTMs.

Adding peepholes to LSTMs

Introduced in 2000, peepholes enable the cell states (from the previous and current cell), which hold the long-term memory of LSTMs, to influence the sigmoid gating mechanisms in the LSTM cell. The intuition is that the long-term memory has information about the past time steps that are not available in the short-term memory held in the hidden state of the previous cell. This allowed for improved predictive performance when compared to a vanilla LSTM. *Figure 4.6* shows the extra peephole connections from the cell state:

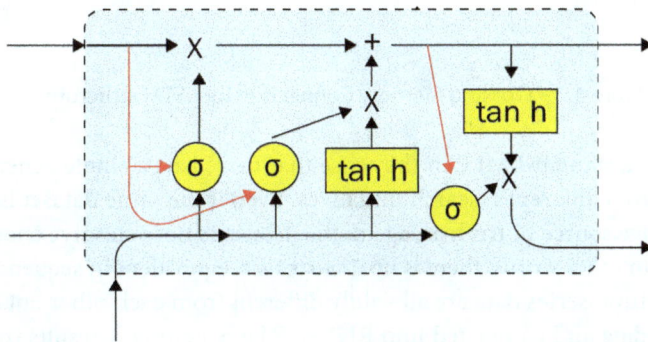

Figure 4.6 – LSTM peephole connections

However, one pitfall of this method is that the cell states can grow to large values over time due to the long-term memory nature of the states, as it is unbounded. This may saturate the gates to always be in an open state and render the gate useless sometimes. This brings us to the last improvement that we will discuss in the next subsection.

Adding working memory to exceed the peephole connection limitations for LSTM

In 2021, an improvement was made on top of peephole connections for LSTMs to enforce a bound to the cell states by using the `tanh` activation. The simple addition proved itself to be better in performance than the LSTM with peepholes unbounded version in experimental benchmarks and was called **Working Memory Connections for LSTM**. *Figure 4.7* shows the Working Memory Connection for LSTM structure:

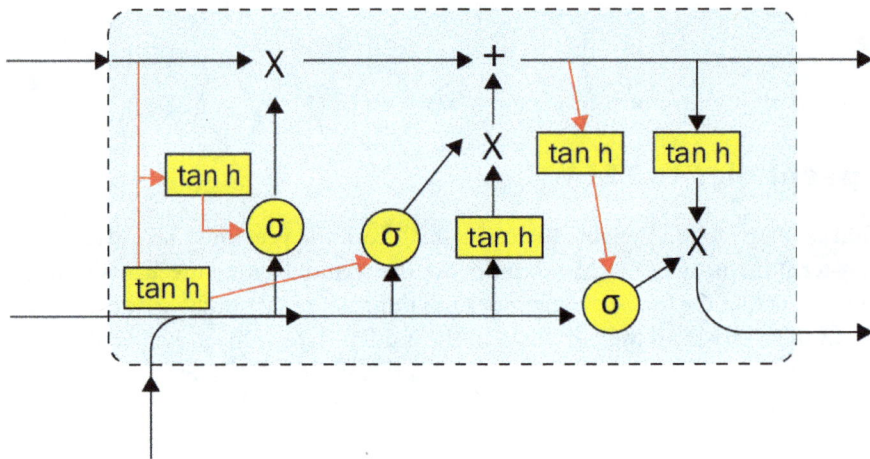

Figure 4.7 – Working Memory Connection for LSTM structure

RNNs are similar to the standard MLP in the sense that there isn't a single generic dataset that is used as a reference across different research initiatives. Even if the same dataset is used, the results might not be a definitive source of truth as, again, the dataset is not extensive enough to generalize across other datasets. In other words, there is no `ImageNet` equivalent in sequence data. Text data, video data, and other time-series data are all wildly different from each other but essentially are all considered sequence data and can be fed into RNNs. Take benchmark results from anywhere on RNNs and MLPs with a pinch of salt as results can vary widely from dataset to dataset. *Figure 4.8*, however, shows one version of benchmarks done with GRU, LSTM, **LSTM peepholes (LSTM-PH)**, and **LSTM Working Memory (LSTM-WM)** on two different model settings out of RNNs, done on an image captioning task using the `COCO` dataset using a metric that considers the naturalism of the produced text – the higher the better.

Model	BLEU-1	BLEU-4	METEOR	ROUGE	CIDEr	SPICE
No Attention, ResNet-152						
LSTM	70.9	27.9	24.4	51.7	92.0	17.6
GRU	69.5	26.2	22.7	50.4	82.3	15.6
LSTM-PH	**71.4**	27.8	24.3	51.7	91.1	17.5
LSTM-WM	**71.4**	**28.3**	**24.6**	**52.4**	**94.0**	**17.8**
Attention, Faster R-CNN						
LSTM	75.9	**36.1**	27.4	56.3	111.9	20.3
GRU	76.0	**36.1**	27.0	56.5	111.0	20.2
LSTM-PH	75.8	35.9	27.3	56.3	111.5	20.2
LSTM-WM	**76.2**	**36.1**	**27.5**	**56.5**	**112.7**	**20.4**

Figure 4.8 – RNN benchmark on image captioning task on COCO dataset

The figure shows that LSTM-WM dominates over the other methods in a single experiment across different evaluation scores. Again, take the results with a pinch of salt as the COCO dataset is by no means a representative dataset of sequential or time-series data.

Try it out on your own dataset to know for sure! With that, we have gone through important concepts of RNN, from basic to advanced levels. Let's summarize the chapter next.

Summary

Recurrent neural networks are a type of neural network that explicitly includes inductive biases of sequential data in its structure.

A couple of variations of RNNs exist but all of them maintain the same high-level concept for their overall structure. Mainly, they provide varying ways to decide which data to learn from and remember along with which data to forget from the memory from the remembering stage.

However, do note that a more recent architecture called transformers, which will be introduced in *Chapter 6*, *Understanding Neural Network Transformers*, demonstrated that recurrence is not needed to achieve a good performance on sequential data.

With that, we are done with RNNs and will dive briefly into the world of autoencoders in the next chapter.

5
Understanding Autoencoders

Autoencoders are a type of model that was built mainly to accomplish **representation learning**. Representation learning is a type of deep learning task that focuses on generating a compact and representative feature to represent any single data sample, be it image, text, audio, video, or multimodal data. After going through some form of representation learning, a model will be able to map inputs into more representable features, which can be used to differentiate itself from other sample inputs. The representation obtained will exist in a latent space where different input samples will co-exist together. These representations are also known as **embeddings**. The applications of autoencoders will be tied closely to representation learning applications, and some applications include generating predictive features for other subsequent supervised learning objectives, comparing and contrasting samples in the wild, and performing effective sample recognition.

Note that autoencoders are not the only way to execute representation learning. The topic of representation learning will be discussed further in *Chapter 8, Exploring Supervised Deep Learning*, and *Chapter 9, Exploring Unsupervised Deep Learning*.

Now, we know that an autoencoder learns to generate distinctive representations or, in other words, embeddings. But what's the architecture like? Let's discover a standard form of the architecture and then discover a couple more useful advancements.

In this chapter, the following topics will be covered:

- Decoding the standard autoencoder
- Exploring autoencoder variations
- Building a CNN autoencoder

Technical requirements

This chapter includes some practical implementations in the **Python** programming language. To complete it, you will need to have a computer with the following libraries installed:

- `pandas`
- `Matplotlib`
- `Seaborn`
- `scikit-learn`
- `NumPy`
- `Keras`
- `PyTorch`

The code files are available on GitHub: `https://github.com/PacktPublishing/The-Deep-Learning-Architect-Handbook/tree/main/CHAPTER_5`.

Decoding the standard autoencoder

Autoencoders are more of a concept than an actual neural network architecture. This is due to the fact that they can be based on different base neural network layers. When dealing with images, you build CNN autoencoders, and when dealing with text, you might want to build RNN autoencoders. When dealing with multimodal datasets with images, text, audio, numerical, and categorical data, well, you use a combination of different layers as a base. Autoencoders are mainly based on three components, called the **encoder**, the **bottleneck layers**, and the **decoder**. This is illustrated in *Figure 5.1*.

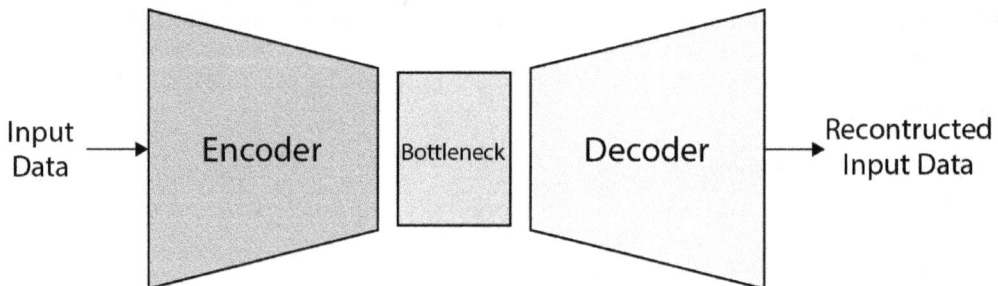

Figure 5.1 – The autoencoder concept

The encoder for a standard autoencoder typically takes in high-dimensional data and compresses it to an arbitrary scale smaller than the original data dimensions, which will result in what is known as a **bottleneck representation**, where it ties itself to the bottleneck, signifying a compact representation without any useless information. The bottleneck component then gets passed into the decoder, where it will expand the dimensionality using the exact opposite of the scale used by the encoder, resulting in output with the same dimensions as the input data.

> **Note**
>
> Encoder and decoder structures are not exclusive to autoencoders but are also utilized in other architectures, such as transformers.

The difference is the bottleneck component that holds the representative features. It is commonly compressed and in smaller dimensions, but sometimes, it can be made larger to hold more representative features for predictive power.

Autoencoders, in general, are trained to reconstruct input data. The training process of the autoencoder model involves comparing the distance between the generated output data and the input data. After being optimized to generate the input data, when the model is capable of reconstructing the original input data completely, it can be said that the bottleneck has a more compact and summarized representation of the input data than the original input data itself. The compact representation can then be used subsequently to achieve other tasks, such as sample recognition, or can even be used generally to save space, by storing the smaller bottleneck feature instead of the original large input data. The encoder and decoder are not constrained to a single layer and can be defined with multiple layers. However, the standard autoencoder only has a single bottleneck feature. This is also known as **code**, or a **latent feature**.

Now, let's explore the different variations of autoencoders.

Exploring autoencoder variations

For tabular data, the network structure can be pretty straightforward. It simply uses an MLP with multiple fully connected layers that gradually shrink the number of features for the encoder, and multiple fully connected layers that gradually increase the data outputs to the same dimension and size as the input for the decoder.

For time-series or sequential data, RNN-based autoencoders can be used. One of the most cited research projects about RNN-based autoencoders is a version where LSTM-based encoders and decoders are used. The research paper is called *Sequence to Sequence Learning with Neural Networks* by Ilya Sutskever, Oriol Vinyals, and Quoc V. Le (https://arxiv.org/abs/1409.3215). Instead of stacking encoder LSTMs and decoder LSTMs, using the hidden state output sequence of each of the LSTM cells vertically, the decoder layer sequentially continues the sequential flow of the encoder LSTM and outputs the reconstructed input in reversed order. An additional decoder LSTM layer was also used to concurrently optimize, to predict future sequences. This structure is shown in *Figure 5.2*. Note that it is also possible to adapt this to the video image modality using raw flattened image pixels as input. The architecture is also called **seq2seq**.

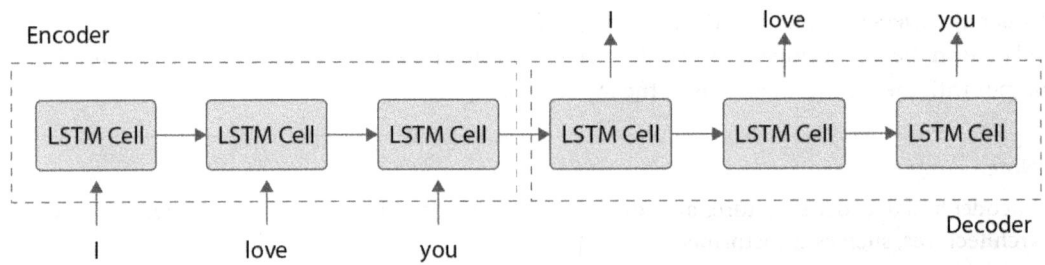

Figure 5.2 – An LSTM-based autoencoder structure

The hidden state outputs of the encoder LSTM can be considered to be the latent feature of the LSTM autoencoder. Transformers, a new architecture that can also deal with sequential data, which will be introduced later in this chapter, have some variations that can also be considered a kind of autoencoder – some transformers can be autoencoders, but not all transformers are autoencoders.

For image data, by using convolutional layers, we can scale down the features gradually with multiple convolutional and pooling layers, until the stage where a global pooling layer is applied and the data becomes a 1-dimensional feature. This represents the encoder of the autoencoder generating the bottleneck feature. This workflow is the same as the one we discussed in the previous section on CNN. However, for the decoder, to scale up the 1-dimensional pooled feature into 2-dimensional image-like data again, a special form of convolution is needed, called the **transpose convolution**.

The variations mentioned in this topic are all about using the standard autoencoder structures but implementing them using different neural network types. There are also two additional variations of the autoencoder, where one variation is based on data input manipulation, and another variation is based on an actual modification of the autoencoder structure to achieve the data generation goal.

For the variation used for data manipulation, the idea is to add noise to the input data during training and maintain the original input data without the added noise, to be used as the target to predictively reconstruct the data. This variation of the autoencoder is called a **denoising autoencoder** because its goal is primarily to *denoise* the data. Since the goal has shifted from compressing the data to denoising the data, the bottleneck features are not restricted to being small in size. The features that are utilized afterward are not constrained to just the single bottleneck feature but, instead, can be the features from multiple intermediate layers in the network, or just simply the denoised reconstructed output. This method takes advantage of the innate capability of neural networks to perform automated feature engineering. The most notable usage of a denoising autoencoder is in the first-place solution of a Kaggle competition based on tabular data, hosted by Porto Seguro, an insurance company, where multiple features from intermediate layers were fed into a separate MLP to predict whether a driver will file an insurance claim in the future (`https://www.kaggle.com/competitions/porto-seguro-safe-driver-prediction/discussion/44629`).

For the variation that uses a modification of the autoencoder structure, the idea is to produce two bottleneck feature vectors that represent a list of standard deviation values and mean values so that different bottleneck feature values can be sampled, based on the mean and standard deviation values. These sampled bottleneck feature values can then be passed into the decoder to generate new random data outputs. This variation of the autoencoder is called a **variational autoencoder**.

Now that we have covered a good overview of autoencoder variations, let's dig further into CNN autoencoders and build a CNN autoencoder using deep learning libraries.

Building a CNN autoencoder

Let's start by going through what a **transpose convolution** is. *Figure 5.3* shows an example transpose convolution operation on a 2x2 sized input with a 2x2 sized convolutional filter, with a stride of 1.

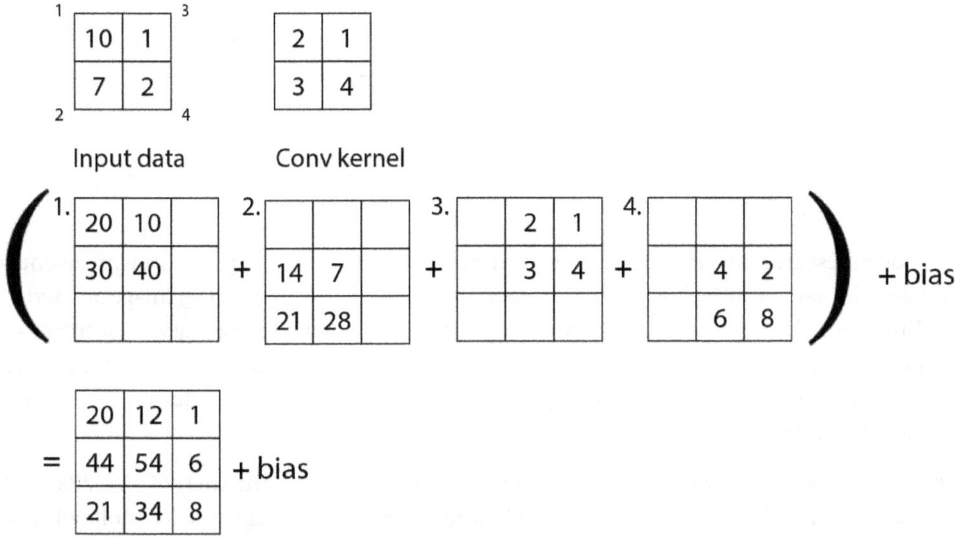

Figure 5.3 – A transposed convolutional filter operation

In *Figure 5.3*, note that each of the 2x2 input data is marked with a number from **1** to **4**. These numbers are used to map the output results, presented as 3x3 outputs. The convolutional kernel applies each of its weights individually to every value in the input data in a sliding window manner, and the outputs from the four convolutional operations are presented in the bottom part of the figure. After the operation is done, each of the outputs will be elementwise added to form the final output and subjected to a bias. This example process depicts how a 2x2 input can be scaled up to a 3x3 data size without relying completely on padding.

Let's implement a convolutional autoencoder model in `Pytorch` below and train it on the `Fashion MNIST` image dataset, an image dataset comprising fashion items such as shoes, bags, and clothes:

1. Let's start by importing the necessary libraries:

```
import torch.nn as nn
import torchvision
from PIL import Image
```

2. Next, we will define the overall convolutional autoencoder structure:

```
class ConvAutoencoder(nn.Module):
  def __init__(self):
    super(ConvAutoencoder, self).__init__()
    self.encoder = None
    self.decoder = None
  def forward(self, x):
    bottleneck_feature = self.encoder(x)
    reconstructed_x = self.decoder(
      bottleneck_feature
    )
    return reconstructed_x
```

The code presented here is a convolutional autoencoder structure in `PyTorch` with an encoder and decoder variable placeholder. The encoder is responsible for taking in an image and reducing its dimensionality until it has a single dimension with a small representation footprint – the bottleneck feature. The decoder will then take in the bottleneck feature and produce a feature map of the same size as the original input image. The encoder and decoder will be defined in the next two steps.

3. The encoder will be designed to take in a grayscale image (one channel) of size 28x28. This image dimension is the default size of the Fashion MNIST image dataset. The following logic shows the code to define the encoder, replacing the placeholder defined in *step 2*:

```
self.encoder = nn.Sequential(
    nn.Conv2d(1, 16, 4),
  nn.ReLU(),
  nn.MaxPool2d(2, 2),
  nn.Conv2d(16, 4, 4),
  nn.ReLU(),
  nn.AvgPool2d(9),
)
```

The defined encoder has two convolutional layers, each followed by the non-linear activation, ReLU, and a pooling layer. The filter sizes of the convolutional layers are `16` and `4` The second pooling layer is a global average pooling layer meant to reduce the 4x9x9 feature map to 4x1x1, where each channel will have only one value to represent itself. This means that the encoder will squeeze the dimensionality of the original 28x28 image, which adds up to around 784 pixels for only four features, which is a 99.4% compression rate!

4. The decoder will then take these four features per image and produce an output feature map of 28x28 again, reproducing the original image size. The entire model has no padding applied. The decoder will be defined as follows, replacing the placeholder decoder defined in *step 2*:

```
self.decoder = nn.Sequential(
  nn.ConvTranspose2d(4, 16, 5, stride=2),
  nn.ReLU(),
  nn.ConvTranspose2d(16, 4, 5, stride=2),
  nn.ReLU(),
  nn.ConvTranspose2d(4, 1, 4, stride=2),
  nn.Sigmoid(),
)
```

Three convolutional transpose layers are used here. Each convolutional layer is followed by a non-linear activation layer, where the first two layers used ReLU (being the standard non-linear activation) and the last layer used sigmoid. sigmoid is used here, as the fashion MNIST data is already normalized to have values between 0 and 1. The convolutional transpose layer defined here adopts a similar number of filter configurations from the encoder, from 16 to 4 and finally, to 1 filter to produce only one channel grayscale image.

5. Now that we have defined the convolutional autoencoder, let's load up the fashion MNIST data from the torchvision library. This tutorial will use the Catalyst library for ease of training, so let's take the fashion MNIST dataset loader and feeder class from torchvision and modify it for usage in the Catalyst library:

```
class FashionMNISTImageTarget(
  torchvision.datasets.FashionMNIST
):
  def __getitem__(self, index):
    img = self.data[index]
    img = Image.fromarray(
      img.numpy(), mode="L"
    )
    if self.transform is not None:
      img = self.transform(img)
    return img, img
```

The class from torchvision already has the necessary logic to download and load the fashion MNIST dataset. However, the data feeder method, getitem, is not in the expected format for image generation and, thus, requires this modification for this experiment to work.

6. Note that the `Pillow` library is used to load the image in *step 5*. This is so that we can easily use the tool from `torchvision` to perform different transformation steps, such as image augmentation. However, in this experiment, we will directly convert the `pillow` image into `Pytorch` tensors, using the transform logic that follows:

```
def transform_image(image):
    return torchvision.transforms.ToTensor()(image)
```

7. Now, let's load the training and validation datasets of `fashion MNIST`:

```
train_fashion_mnist_data = FashionMNISTImageTarget(
    'fashion_mnist/', download=True, train=True,
    transform=transform_image,
)
valid_fashion_mnist_data = FashionMNISTImageTarget(
    'fashion_mnist/', download=True, train=False,
    transform=transform_image,
)
loaders = {
    "train": DataLoader(
        train_fashion_mnist_data, batch_size=32,
        shuffle=True
    ),
    "valid": DataLoader(
        valid_fashion_mnist_data, batch_size=32
    ),
}
```

The preceding code downloads the dataset into the `fashion_mnist` folder if it doesn't already exist. Additionally, the `loaders` variable is here so that it can be consumed by the `Catalyst` library.

8. Since the optimization goal is to reduce the different of reproduced pixel values compared to the target pixel values, we will use the mean squared error as the reconstruction loss here:

```
criterion = nn.MSELoss()
```

It's important to note that while reconstruction loss is a common objective in unsupervised representation learning, there may be other metrics or objectives used, depending on the specific algorithm or approach. For example, in **variational autoencoders (VAEs)**, the objective is to maximize the **evidence lower bound (ELBO)**, which consists of both the reconstruction loss and the KL divergence that encourages the learned latent space to follow a specific probability distribution. Another example is perceptual loss, which can be used as a loss function for autoencoders when the goal is to preserve high-level semantic features, rather than achieving pixel-wise accuracy.

9. Now, let's define the supervised runner instance from `Catalyst` so that we can train our model:

```
runner = dl.SupervisedRunner(
     input_key="features", output_key="scores", target_
key="targets", loss_key="loss"
)
```

10. Next, we will define a generally usable function that can make it easy to perform multiple training and validation experiments through code:

```
def train_and_evaluate_mlp(
  trial_number, net, epochs,
  load_on_stage_start=False, best_or_last='last',
  verbose=False
):
  model = net
  optimizer = optim.Adam(
    model.parameters(), lr=0.02
  )
  checkpoint_logdir = "logs/trial_{}_autoencoder".format(
trial_number)
  runner.train(
    model=model,
    criterion=criterion,
    optimizer=optimizer,
    loaders=loaders,
    num_epochs=epochs,
    callbacks=[
            dl.CheckpointCallback(
                logdir=checkpoint_logdir,
                loader_key="valid",
                metric_key="loss",
                load_on_stage_end='best',
            )
    ],
    logdir="./logs",
    valid_loader="valid",
    valid_metric="loss",
    minimize_valid_metric=True,
    verbose=verbose,
  )
  with open(
    os.path.join(checkpoint_logdir, '_metrics.json'),
    'r'
  ) as f:
```

```
      metrics = json.load(f)
      if best_or_last == 'last':
        valid_loss = metrics['last']['_score_']
      else:
        valid_loss = metrics['best']['valid']['loss']
    return valid_loss
```

These are just basic training and evaluation boiler code without any tricks applied. We will explore in depth the tricks to train supervised models in *Chapter 8, Exploring Supervised Deep Learning*.

11. Now, we are ready to train and evaluate the CNN autoencoder model through the following logic:

```
cnn_autoencoder = ConvAutoencoder()
best_valid_loss = train_and_evaluate_mlp(
    0, cnn_autoencoder, 20, load_on_stage_start=False, best_or_
last='last', verbose=True
)
```

The best performing `cnn_autoencoder` weights based on the validation loss will be automatically loaded after training on 20 epochs is complete.

12. After training the preceding model with the provided training code, on the Fashion MNIST dataset with 1x28x28 image dimensions, you should get something similar to the example input/output pairs shown in *Figure 5.4* through the following code:

```
input_image = valid_fashion_mnist_data[0][0].numpy()
predicted_image = cnn_autoencoder(
   torch.unsqueeze(valid_fashion_mnist_data[0][0], 0)
)
predicted_image = predicted_image.detach().numpy(
).squeeze(0).squeeze(0)
f, axarr = plt.subplots(2,1,  figsize=(5, 5))
axarr[0].imshow(predicted_image, cmap='gray')
axarr[1].imshow(input_image.squeeze(0), cmap='gray')
```

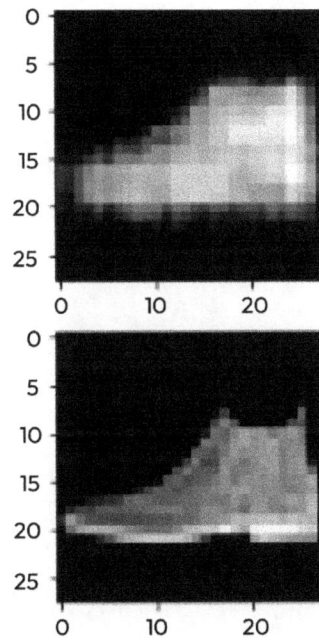

Figure 5.4 – Autoencoder sample results on the Fashion MNIST dataset; the bottom represents the original image, and the top represents the reproduced image

With the results, it's clear that the bottleneck features are capable of reproducing the entire picture somewhat completely, and they can be treated as more representative and compact features for use instead of the original input data. Increasing the number of representation values of the bottleneck feature from 4 to something like 10 features should also increase the quality of the reproducible image. Feel free to try it out and experiment with the parameters!

Summary

Autoencoders are considered a fundamental method to achieve representation learning across data modalities. Consider the architecture as a shell that you can fit in a variety of other neural network components, allowing you to ingest data of different modalities or benefit from more advanced neural network components.

However, do note that they are not the only method to learn representative features. There are many more applications for autoencoders that primarily revolve around different training objectives using the same architecture. Two of these adaptations that were briefly introduced in this chapter are denoising autoencoders and variational autoencoders, which will be introduced properly in *Chapter 9, Exploring Unsupervised Deep Learning*. Now, let's shift gears again to discover the model family of transformers!

6

Understanding Neural Network Transformers

Not to be confused with the electrical devices that are also called transformers, neural network transformers are the jack-of-all-trades variant of NNs. Transformers are capable of processing and capturing patterns from data of any modality, including sequential data such as text data and time-series data, image data, audio data, and video data.

The transformer architecture was introduced in 2017 with the motive of replacing RNN-based sequence-to-sequence architectures and primarily focusing on the machine translation use case of converting text data from one language to another language. The results performed better than the baseline RNN-based model and proved that we don't need inherent inductive biases on the sequential nature of the data that the RNNs employ. Transformers then became the root of a family of neural network architectures and branched off to model variants that are capable of capturing patterns in other data modalities while continually raising the bar of the performance of the original text sequence data-based tasks. This showed that we don't really need to have inherent inductive bias built into a model in general, regardless of the data modality, and instead, we can allow the model to *learn* these inherent structures and patterns. Do you have sequential data such as video, text, or audio? Let the neural network learn its sequential nature. Do you have image data? Let the neural network learn the spatial and depth relationships between the pixels. You get the picture.

Before we explore a more in-depth overview of transformers, take a breather to check out the topics that will be covered in this chapter:

- Exploring neural network transformers
- Decoding the original transformer architecture holistically
- Uncovering transformer improvements using only the encoder
- Uncovering transformer improvements using only the decoder

Exploring neural network transformers

Figure 6.1 provides an overview of the impact transformers have had, thanks to the plethora of transformer model variants.

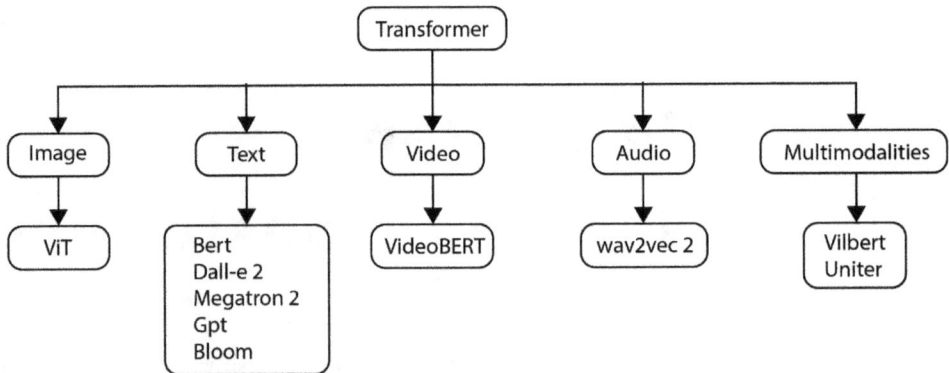

Figure 6.1 – Transformers' different modality and model branches

The transformer does not have inherent inductive bias structurally designed into its architecture. Inductive bias refers to the pre-assumptions made by a learning algorithm on the data. This bias can be built into the model architecture or the learning process, and it helps to guide the model toward learning specific patterns or structures in the data. Traditional models, such as RNNs, incorporate inductive bias through their design, for instance, by assuming that data has a sequential structure and that the order of elements is important. Another example is CNN models, which are specifically designed for processing grid-like data, such as images, by incorporating inductive bias in the form of local connectivity and translation invariance through the use of convolutional layers and pooling layers. In this context, the model architecture itself enforces certain constraints on the patterns that can be learned.

Transformers were designed with the idea that we should allow the model to decide how and where to focus, based on all the input data provided. This is executed technically using an aggregate of multiple mechanisms that each decide where to focus in different ways. Thus, transformers depend entirely on the information provided by the input data to determine any form of inductive bias your data has, if any. These focusing components are formally called attention layers. The improvements on top of transformers that form the branches you see in *Figure 6.1* do not deviate far from the base structure of transformers. Usually, the improvements to add different modalities are done by data setup variations to adapt the input data structure to the structure of the transformer along with the different prediction output details specific to variations in the target task application.

At the moment, transformers are the architecture family with the biggest capacity for learning and identifying highly complex patterns that exist in the real world. To add to that, for a period of almost a year from mid-2022 to 2023, transformers demonstrated that their informational capacity is bounded only by the hardware resource capacity. Currently, one of the largest transformer models, scaled up many times from the base transformer, with an astronomical *540 billion* parameters, is a model variant called **PaLM** provided by Google. Additionally, it is rumored that the GPT-4 multimodal text generation model by OpenAI, which is not available as an open sourced model but as a service, consists of either multiple models or a single model that adds up more than a trillion parameters. These huge models usually take months of training to allow the model to achieve peak performance, along with the need to have highly performant and state-of-the-art GPUs.

You might wonder why we would want to train such a big model and wonder whether the value it provides is worth it. Let's evaluate one of the transformer models, called **GPT-3**, developed by OpenAI. GPT-3 is a type of language model; it takes in input text and outputs text-based predictions based on what it thinks is the most appropriate and useful response. Now, this spans many different tasks in the NLP space that conventionally would have been accomplished by individual models for each task, making it a **task-agnostic model**. The tasks that can be accomplished are machine language translation, reading comprehension, reasoning, arithmetic processing, and in general, demonstrating a wide variety of language understanding capabilities, providing results in different formats, depending on the query input text. For example, end user applications include generating code in any specified languages capable of achieving the described objectives, writing fiction with a specified theme, obtaining any requested information on any publicly known person, summarizing customer feedback, in general or with a focus on certain topics, such as "what's frustrating the customer," and adding realism to the conversations of characters in a virtual world. To date, there are more than 300 applications of GPT-3 being used in a variety of categories and industries, a number that will seem small but will pave the way for more innovation and adoption of the immense capabilities of NNs.

Transformers have gone through multiple years of gradual improvements through rigorous research but have not deviated a lot from the base architecture. This means that understanding the original architecture is key to understanding all the latest and greatest improved transformers such as GPT-3. With that, let's dive into the original transformer architecture from 2017 and discuss which components have been changed or adapted in the past five years of research that gave birth to new model architectures.

Decoding the original transformer architecture holistically

Before we look into the structure of the model, let's talk about the basic intent of transformers.

As we covered in the previous chapter, transformers are also a family of architectures that utilize the concept of encoder and decoder. The encoder encodes data into what is known as the code and the decoder decodes the code into a data format that looks similar to raw, unprocessed data. The very first transformer used both the encoder and decoder concepts to build the entire architecture and demonstrated its application in text generation. The subsequent adaptations and improvements either used only the encoder or only the decoder to achieve different tasks. In a transformer, however, the encoder's goal is not to compress the data to achieve a smaller and more compact representation of the data, but instead mainly to serve as a feature extractor. Additionally, the decoder's goal for transformers is not to reproduce the same input.

> **Note**
>
> It is possible to build an actual autoencoder structure with the transformer component instead of using CNN or RNN components, but this will not be covered in this book.

The original transformer fixed the data/weight dimensions in the encoder and decoder to a unified single size so that residual connections could be made, using a dimension of 512.

Transformers also utilize logical block structures to define their novelties, which allows you to scale their size easily. The core operation of the transformer is the mechanism to identify which part of the entire input data to focus on for each input data unit in an input data sample. The focusing mechanism is achieved technically by a type of neural network layer called the **attention layer**. There are many types of attention variants that will not be covered in this book.

Transformers utilize the simplest variant of attention that utilizes the `softmax` activation to achieve **dot-product**-based attention. Since `softmax` forces values to add up to `1.0`, there is usually one strongest focal point for each layer. Take this as a form of gate similar to the gates in RNN, for which you can refer to *Chapter 4, Understanding Recurrent Neural Networks*, again for the full context on what an RNN is and how it functions. Transformers use multiple attention layers where each input will be focused on in multiple ways, and the aggregated name to simplify referencing and scaling is called the **multi-head attention** layer.

The inputs and outputs of transformers are in the form of tokens. Tokens are a way to refer to the individual units of input data that can be passed into and out of a transformer. These tokens can be sequential or non-sequential, grouped or non-grouped, although the structure of transformers is not explicitly designed with these assumptions. Through research and advancements since they were devised, tokens for text can be words or sub-words, tokens for images can be image patches, tokens for audio can be sequential time windows of audio data, for example, one-second windows, and tokens for videos can be single image frames or a group of image frames. As the model does not have an inherent inductive bias toward any type of data modalities, the positional or sequential identification of the data is encoded into the data explicitly by tokens. These can range from incremental discrete integer positions (1, 2, 3 ...) to continuous floating-point positions in between discrete integer values (1.2, 1.4556, 2.42325 ...), from absolute positions to relative positions, and finally, single-valued positions for each token or embedding that can be learned. The original transformer used a mapping function that maps absolute integer positions to relative floating-point positions that consider the data dimensions of the tokens and the intended unified data dimension size, but the state-of-the-art models commonly adopt instead embeddings that can be learned. Embeddings are a look-up table to encode any discrete categorical data into a higher-dimensional and more complex representation of the original category that can accurately discern categories among each other. Note that the actual input data token itself is also usually applied with embeddings depending on the input data (image frames or patches usually don't use embeddings as they are already high in dimensionality, but text tokens are categorical and use embeddings).

Figure 6.2 shows a high-level overview of the architecture:

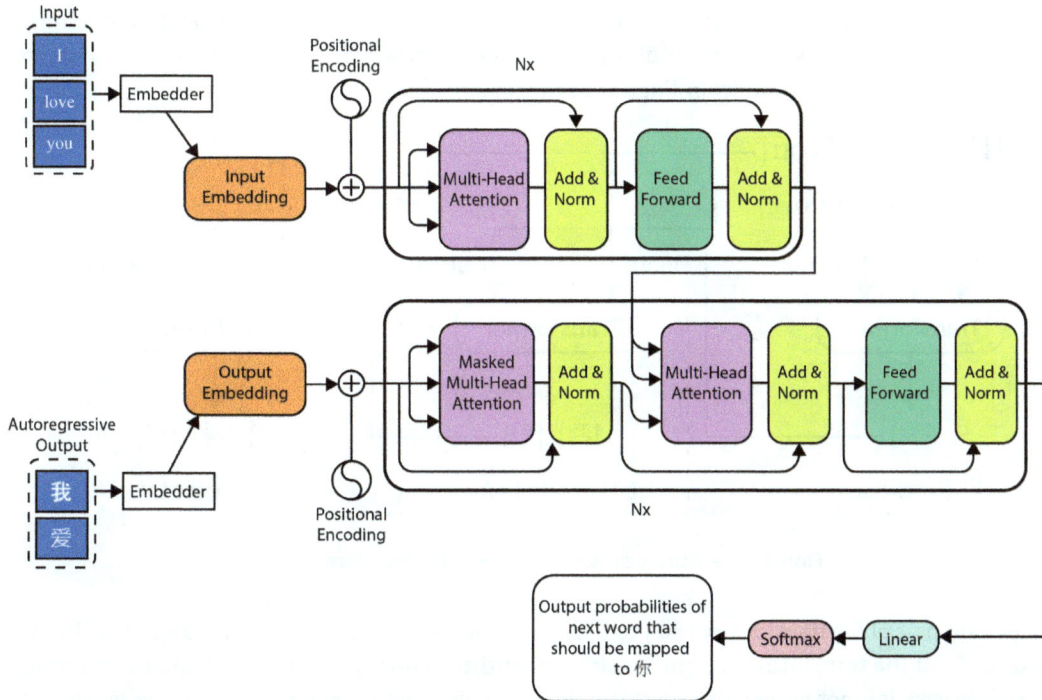

Figure 6.2 – Transformers viewed with input and output visualizations with both the encoder and decoder of the task to translate an English sentence to its Mandarin counterpart

The encoder and decoder have somewhat similar structures, with the exception that the decoder has an extra middle **multi-head attention** layer that connects to the output of the encoder and a masked version of the multi-head attention block. The purpose of masking is to prevent the use of irrelevant data, which will be apparent in the next paragraph of the input text. The encoder's purpose in the original transformer is to transfer the input data information to the middle multi-head attention layer for multiple blocks of the decoder. The decoder's purpose is to finally produce the output that gets fed into a token-wise fully connected layer (formally known as the pointwise linear layer) for supervised target predictions. Predictions can be for regression or classification-based targets in which the latter employs a softmax activation function.

> **Note**
> The decoder also takes in other data, aside from the encoder outputs.

To achieve data generation without a constraint on the number of outputs, the transformer was designed into an autoregressive model. An autoregressive model uses the result of the first prediction as input for the second prediction and is subsequently used as the input for the next predictions. The following figure shows an example of this prediction process with text-based data that aims to accomplish machine translation from English to Mandarin.

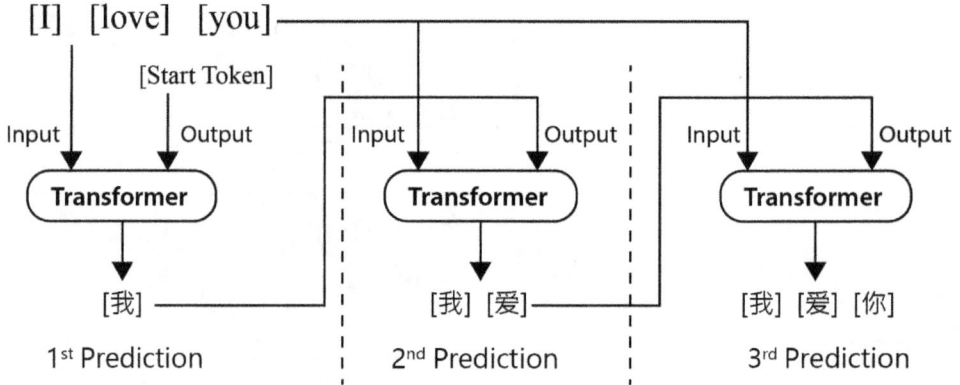

Figure 6.3 – Autoregressive workflow for transformers

Each forward pass of the transformer will predict a single output. An end-of-sentence token prediction is factored in at the transformer output to signal that the prediction is done. During the training process, however, it is not necessary to perform this autoregressive loop; instead, masks are generated at random positions to nullify all the future tokens from the target tokens to prevent the model from copying the single target token directly and taking in future token information. Plainly speaking, the masking mechanism is not used during the prediction or inference stage.

The multi-head attention layer has a few other key operations besides the actual attention mechanism. *Figure 6.4* shows the unraveled multi-head attention layer along with the previously mentioned attention mechanism in the scaled dot-product attention layer.

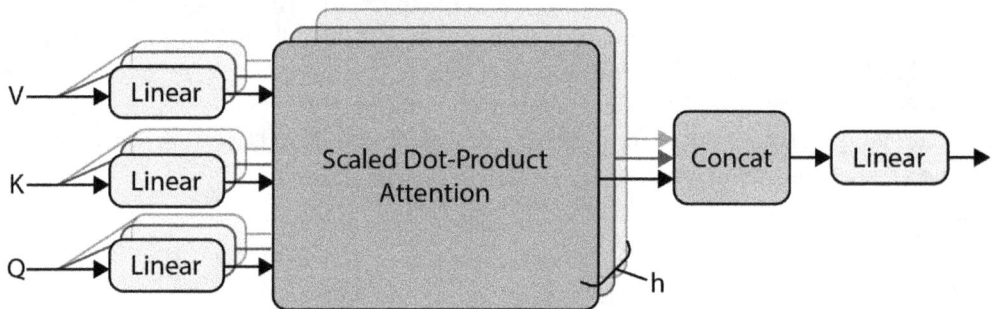

Figure 6.4 – Multi-head attention layer with the scaled dot-product attention structure

Take the multiple attention layer as multiple humans contributing to different focusing patterns on the same data. **Q**, **K**, and **V** represent the query, key, and value respectively where the query represents the individual tokens, and the key and value are just the entire lists of tokens. The linear layers are the only place where the weights live in the multi-head attention layer where each query, key, and value component has its own linear layer. The scaled dot-product attention layer is where the mask is used to nullify the future tokens during training. One unique part of this layer is that it is scaled by the square root of the linear layer output dimensions to prevent `softmax` from focusing on unimportant areas. The attention mechanism is meant to find the degree of relevance of each token to the entire set of input tokens, in other words, to find out what tokens interacted with what tokens and figure out how much they depend on each other. Recall that the model's dimension has to stay a fixed sized throughout the architecture to ensure values can be added together without additional processing. Since the output of the multiple attention heads will be concatenated instead of added, the dimensions of the linear layer have to be evenly distributed among the heads. With 8 heads, for example, and a data/weight dimension of 512, the linear layer dimension for each query, key, and value for each head would then be 512/8=64 neurons. *Figure 6.5* shows a simple example of the outputs and their shapes in the workflow with the number of fixed data/weight dimensions per token being 3, using the same input text data as *Figure 6.4*.

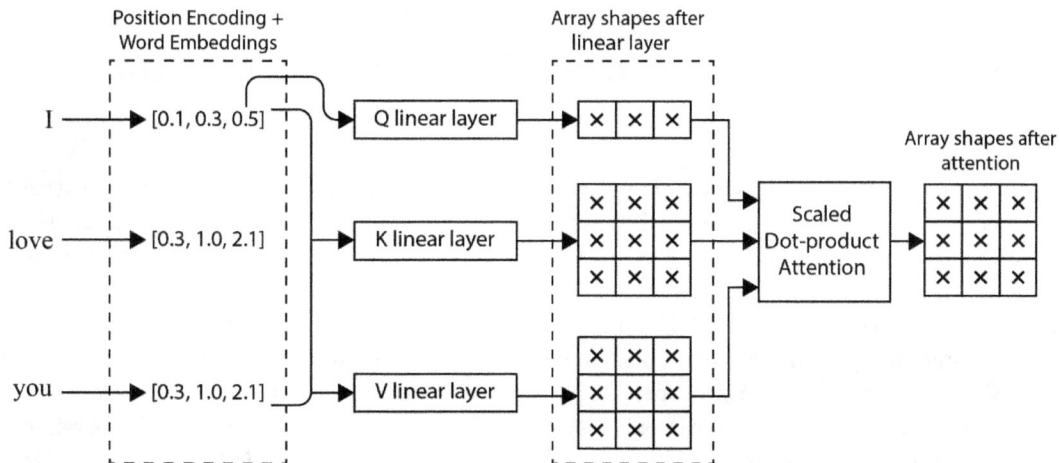

Figure 6.5 – Part operation of the multi-head attention before concatenating using the query of the word "I" in the context of "I love you" with a linear layer with an output dimension of 3

The same operation will be done for the other two query words, "love" and "you." An operation from a single head focuses on the input data in one way while the other heads focus on the input data in a different way. In the encoder-decoder layer, the query comes from the decoder layer while the keys and value come from the encoder layer. This allows the decoder to choose where to focus in the given input text to produce the next output. *Figure 6.6* shows a good way to think about the output of multi-head attention with a four-headed multi-head attention focusing on different parts of the text.

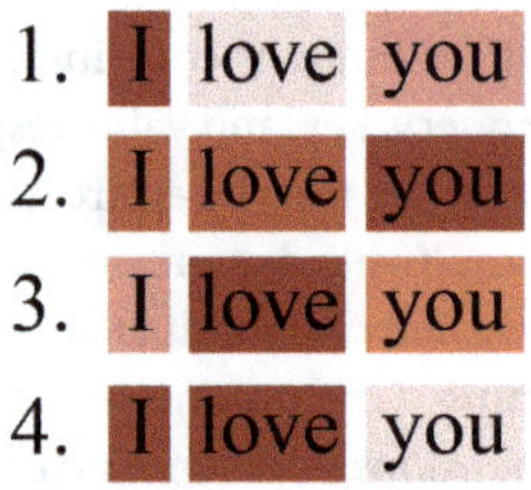

Figure 6.6 – Multi-head attention example output with four heads

The regularization mechanisms part of the transformer has the same purpose as described in other previous architectures such as CNNs, RNNs, and MLPs, and won't be discussed again here. This sums up all the components of the transformer. This base architecture allows us to achieve better data generation results compared to older architectures such as sequence-to-sequence or autoencoders. Although the original architecture focused on applications for text data, the concept can and has been adapted to handle other types of data modalities such as image, video, and audio data. Recall that the improvements and adaptations that came after the original model used only the encoder or only the decoder. In the next two topics, we will dive into the two different concepts separately.

Uncovering transformer improvements using only the encoder

The first type of architectural advancements based on transformers we will discuss are transformers that utilize only the encoder part of the original transformer using the same multi-head attention layer. The encoder-only line of transformers is adopted generally because there is no masked multi-head attention layer since the next token prediction training setup is not used. In this line of improvements, training goals and setups vary across different data modalities and vary slightly for sequential improvements under the same data modality. However, one concept that stays pretty much constant across different data modalities is the fact that a semi-supervised learning method is used. In the case of transformers, this means that a form of unsupervised learning is executed first and then the straightforward supervised learning method is executed next. Unsupervised learning offers transformers a way to initialize their state based on a wider understanding of the nature of the data. This process is also known as **pre-training** a model, which is just one form of unsupervised learning. Unsupervised learning will be discussed more extensively in *Chapter 8, Exploring Unsupervised Deep Learning*.

Some improvements are specific to the nature of the data modality, which mostly consists of text-specific improvements and involves a change in the type of task to optimize. Other improvements are general transformer architectural improvements that can generalize to different data modalities. We will start by focusing on the general architectural improvements before diving into the improvements specifically crafted for text to date, which might or might not be adaptable to other data modalities. To start off, we will go through the base architecture of an encoder-only structure called BERT.

The **Bidirectional Encoder Representations from Transformers (BERT)** architecture, introduced in 2018, is simply a decoder-only transformer that utilizes the same multi-head attention layer and introduces a pre-training task that can be generalized to other data modalities, but was built for text data. This task was called **masked language modeling (MLM)** where the objective is to predict the randomly masked input tokens and only learn from these token outputs. Note that the encoder of the transformer outputs token-based outputs and thus token classification can be achieved by simply adding a token-wise `softmax` layer. This specific task was proven to be highly effective as a pretraining method that allows downstream subsequent supervised tasks to achieve better performance. Since the tokens are masked out in different positions from the original sequence, the architecture is said to be bidirectional, as information from the future tokens can be attended to predict tokens in the past in addition to the past tokens. 15% of tokens were masked, for example, and from that percentage, 10% were randomly replaced with random tokens to act as noise, making the model similar to a denoising autoencoder and allowing it to be robust to noise. Standard autoregressive structures are naturally uni- and forward-directional and can't take advantage of future tokens. BERT also introduced ways to encode multiple sentences in a single input representation. This is achieved by adding a special separator token that has its own learned embeddings to signify a separation between text along with an extra segment embedding that signals to the transformer which sentence number the token is part of. This allows another task called **next sentence prediction (NSP)** to be optimized along with the MLM objectives for pre-training purposes. This is achieved by reserving the first token to be a special class token, shown as **CLS** in *Figure 6.7*, and the first output token to be used as the prediction for the next sentence, where positive examples are two consecutive sentences from a single document and negative examples are created by pairing sentences from different documents. The pre-training mechanism is a way to allow the model to learn generalizable patterns of the nature of the data, and these two tasks proved to improve the performance of downstream supervised learning tasks. Downstream supervised learning simply adds a global fully connected layer across all the tokens along with target-specific layers such as `softmax`. The three embeddings (positional embeddings, segment embeddings, and token embeddings) are added together before passing the result into the transformer. *Figure 6.7* depicts this architecture.

Output   [NPS] [Token 1] ... [Token N] [Separation Token] [Token 1] ... [Token N]

Add & Normalize

Fully Connected Layer

Add & Normalize

Multi-head
Attention Layer

Position Embeddings

+

Segment Embeddings

+

Token Embeddings

Input   [CLS] [Token 1] ... [Token N] [Separation Token] [Token 1] ... [Token N]

Sentence 1   Sentence 2

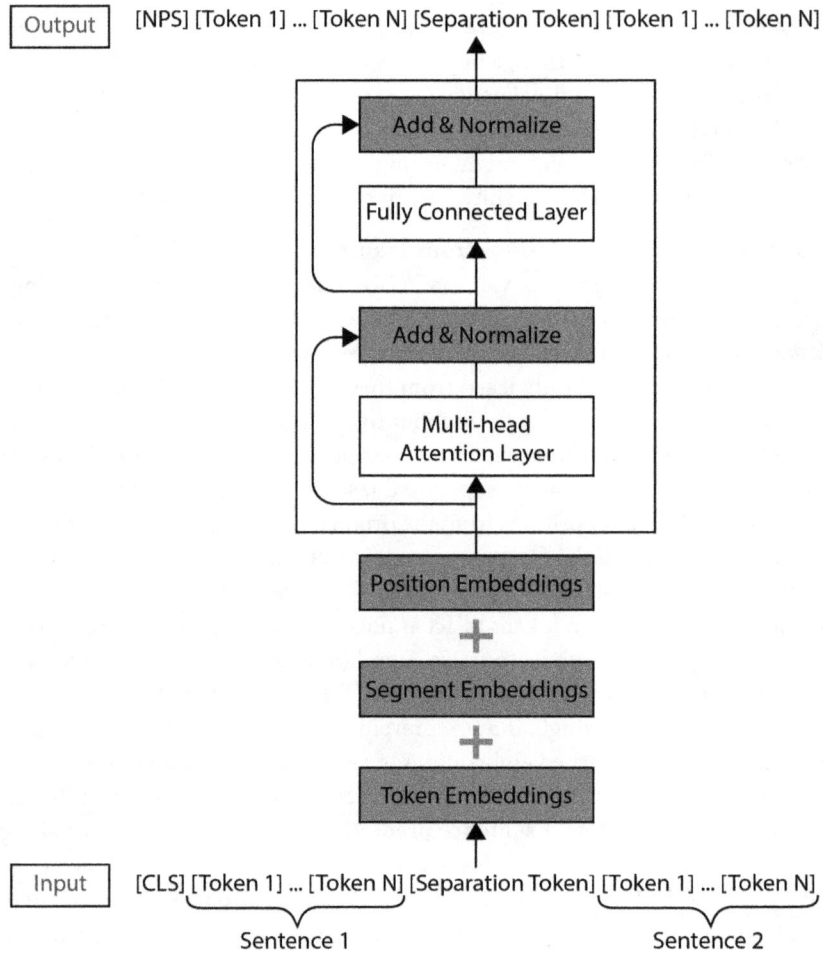

Figure 6.7 – BERT architecture

The architecture focused on ways to represent extra input and methods to increase model understanding of the nature of the data modality through multiple-task learning and exceeded prior performance on multiple downstream text-based tasks.

Now that we have introduced the base encoder-only model, we are ready to explore the different categories of advancements specifically for an encoder-only transformer. Improvements will be explored in terms of advancement categories instead of by model this time, as the different models combine many advancements, which makes it hard to compare actual advancements. The three types of advancements are as follows:

- Improved data-modality-based tasks used for pre-training
- Architectural improvements in terms of compactness, speed, and efficiency
- Core/functional architectural improvements

We will skip the details of data-based improvements such as using multilingual data to build a multilingual BERT or the fact that sub-word-based tokens are used to reduce vocabulary and increase vocabulary reuse between words.

Improving the encoder only pre-training tasks and objectives

Recall that MLM and NSP are used in the base BERT model and that MLM was mentioned to be a robust task for the understanding of text language but can be easily adapted to tokens from other data modalities. NSP, however, has been proven to be unstable and doesn't concretely help models improve performance in every case. One improvement in this direction is to use **sentence order prediction** (**SOP**). SOP demonstrated higher consistency in improving the downstream supervised task performance compared to simply utilizing the inverted sequence of positive consecutive sentences from the same document as a negative sentence. Conceptually, it learns the cohesion between sentences instead of trying to simultaneously predict whether two sentences are from the same topic or not. This method was introduced in the **ALBERT** model in 2020, which mainly focused on transformer efficiency.

Another notable improvement is called **replaced token detection** (**RTD**). RTD, from the **ELECTRA** model, predicts whether a token is replaced by a random token or not and requires that all token positions are learned from, compared to the MLM process of only learning from the masked token positions. This was introduced as an improvement on the MLM objective.

Next, let's dive into architectural improvements in terms of compactness, speed, and efficiency.

Improving the encoder-only transformer's architectural compactness and efficiency

Transformers have proved themselves to hold an immense structural capacity to take in information that grows along with the model. Some research on transformers has been about increasing and scaling the transformer model to the level of hundreds of billions of parameters in both decoder-only and encoder-only models. The results achieved using these huge models are astounding, but they are unsuitable for practical use due to the need to have state-of-the-art GPUs and machines that are not readily available or affordable for most people in the world. This is where the line of architectural compactness and efficiency improvements can help to level the field. Since we are mostly limited by the hardware resource that we can use, if the model can be efficiently reduced in size architecturally while maintaining the same performance, this would allow a more performant model when you scale it up to the limits of your hardware resource.

One of the notable improvements in this line comes again from the ALBERT model, a lite BERT model. A key improvement made here was to factorize the embeddings layer in the same way as convolutional layers were factorized to make depthwise convolutions. A smaller embedding dimension is used with a fully connected layer that maps the small dimensional embedding to the same desired dimensions. As embeddings are essential weights and contribute to the number of parameters, factorizing the embeddings layer allows for a transformer model with fewer parameters. This allows BERT to be more performant at a lower number of parameters. Note that some improvements add model parallelization and use clever memory management, but these will not be covered extensively here. However, since there are some methods that speed up the model in general for both inference and training stages, they will be discussed and covered in *Chapter 15, Deploying Deep Learning Models in Production*, which is all about deployment.

Improving the encoder-only transformers' core functional architecture

The most notable improvement made in this line is from the model called **Decoding-enhanced BERT with Disentangled Attention (DeBERTa)**, introduced in 2021, which is considered to be the current architectural SOTA. The main idea here is that two separate positional encoded embeddings are used, which are relative positions and absolute positions. These positional embeddings, however, are not added along with the input token embeddings but instead are treated as separate queries and keys to be fed into the layers of the transformer, with their own fully connected layers. This allows three explicit attention maps to be obtained, specifically, content-to-position attention, content-to-content attention, and position-to-content attention. The different attention maps are then summed together. The relative positions are specifically used by each intermediate multi-head attention layer's internal scaled dot-product attention layer by acting either as the key or query, depending on whether it is being attended to by the content or attending to the content.

This allows the raw relative positional information to be explicitly considered in every attention layer instead of possibly being forgotten after going through many layers. We are essentially applying a form of skip connections for relative position data. Relative encoding allows the model to learn more generalizable representations by conceptually allowing pattern identifiers to be reused at each relative position. This solution is similar to how the same convolutional filters are used at different partitions of a single-image data instead of using one fully connected layer on an entire image, which doesn't allow the reuse of pattern identifiers across the entire input space. As for the usage of absolute positions, it was only added due to the nature of text data where the absolute position is needed to discern the absolute importance of different things, such as police holding more power than most people and trucks being bigger than motorcycles. Absolute positions, however, are only added to the last few layers once, using a mechanism called **enhanced mask decoder** (**EMD**). EMD applies absolute positional embeddings as the query component of the scaled dot-product attention. There are no concrete reasons why it is only applied in the final few layers and it can probably be extended to be applied in every layer.

> **Bucketing**
> A bucketing mechanism to group multiple positions into a single number is used to reduce the number of embeddings and thus the number of parameters. Consider this as grouped position information that again increases the compactness of the model.

Uncovering encoder-only transformers' adaptations to other data modalities

Transformers are capable of handling other data modalities. First, the input data just needs to be structured properly in a sequence format. After that, since pre-training is at the core of a transformer, suitable task objectives need to be crafted for the encoder-only transformers to achieve a pre-trained state that captures an understanding of the specific data modality. Even though the introduction to transformers using text is focused on an unsupervised pre-training method, it really can be either supervised, unsupervised, or semi-supervised, where the key is to set the weights of the model to be in a state that the subsequent supervised downstream task can leverage.

For images, **Vision Transformer (ViT)**, introduced in 2021, forms the base of an image-based transformer and has proved that transformers can achieve competitive performance for image-based tasks when compared to convolutional-based models. One of the most problematic issues for transformers to handle images is that images are big in size and attending to the entire large input space bloats up the model size of transformers very quickly. ViT solves this by splitting images into systematic patches where each patch would subsequently be fed into a single fully connected layer that reduces the dimension of each image patch substantially. The patch data with reduced dimensions will be the token embeddings for the transformer. The patch conversion is an essential process to represent an image input in a format that a transformer can process. Additionally, this patch-based mechanism introduces some form of image domain inductive bias, but the rest of the architecture remains practically free of inductive bias for image data. As transformers are almost always coupled with a pre-training method for performance improvement reasons, ViT adopted a similar style of pre-training as the CNN model families, using supervised classification pre-training.

For audio, **wav2vec 2.0**, created in 2020, utilized similar concepts to its predecessor, **wav2vec**, but instead utilized a decoder-only transformer as part of the architecture. A brief overview is that wav2vec 2.0 first encodes audio of a specific time window length using 1D convolutions into a lower-dimensional space as a feature vector. The encoder feature vectors for each window are then fed into the decoder transformer and act as tokens. The unsupervised pre-training method is applied here, similarly to MLM from BERT, where a certain portion of the input encoded feature vectors is masked, and the task is to predict the masked encoded feature vector. After pre-training, the model is then fine-tuned on downstream tasks, using a linear layer right after the transformer outputs for tasks such as speech recognition. wav2vec 2.0 achieved a new SOTA on a downstream speech recognition dataset and proved that audio works extremely well with transformers.

VideoBERT, created in 2019 by Google, not only demonstrates methods to model video data for transformers but also the capability of the transformer model to learn joint representations in a multimodal way using both video and text. Note that a video by itself is also inherently multimodal since video data contains multiple image frames as well as an audio data component. In VideoBERT, a cooking video sentence pair dataset with clear annotations for the starting and ending time stamps of each cooking action is used to pre-train the model. The model takes in the sequence of image frames by encoding a preset number of frames into a feature vector using a pre-trained 3D convolutional model and treats the feature vector as a token. These encoded video tokens are then paired with the text data and use a similar objective to MLM to predict masked tokens of videos and text tokens. The model then can either be used as a featurizer for video and text-related data or similarly be fine-tuned with an additional linear layer on the outputs of the transformer model.

The preceding summaries of different approaches to different data modalities are meant to serve as a reminder that transformers can be applied to more than just text data and not as an introduction to the current state of the art for these modalities. A key takeaway here is that the different data modalities need to be arranged into a sequence of token formats with reduced dimensions and suitable task objectives need to be crafted for the encoder-only transformers to achieve a pre-trained state that captures an understanding of the main nature of the data modality. Moreover, the same pre-training method from BERT can be easily adapted to other data modalities. So instead of naming it "masked language modeling," maybe it would be better to name it "masked token modeling." Next, let's uncover the transformer's improvements using only the decoder.

Uncovering transformer improvements using only the decoder

Recall that the decoder block of the transformer focuses on an autoregressive structure. For the decoder-only transformer line of models, the task of predicting tokens autoregressively remains the same. With the removal of the encoder, the architecture has to adapt its input to accept more than one sentence, similar to what BERT does. Starting, ending, and separator tokens are used to encode input data sequentially. Masking is still performed to prevent the model from depending on the current token to predict future tokens from the input data during predictions, which is similar to the original transformer along with positional embeddings.

Diving into the GPT model family

All these architectural concepts were introduced by the GPT model in 2018, which is short for **generative pre-training**. As the name suggests, GPT also adopts unsupervised pre-training as the initial stage and subsequently moves into the supervised fine-tuning stage for any downstream task. GPT is focused on text data, but the same concepts can be applied to other data modalities. GPT uses the basic language modeling task for pre-training to predict the next token given the prior tokens as context with a fixed window to limit the number of context tokens taken as input. After the weights are pre-trained, it is subsequently fine-tuned to the objectives defined by any of the supervised task types. GPT-1 utilized the same language modeling task objective as an auxiliary side objective along with the main supervised task as an attempt to boost the performance on downstream tasks and showed that it boosts performance only on larger datasets. Furthermore, the model showcases first-hand the generalizability of using language modeling as a pre-training task to other tasks without actually fine-tuning it and still achieving good scores. This line of behavior is termed **zero-shot** and will be discussed more extensively in *Chapter 9, Exploring Unsupervised Deep Learning*. The model was pre-trained on 7,000 unpublished books and had 117 million parameters with 12 repeated decoder blocks.

It's important to clarify that the GPT model's ability to perform without fine-tuning (few-shot or zero-shot learning) doesn't negate the benefits that fine-tuning can provide in domain-specific tasks. While the GPT line of models has shown impressive capabilities in various tasks, fine-tuning may still be advantageous for specific applications, leading to improved performance and better adaptation to the unique requirements of a given task.

Proceeding to the next advancement, GPT-2 had no concrete architectural changes as the main idea but emphasized the capability of zero-shot learning with the concept of task conditioning along with scaling the GPT-1 model by 10 times in the number of layers, embedding dimensions, the context window size, and vocabulary size. Task conditioning is the idea that you tell the model what kind of outputs to generate instead of having a fixed, single, known prediction output that is determined by the type of task you train it for. To allow true flexibility similar to humans, they didn't add any new architecture to accommodate the task conditioning idea but instead decided that the textual input for the transformer could be directly used as the task conditioning method. For example, free-form text model input such as "I love you = 我爱你, I love deep learning=" naturally directs the transformer to perform a translation from English to Mandarin. This can also be done differently with an example that specifies both the task and the context such as "(translate to mandarin), I love deep learning =". GPT-2 achieves this by pre-training with the language modeling objective on a wide variety of textual domains such as news articles, fiction books, Wikipedia, and blogs, essentially a much larger dataset compared to the training dataset in GPT-1. This work exemplifies the huge potential of true AI generalization in general even though at the moment only text and image data have been utilized. For image data, this is done in a model called **DALL-E**, which is capable of generating images according to the provided input text.

Finally, the GPT-3 model was introduced in 2020. It is similar to the previous advancement, with the main changes being increasing the size of the model in all components and characterizing different types of text examples with different ways of specifying the task conditions. We won't go into this too much here. Although GPT-2 performed well on zero-shot settings without fine-tuning for the downstream tasks, only by fine-tuning could the model exceed previous benchmarks on multiple text language datasets. The sheer size of GPT-3 completely removed the need for fine-tuning by surpassing previous benchmarks without any form of fine-tuning. This model, however, requires specialized machines to train and is not readily accessible to individuals or small organizations. It is currently part of OpenAI's API offering. GPT-3 demonstrates a wide range of applicability to many different tasks without any fine-tuning, where some of the most notable uses are code generation for any programming language, storytelling, and producing amazingly human-seeming chatbots. The three improvements pushed the performance limitations on many datasets with language-based tasks to higher levels, and serve as an example of the generalizability component of machine learning, as well as the immense learning potential of transformers, limited only by the amount of hardware resources you have.

Next, we will discuss a slightly different form of a decoder model that is worth mentioning, called the **XLNet** model.

Diving into the XLNet model

If you and a friend were given the prompt "_ _ is a city," would you both separately predict "New York" to fill in the blanks? This is one of the flaws of MLM objectives of the encoder-only transformers as they learn by predicting on multiple tokens at once. Standard autoregressive models, however, are not susceptible to this issue due to their autoregressive nature of only predicting the future and only one token at a time. Encoder-only transformers, however, have a bidirectional sequence support that leads to improved performance. Mainly, XLNet works on the idea of making an autoregressive model attend to data in a bidirectional way like encoder-only models, while maintaining the benefits of an autoregressive model of predicting a single output.

The way XLNet achieves the best of both worlds is by conditioning the decoder-only transformer during MLM pre-training. The idea is to pre-train the model on all permutations of the token's sequence order by using the masking mechanism. Standard autoregressive models use the order of "1-2-3-4" sequentially to predict the fifth token during pre-training, but in XLNet, the order can be changed into "3-4-1-5" to predict the second token. XLNet also introduced another extra query linear layer hidden state along with an additional single fixed learnable embedding that will be traversed through the attention layer along with each token embedding, as an extra stream along with the original token-based content stream. The extra path is called the query stream. The entire process is called the masked two-stream attention layer, depicted in *Figure 6.8*.

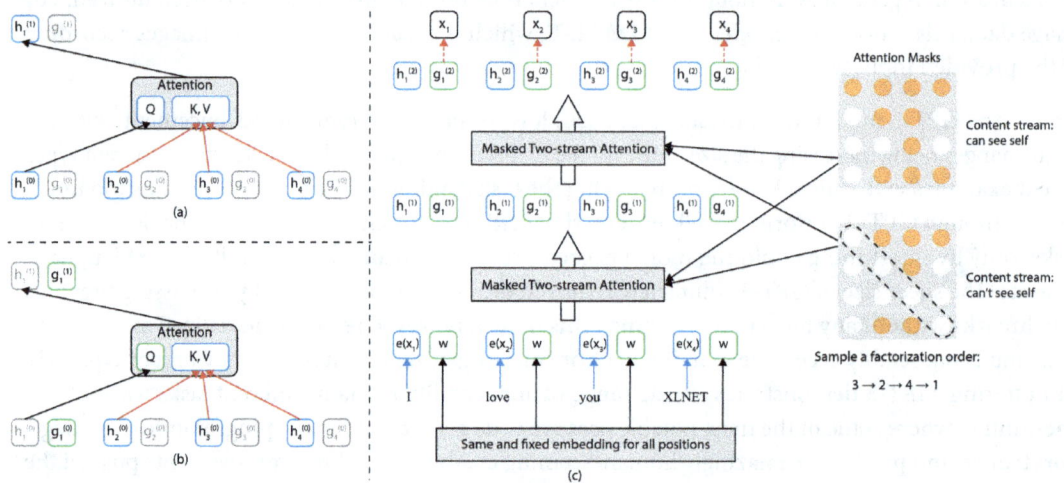

Figure 6.8 – XLNet workflow

The forward propagation mechanism of the content-based stream for any given position of a sequence ordering has the capability to attend to the content of its own position and the positions before it while the query-based stream for any given position of a sequence ordering only has the capability to attend to the content of positions before it. *Figure 6.8* (**a**) shows how the attention operation for the content stream for the actual position 1 in the model can attend to all tokens in other positions due to the ordering of "3-2-4-1" where position 1 is permuted to be the last token. *Figure 6.8* (**b**) shows how the token content at position 1 is not used to update the additional query stream path data denoted as **g** and that the **g** component acts as the query to obtain the next **g** as the output of the attention layer. If the fourth position of the model needs to be updated, only the second and third position content will be attended to in the content stream, as the first position is the future token according to the permuted sequence order. The masking mechanisms are applied dynamically based on this concept to prevent each position of **h** and **g** from attending to the content positions before it is defined from the permuted sequence order. Note the **w** component in *Figure 6.8* (**c**) denotes the single-weight embedding shared across all positions for the query stream. During pre-training, the query stream acts as the path for the final classification output for the MLM objective where the final **h** components are ignored. When passed on to the fine-tuning stage, however, the query stream components and query-based linear layer hidden states are discarded or ignored completely.

The novel way of enabling the bidirectional context of the data while leveraging the benefits of a single-word prediction-based language modeling objective during pre-training led to improved performance on downstream supervised learning tasks.

Discussing additional advancements for a decoder-only transformer model

Since the advent of GPT, there have not been any highly impactful changes to the base decoder-only architecture. Most of the work focuses on three things:

- Applying transformers to different problem types

- Scaling up the model size to hundreds of billions of parameters

- Focusing on engineering solutions so that a huge model is feasible to be trained on the available hardware resources

One of the most notable works utilizing decoder-only transformers for other problem types is the DALL-E model from the OpenAI team. Transformers were used in both the first and second versions of DALL-E to autoregressively predict a compact image embedding given a text input with some other implementation-specific inputs. These image embeddings are then transformed into actual image embeddings through other mechanisms.

For model scaling, GPT-3 showed that the model capacity can still be increased along with the performance. First to mention is **Megatron**, which focused on making the transformer architecture more efficient with engineering strategies made to achieve parallelism and capable of being trained with 512 GPUs, making it 5.6 times larger than GPT-2. GPT-3 outsized Megatron, but another notable model called **BLOOM** scaled it to 176B using Megatron as a base to match the GPT-3 model size of 175B. What's most notable is not that it is bigger than GPT-3 but the fact that BLOOM gives birth to a new paradigm – open source – which means that the collective work and contributions of the community of researchers and institutions can also train highly useful and performant models on a massive scale. These huge models in the past have always been exclusive to big corporations due to their immense hardware resources. Other than that, another model, subsequently released in 2022, that scaled up GPT to 540 billion parameters is **PaLM**, which exceeded the performance of every other smaller GPT variant model. **PaLM 2** outperforms its predecessor by offering enhanced performance across various tasks, including English and multilingual understanding, reasoning, and code generation. It achieves this by optimizing computing scaling laws, utilizing diverse multilingual datasets, and implementing architectural improvements. PaLM 2 excels in multilingual proficiency, classification, question-answering, and translation tasks, demonstrating its broad applicability. The model also incorporates control tokens to mitigate toxicity and provides guidelines for responsible development and deployment.

When selecting a transformer model for a specific task, it's crucial to take a balanced approach, considering several factors in addition to performance metrics. Model size and resource requirements play a significant role in determining the feasibility and applicability of the chosen model. While larger models such as GPT-3 have shown impressive capabilities, their resource demands may not be suitable for all situations, especially for individuals or small organizations with limited computational resources.

It's essential to weigh the trade-offs between model size, performance, and resource requirements when choosing a transformer for a particular task. In some cases, smaller models may provide adequate performance while being more efficient and environmentally friendly. Additionally, fine-tuning can often improve the performance of a model in domain-specific tasks, even if it is not as large as some of the most prominent models such as GPT-3. Ultimately, the best choice depends on the unique requirements and constraints of the specific task, as well as the resources available for model training and deployment.

Summary

Transformers are versatile NNs capable of capturing relationships of any data modality without explicit data-specific biases in the architecture. Instead of a neural network architecture capable of ingesting different data modalities directly, careful considerations of the data input structure along with crafting proper training task objectives are needed to successfully build a performant transformer. The benefits of pre-training still hold true even for the current SOTA architecture. The act of pre-training is part of a concept called transfer learning, which will be covered more extensively in the supervised and unsupervised learning chapters. Transformers can currently perform both data generation and supervised learning tasks in general with more and more research experimenting with using transformers in unexplored niche tasks and data modalities. Look forward to more deep learning innovations in the coming years with transformers being at the forefront of the advancement.

By now, you have gained the knowledge needed to appropriately choose and design a neural network architecture according to your data and requirements. Most of the architectures and concepts presented in *Chapters 2* to *6* are tricky to implement from scratch but with the help of open source work such as `https://github.com/rwightman/pytorch-image-models` and the help of many deep learning frameworks, using a model is a matter of importing libraries and adding a few lines of code. Truthfully, understanding what goes into each of the architectures under the hood is not actually needed to utilize these models, due to the ease of adopting publicly available work complete with pre-trained weights on big datasets. More often than not, understanding architectural concepts such as CNNs at a high level and knowing some bits and pieces about training a model is really all it takes today to benefit practically from these mostly ready-off-the-shelf architectures/models. However, understanding the implementation details of these architectures is essential and will prove to be beneficial when it comes to the following cases:

- When things fail or don't work as expected (the model does not converge to an optimum solution or diverges, model errors with the prepared input with unexpected data shapes, and so on).

- When it's necessary to choose a more appropriate architecture based on your dataset, runtime requirements, or performance requirements.

- When you are inventing a shiny new neural network layer.

- When you want to decode what the neural network actually learns and when you want to understand why the neural network makes its predictions. This will be explored in more depth in *Chapter 11, Explaining Neural Network Predictions*, and *Chapter 12, Interpreting Neural Networks*.

- To adopt concepts from one architecture domain to another domain. For example, skip connections from DenseNet and ResNet can be easily transferable to MLPs.

Since you've read to the end of this chapter, which completes the neural network architecture specific content, give yourself a pat on the back. You now have knowledge about models that deal with images (CNNs), time-series or sequence data (RNNs), models that deal with tabular data (MLPs), and models that are a jack of all trades (transformers). The performance of architectures, however, is closely coupled with the data type, the data structure, the data preprocessing method, the weights learning and optimizing method, the loss function, and the optimization task. More details about these components will be decoded in *Chapter 8, Exploring Supervised Deep Learning*, and *Chapter 9, Exploring Unsupervised Deep Learning*.

In the next chapter, we will explore an emerging method to design neural network models in an automated way called neural architecture search.

7

Deep Neural Architecture Search

The previous chapters introduced and recapped different **neural networks** (**NNs**) that are designed to handle different types of data. Designing these networks requires knowledge and intuition that can only be gained by consuming years of research in the field. The bulk of these networks are hand-designed by experts and researchers. This includes inventing completely novel NN layers and constructing an actually usable architecture by combining and stacking NN layers that already exist. Both tasks require a ton of iterative experimentation time to burn to actually achieve success in creating a network that is useful.

Now, imagine a world where we can focus on inventing useful novel layers while the software takes care of automating the final architecture-building process. Automated architecture search methods help to accomplish exactly that by streamlining the task of designing the best final NN architecture, as long as appropriate search spaces are selected based on deep domain knowledge. In this chapter, we will focus on the task of constructing an actual usable architecture from already existing NN layers using an automated architecture creation process called **neural architecture search** (**NAS**). By understanding the different types of NAS, you will be able to choose the most straightforward automated search optimization approach based on your current model-building setup, which ranges from simple to efficiently complicated. Specifically, the following topics will be introduced:

- Understanding the big picture of NAS
- Understanding general hyperparameter search-based NAS
- Understanding **reinforcement learning** (**RL**)-based NAS
- Understanding non-RL-based NAS

Technical requirements

This chapter includes practical implementation in the Python programming language. These simple methods will need to have the following libraries installed:

- `numpy`
- `pytorch`
- `catalyst == 21.12`
- `scikit-learn`

You can find the code files for this chapter on GitHub at `https://github.com/PacktPublishing/The-Deep-Learning-Architect-Handbook/tree/main/CHAPTER_7`.

Understanding the big picture of NAS

Before we dive into the details of the big picture of NAS methods, it's important to note that although NAS minimizes the manual effort necessary for shaping the final architecture, it doesn't completely negate the need for expertise in the field. As we discussed earlier, foundational knowledge in **deep learning (DL)** is crucial for selecting appropriate search spaces and interpreting the results of NAS accurately. Search spaces are the set of possible options or configurations that can be explored during a search. Furthermore, the performance of NAS heavily relies on the quality of the training data and the relevance of the search space to the task at hand. Therefore, domain expertise is still necessary to ensure that the final architecture is not only efficient but also accurate and relevant to the problem being solved. By the end of this section, you will have a better understanding of how to leverage your domain expertise to optimize the effectiveness of NAS.

The previous chapters on NNs have only introduced a few prominent NN layer types and only scratched the surface of the entire library of neural layers out there. Today, there are too many variations of NN layers, which makes it hard to design precisely which layers get used at which point in the architecture. The main problem is that the space of possible NN architectures is infinitely big. Additionally, evaluating any possible architectural design is slow and expensive in terms of resources. These are the reasons that make it impossible to evaluate all the possible NN architectures. Let's take the training of a **convolutional NN (CNN)** ResNet50 architecture on ImageNet, for example, to get a sense of how impossible this is. This would take around 3-4 days with a single RTX 3080 Ti Nvidia GPU, which is a GPU meant for normal consumers and available to be procured off-the-shelf. Business consumers, on the other hand, usually obtain industrial-grade GPU variants that have much greater processing power, which can bring down the runtime to under a day.

Typically, researchers will hand-design architectures with already available NN layers and operations by intuition. This manual method is a one-off effort, and doing so repeatedly when newer and better core NN layers are invented is not scalable. This is where NAS comes into play. NAS leverages already invented NN layers and operations to build a more performant NN architecture. The core of NAS lies in using a smarter way to conceptually search through different architectures. The searching mechanism of NAS can be implemented in three ways, namely: **general hyperparameter search optimization**, RL, and NAS methods that do not use RL.

General hyperparameter search optimization pertains to methods that can be applied to any **machine learning** (ML) algorithm hyperparameter optimizations. RL is another high-level ML method, alongside **supervised learning** (SL) and **unsupervised learning** (UL), that deals with some form of optimizing actions taken in an environment that produces states with a quantifiable reward or punishment. Non-RL-based NAS can be further broken down into three distinctive types: progressive architecture growing from a small architecture baseline, progressive architecture downsizing from a complex fully defined architecture graph, and evolutionary algorithms. The progressive architecture-growing method includes all algorithms that slowly grow a simple network to be a larger network with increasing depth or width. Vice versa, there are methods that first define an architecture with all the possible connections and operations and slowly drop these connections. Finally, **evolutionary algorithms** are a branch of algorithms that are based on biological phenomena such as mutation and breeding. In this chapter, we will only cover some general hyperparameter search optimization methods, RL methods, a simple form of progressive growing-based NAS, and a competitive version of progressive downsizing-based NAS. Technical implementations will be available for the progressive growing-based NAS methods but not for the other more complicated methods. Open sourced implementations from the authors of the more complicated methods will be referred to instead.

Before diving into any of the mentioned NAS methods, you need to first understand the notion of **microarchitecture** and **macroarchitecture**. Microarchitecture refers to the details of the exact combination of layers being used in a logical block. As introduced in *Chapter 3, Understanding Convolutional Neural Networks*, some of these logical blocks can be repeatedly stacked onto each other to generate the architecture that will be actually used. There can also be different logical blocks with different layer configurations in the final created architecture. Macroarchitecture, in comparison, refers to a higher-level overview of how the different blocks are combined to form the final NN architecture. The core idea behind NAS methods always revolves around reducing the search space based on already curated knowledge about which layer or which layer configurations work the best. The methods that will be introduced in this chapter will either keep the macroarchitecture setup fixed while only searching in the microarchitecture space or have the flexibility to explore both the micro- and macroarchitecture space with creative tricks to make searching feasible.

First, let's start with the simplest NAS method, which is general hyperparameter search optimization algorithms.

Understanding general hyperparameter search-based NAS

In ML, parameters typically refer to the weights and biases that a model learns during training, while **hyperparameters** are values that are set before training begins and influence how the model learns. Examples of hyperparameters include learning rate and batch size. General hyperparameter search optimization algorithms are a type of NAS method to automatically search for the best hyperparameters to use for constructing a given NN architecture. Let's go through a few of the possible hyperparameters. In a **multi-layer perceptron (MLP)**, hyperparameters could be the number of layers that control the depth of the MLP, the width of each of the layers, and the type of intermediate layer activation used. In a CNN, hyperparameters could be the filter size of the convolutional layer, the stride size of each of the layers, and the type of intermediate layer activation used after each convolutional layer.

For NN architectures, the available types of hyperparameters that you can configure depend heavily on the capabilities of the helper tools and methods used to create and initialize the NN. For instance, consider the task of configuring the hidden layer size of three layers individually. Having a method that produces an MLP with a fixed number of layers of three makes it possible to perform a hyperparameter search only on the hidden layer sizes. This is achievable by simply adding three hyperparameters to the function, which sets the three hidden layer sizes respectively. However, to enable the flexibility to perform a hyperparameter search for both the number of layers and the hidden layer size, you have to build a helper method that can dynamically apply these hyperparameters to create an MLP.

The simplest form of NAS leverages these tools to perform a slightly smarter search of the defined hyperparameters. Three well-known variations of hyperparameter search will be covered here; these include **successive halving**, **Hyperband**, and **Bayesian hyperparameter optimization**. We will go through these three algorithms using an MLP from `pytorch`, as the implementation is short enough to fit in a chapter. Let's start with successive halving.

Searching neural architectures by using successive halving

The most basic method to search with a reduced search space to optimize runtime is to randomly sample a few hyperparameter configurations and execute the full training and evaluation only of the sampled configurations. This method is simply called **random search**. What if we know that certain configurations are almost certain to perform badly after a certain quantity of resources are consumed?

Successive halving is an extension of random search that helps to save resources while searching for the best neural architecture. The idea behind successive halving is to eliminate half of the poorly performing configurations at each step, allowing us to focus on the more promising ones. This way, we don't waste time on configurations that are less likely to yield good results.

Let's break down the concept using a simple example. Imagine you are trying to find the best configuration for an MLP with varying hyperparameters such as the number of layers and layer sizes. You start by randomly sampling 100 different configurations. Now, instead of training all 100 configurations to completion, you apply successive halving. You train each of the 100 configurations for a short period (for example, 5 epochs) and then evaluate their performance on a validation dataset. At this point, you eliminate the 50 worst-performing configurations and continue training the remaining 50 configurations for another 5 epochs.

After this second round of training, you again evaluate the performance of the remaining configurations and eliminate the 25 worst-performing ones. The top 25 configurations can then continue to train until convergence. By applying successive halving, you save resources and time by focusing on the most promising configurations while discarding the poorly performing ones early in the process. This allows you to more efficiently search for the best neural architecture for your problem.

Let's dive into the technical implementation of successive halving that will also set the stage for all the other methods under general hyperparameter tuning-based NAS:

1. Let's start by importing the relevant libraries and setting the `pytorch` library seed to ensure reproducibility:

```
import json
import os
import numpy as np
import torch
import torch.nn as nn
import torch.nn.functional as F
from catalyst import dl, utils
from catalyst.contrib.datasets import MNIST
from sklearn import datasets
from sklearn.metrics import log_loss
from sklearn.model_selection import train_test_split from
sklearn.preprocessing import MinMaxScaler from torch import
nn as nn from torch import optim from torch.utils.data import
DataLoader, TensorDataset torch.manual_seed(0)
```

2. Next, let's define a `pytorch` MLP class that has the functionality to build an MLP dynamically based on the number of hidden layers and hidden sizes of each layer:

```
class MLP(nn.Module):
    def __init__(self, input_layer_size, output_layer_size, layer_
configuration, activation_type='relu'):
        super(MLP, self).__init__()
        self.fully_connected_layers = nn.ModuleDict()
        self.activation_type = activation_type
        hidden_layer_number = 0
        for hidden_layer_idx in range(len(layer_configuration)):
        if hidden_layer_idx == 0:
            self.fully_connected_layers[
                str(hidden_layer_number)
            ] = nn.Linear(
                input_layer_size,
                layer_configuration[hidden_layer_idx]
            )
            hidden_layer_number += 1
        if hidden_layer_idx == len(layer_configuration) - 1:
            self.fully_connected_layers[
                str(hidden_layer_number)
```

```
        ] = nn.Linear(
            layer_configuration[hidden_layer_idx],
            output_layer_size
        )
    else:
        self.fully_connected_layers[
            str(hidden_layer_number)
        ] = nn.Linear(
            layer_configuration[hidden_layer_idx],
            layer_configuration[hidden_layer_idx+1]
        )
        hidden_layer_number += 1
def forward(self, x):
    for fc_key in self.fully_connected_layers:
        x = self.fully_connected_layers[fc_key](x)
        if fc_key != str(len(self.fully_connected_layers) -1):
            x = F.relu(x)
    return x
```

3. Next, we need the logic that trains this MLP when provided with a specific layer configuration list. Here, we will use the `pytorch` abstraction library called `catalyst` to train the model and save the best and last epoch model with a few convenient methods:

```
def train_and_evaluate_mlp(
    trial_number, layer_configuration, epochs,
    input_layer_size, output_layer_size ,
    load_on_stage_start=False, best_or_last='last',
    verbose=False
):
    criterion = nn.CrossEntropyLoss()
    runner = dl.SupervisedRunner(
        input_key="features", output_key="logits",
        target_key="targets", loss_key="loss"
    )
    model = MLP(
        input_layer_size=input_layer_size,
        layer_configuration=layer_configuration,
        output_layer_size=output_layer_size,
    )
    optimizer = optim.Adam(model.parameters(), lr=0.02)
    checkpoint_logdir = "logs/trial_{}".format(
        trial_number
    )
    runner.train(
```

```
        model=model, criterion=criterion,
        optimizer=optimizer, loaders=loaders,
        num_epochs=epochs,
        callbacks=[
            dl.CheckpointCallback(
                logdir=checkpoint_logdir,
                loader_key="valid",
                metric_key="loss",
                mode="all",
                load_on_stage_start="last_full" if load_on_stage_
start else None,
            )
        ], logdir="./logs", valid_loader="valid",
        valid_metric="loss", minimize_valid_metric=True,
        verbose=verbose
    )
    with open(os.path.join(checkpoint_logdir, '_metrics.json'),
'r') as f:
        metrics = json.load(f)
    if best_or_last == 'last':
        valid_loss = metrics['last']['_score_']
    else:
        valid_loss = metrics['best']['valid']['loss']
    return valid_loss
```

4. Next, we need a method that generates a random number of hyperparameters for the MLP. The hyperparameter is structured to be a list of hidden layer size specifications where the number of items in the list determines the number of layers. We fix the range number of hidden layers to be between 1 and 6 layers and hidden layer sizes to be between 2 and 100:

```
def get_random_configurations(
    number_of_configurations, rng
):
    layer_configurations = []
    for _ in range(number_of_configurations):
        layer_configuration = []
        number_of_hidden_layers = rng.randint(
            low=1, high=6
        )
        for _ in range(number_of_hidden_layers):
            layer_configuration.append(
                rng.randint(low=2, high=100)
            )
        layer_configurations.append(layer_configuration)
```

```
    layer_configurations = np.array(
        layer_configurations
    )
    return layer_configurations
```

5. Now, with the helpers defined, let's set up our tabular dataset to apply MLP with. The `iris` dataset from `scikit-learn` will be used here. We will load it, scale the values, split the dataset into train and validation partitions, and prepare it to be consumed by the `catalyst` library. Note that the code up until this step will be reused for the next two methods:

```
iris = datasets.load_iris()
iris_input_dataset = iris['data']
target = torch.from_numpy(iris['target'])
scaler = MinMaxScaler()
scaler.fit(iris_input_dataset)
iris_input_dataset = torch.from_numpy(
  scaler.transform(iris_input_dataset)
).float()
(
  X_train, X_test, y_train, y_test
) = train_test_split(
  iris_input_dataset, target, test_size=0.33,
  random_state=42
)
training_dataset = TensorDataset(X_train, y_train)
validation_dataset =  TensorDataset(X_test, y_test)
train_loader = DataLoader(
  training_dataset, batch_size=10, num_workers=1
)
valid_loader = DataLoader(
  validation_dataset, batch_size=10, num_workers=1
)
loaders = {"train": train_loader, "valid": valid_loader}
```

6. The approach we are going to take here with successive halving is to use epochs as the resource component where we will execute three iterations of successive halving once a predefined number of epochs for that iteration has been executed. Half of the top-performing configurations will continue to be trained in the next iteration. Here, we use 20 initial configurations and 3 iterations of successive halving with 5 epochs each. Let's start by defining these values along with the seeded random number generator that controls the randomness of the generated configurations:

```
rng = np.random.RandomState(1234)
number_of_configurations = 20
layer_configurations = get_random_configurations(
```

```
    number_of_configurations, rng
)
successive_halving_epochs = [5, 5, 5]
```

7. Finally, we will define the execution logic for successive halving. Note that the last trained epoch weights are used here, at the next iteration, instead of the epoch with the best validation score:

```
for succesive_idx, successive_halving_epoch in
enumerate(successive_halving_epochs):
    valid_losses = []
    for idx, layer_configuration in enumerate(layer_
configurations):
        trial_number = trial_numbers[idx]
        valid_loss = train_and_evaluate_mlp(
            trial_number, layer_configuration,
            epochs=successive_halving_epoch,
            load_on_stage_start=False if succesive_idx==0 else True
        )
        valid_losses.append(valid_loss)
        if succesive_idx != len(successive_halving_epochs) - 1:
            succesive_halved_configurations = np.argsort(
                valid_losses
            )[:int(len(valid_losses)/2)]
            layer_configurations = layer_configurations[
                succesive_halved_configurations
            ]
            trial_numbers = trial_numbers[
                succesive_halved_configurations
            ]
```

8. The best configuration can then be found via the following logic:

```
best_loss_idx = np.argmin(valid_losses)
best_layer_configuration = layer_configurations[best_loss_idx]
best_loss_trial_number = trial_numbers[best_loss_idx]
```

In successive halving, some configurations might only be performant at the later stages of the training process, while some configurations can be performant from the early stages of the training process. Choosing either a longer wait time or a faster wait time will put some models at a disadvantage and requires finding a balance that we might not know the truth about. The Hyperband method that will be introduced next is an attempt to solve this issue.

Searching neural architectures by using Hyperband

Hyperband improves upon the caveats in successive halving by executing multiple separate end-to-end iterations of successive halving called brackets. Each consecutive bracket would have smaller original sample configurations but has a higher number of resources allocated. This algorithm essentially allows some randomly sampled configurations to be trained longer, increasing the probability that the inherent potential for good performance is shown so that abandoning these configurations won't be a waste at the later brackets. The full algorithm is shown in *Figure 7.1*:

Algorithm 1: HYPERBAND algorithm for hyperparameter optimization.

input $: R, \eta$ (default $\eta = 3$)
initialization$: s_{\max} = \lfloor \log_\eta(R) \rfloor$, $B = (s_{\max} + 1)R$

1 **for** $s \in \{s_{\max}, s_{\max} - 1, \ldots, 0\}$ **do**

2 $\quad n = \lceil \frac{B}{R} \frac{\eta^s}{(s+1)} \rceil$, $r = R\eta^{-s}$

$\quad$ // begin SUCCESSIVEHALVING with (n, r) inner loop

3 $\quad T = $get_hyperparameter_configuration(n)

4 $\quad$ **for** $i \in \{0, \ldots, s\}$ **do**

5 $\quad\quad n_i = \lfloor n\eta^{-i} \rfloor$

6 $\quad\quad r_i = r\eta^i$

7 $\quad\quad L = \{$run_then_return_val_loss$(t, r_i) : t \in T\}$

8 $\quad\quad T = top_k(T, L, \lfloor n_i/\eta \rfloor)$

9 $\quad$ **end**

10 **end**

11 **return** *Configuration with the smallest intermediate loss seen so far.*

Figure 7.1 – Hyperband algorithm pseudocode

Two user input configurations are needed for this algorithm: specifically, R, the maximum amount of resources to train and evaluate a single configuration, and η, the divider number that decides the number of configurations to keep at the end of every successive halving iteration. The total number of brackets, s_{max}, the total resource allocated for each bracket, B, the total number of configurations by bracket and iteration n and n_i, and the resource allocated by brackets r_i, are all computed by formula. To make this easier to digest, *Figure 7.2* shows the example Hyperband resulting configuration results in each bracket when $R = 81$, and $\eta = 3$:

i	$s=4$		$s=3$		$s=2$		$s=1$		$s=0$	
	n_i	r_i	n_i	r_i	n_i	r_i	n_i	r_i	n_i	r_i
0	81	1	27	3	9	9	6	27	5	81
1	27	3	9	9	3	27	2	81		
2	9	9	3	27	1	81				
3	3	27	1	81						
4	1	81								

Figure 7.2 – Example Hyperband resulting configuration results in each bracket

These settings produce a total of 5 brackets and produce a total of 10 final models. The best model out of these 10 models would then be used as the final produced model from the search operation.

Note that, in this method, expert knowledge can be explicitly injected into the process of the two search methods by, for example, fixing the macroarchitecture of the model and only searching the hyperparameters for a microarchitecture logical block. Let's go through an implementation of Hyperband using the methods and dataset defined in the successive halving topic:

1. First, let's define the additional library needed here to compute logarithms:

   ```
   import math
   ```

2. Next, we will define the two input parameters needed for the Hyperband implementation, the maximum resource we want to run per configuration in terms of epochs, and the divisor of configurations, N, after every successive halving operation in the Hyperband algorithm:

   ```
   resource_per_conf = 30  # R
   N = 3
   ```

3. Now, we will define the main logic of Hyperband according to *Figure 7.1*:

   ```
   s_max = int(math.log(resource_per_conf, N))
   bracket_resource = (s_max + 1) * resource_per_conf  bracket_
   best_valid_losses = []
   bracket_best_layer_configuration = []
   for s in range(s_max, -1, -1):
     number_of_configurations = int(
         (bracket_resource / resource_per_conf) *
         (N**s / (s+1))
     )
     r = resource_per_conf * N**-s
     layer_configurations = get_random_configurations(
         number_of_configurations, rng
     )
   ```

```
trial_numbers = np.array(
    range(len(layer_configurations))
)
valid_losses = []
for i in range(s+1):
    number_of_configurations_i = int(
        number_of_configurations * N**-i
    )
    r_i = int(r * N**i)
    valid_losses = []
    for idx, layer_configuration in enumerate(layer_
configurations):
        trial_number = '{}_{}'.format(
            s, trial_numbers[idx]
        )
        valid_loss = train_and_evaluate_mlp(
            trial_number, layer_configuration,
            epochs=r_i, load_on_stage_start=False if
            s==s_max else True
        )
        valid_losses.append(valid_loss)
        if succesive_idx != len(successive_halving_epochs) - 1:
            succesive_halved_configurations = np.argsort(
             valid_losses
            )[:int(number_of_configurations_i/N)]
            layer_configurations = layer_configurations[
                succesive_halved_configurations
            ]
            trial_numbers = trial_numbers[
             succesive_halved_configurations
            ]
            best_loss_idx = np.argmin(valid_losses)
            best_layer_configuration = layer_configurations[best_
loss_idx]
            best_loss_trial_number = trial_numbers[
                best_loss_idx
            ]
            bracket_best_valid_losses.append(
                valid_losses[best_loss_idx]
            )
            bracket_best_layer_configuration.append(
                best_layer_configuration
            )
```

The two methods utilize random search heavily without being very smart about choosing the hyperparameters' configuration. Next, we will explore a search method that optimizes the next choice of hyperparameters after some initial searching has been done.

Searching neural architectures by using Bayesian hyperparameter optimization

Bayesian hyperparameter optimization is a method that utilizes a surrogate performance estimation model to choose an estimated best set of configurations to sample and evaluate from. The act of sampling configurations to train and evaluate is formally called the **acquisition function**. Instead of random sampling and hoping that it'll perform well, Bayesian optimization attempts to leverage the prior information gained from an initial random configuration sampling and actual training and evaluation to find new configurations that are estimated to perform well. Bayesian optimization takes the following steps:

1. Sample a number of hyperparameter configurations.

2. Perform full training and evaluation with these configurations to obtain performance scores.

3. Train a surrogate regression model (typically, a **Gaussian processes** (**GP**) model is used) with all the available data to estimate the performance scores based on the hyperparameters.

4. Either use all possible hyperparameter configurations or randomly sample a good amount of hyperparameter configurations and predict the performance score using the surrogate model.

5. Get the k hyperparameter configurations that have the minimum estimated performance scores from the surrogate model.

6. Repeat *step 1* to *step 5* either a predetermined number of times, until a good enough result is obtained, or until your resource budget is all used up.

The process essentially attempts to speed up the actual training and evaluation process by estimating the scores it will produce and only actually training the estimated top few configurations. The optimization only works if the surrogate model performance score estimation function is considerably faster than the actual training and evaluation of the main model. Note that the standard Bayesian optimization works only in the continuous space and can't deal with discrete hyperparameters properly. Let's go through the technical implementation of Bayesian optimization-based NAS using the same methods and dataset defined earlier:

1. Let's start by importing the main powerhouse behind the Bayesian optimization approach here, which is the GP regressor from `scikit-learn`:

```
from sklearn.gaussian_process import GaussianProcessRegressor
```

2. Next, let's define the method that creates a structured fixed-sized column of 6, which is the maximum number of possible layers defined earlier. When there are fewer than 6 layers, the later columns will just have 0 layers as a feature:

```
def get_bayesian_optimization_input_features(
  layer_configurations
):
  bo_input_features = []
  for layer_configuration in layer_configurations:
    bo_input_feature = layer_configuration + [0] * (6 -
len(layer_configuration))
    bo_input_features.append(bo_input_feature)
  return bo_input_features
```

3. Next, let's define three important parameters for Bayesian optimization-based NAS using MLP. The first parameter is the number of configurations. The approach we are taking here is to initially train 100 configurations in the first iteration according to the specified epochs per configuration. After that, we build a GP regressor to predict the validation loss. Then, we will sample configurations in the next few iterations and use the model to predict and pick the top five configurations to perform full training. In every iteration, a new regressor model is built with all the available validation loss data:

```
number_of_configurations = [100, 2000]
epochs_per_conf = 15
topk_models = 5
```

4. Finally, let's define the main logic that accomplishes a version of Bayesian optimization-based NAS with MLP:

```
Trial_numbers = np.array(
  range(len(layer_configurations))
)
trial_number = 0
model = None
best_valid_losses_per_iteration = []
best_configurations_per_iteration = []
overall_bo_input_features = []
overall_bo_valid_losses = []
for number_of_configuration in number_of_configurations:
  valid_losses = []
  layer_configurations = get_random_configurations(
    number_of_configuration, rng
```

```
    )
    if model:
        bo_input_features = get_bayesian_optimization_input_
features(layer_configurations)
        predicted_valid_losses = model.predict(
            bo_input_features
        )
        top_k_idx = np.argsort(
            predicted_valid_losses
        )[:topk_models]
        layer_configurations = layer_configurations[
            top_k_idx
        ]
    for idx, layer_configuration in enumerate(layer_
configurations):
        trial_identifier = 'bo_{}'.format(trial_number)
        valid_loss = train_and_evaluate_mlp(
            trial_number, layer_configuration,
            epochs=epochs_per_conf,
            load_on_stage_start=False, best_or_last='best'
        )
        valid_losses.append(valid_loss)
        trial_number += 1
    best_loss_idx = np.argmin(valid_losses)
    best_valid_losses_per_iteration.append(
        valid_losses[best_loss_idx]
    )
    best_configurations_per_iteration.append(
        layer_configurations[best_loss_idx]
    )
    bo_input_features = get_bayesian_optimization_input_
features(layer_configurations)
    overall_bo_input_features.extend(bo_input_features)
    overall_bo_valid_losses.extend(valid_losses)
    model = GaussianProcessRegressor()
    model.fit(
        overall_bo_input_features,
        overall_bo_valid_losses
    )
```

With that, we've achieved MLP hyperparameter search with Bayesian optimization!

In addition to the hyperparameter search-based NAS methods discussed in this chapter, it is worth mentioning three other approaches, which are hierarchical search, proxy models, and evolutionary algorithms:

- Hierarchical search focuses on optimizing architectures at different levels of granularity, allowing for a more efficient exploration of the search space

- Proxy models serve as lightweight approximations of the target models, reducing the computational cost of evaluating candidate architectures during the search

- Lastly, evolutionary algorithms are inspired by natural selection processes and can be applied to the NAS problem, enabling the exploration and optimization of architectures through mutation, crossover, and selection operations

These methods can also be considered when choosing among hyperparameter search-based NAS techniques.

NAS with general hyperparameter search methods provides a simple way to search different configurations in a smarter way than just plain random or brute-force search. It provides the most help when you already have the infrastructure ready to choose different hyperparameters easily, along with the expert knowledge from the field already built in under the hood.

However, NAS with general hyperparameter search generally requires a lot of out-of-algorithm tooling for building the model and formalizing the helper methods that can reliably be controlled by hyperparameters. On top of that, it is still required to have quite a bit of knowledge of which types of layers to use along with out-of-algorithm crafting of the macro- and microarchitecture of the NN model.

In the next section, we will go through a line of NAS methods that covers more extensively all the steps needed to achieve NAS for any NN, called RL-based NAS.

Understanding RL-based NAS

RL is a family of learning algorithms that deal with the learning of a policy that allows an agent to make consecutive decisions on its actions while interacting with states in an environment. *Figure 7.3* shows a general overview of RL algorithms:

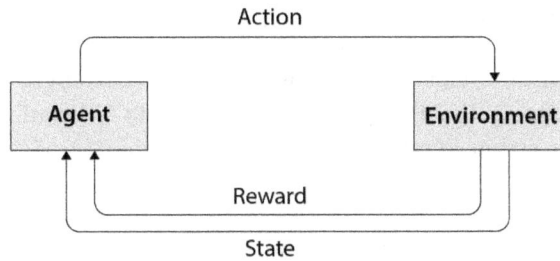

Figure 7.3 – General overview of RL algorithms

This line of algorithms is most popularly utilized to create intelligent bots for games that can act as offline players against real humans. In the context of a digital game, the environment represents the entire setting in which the agent operates, including aspects such as the position and status of the in-game character, as well as conditions of the in-game world. The state, on the other hand, is a snapshot of the environment at a given time, reflecting the current conditions of the game. One key component in RL is the environment feedback component that can provide either a reward or punishment. In digital games, examples of rewards and punishments are some forms of a competitive scoring system, in-game cash, the leveling system, or sometimes negatively through death. When applied to the realm of NAS, the state will then be the generated NN architecture, and the environment will be the evaluation of the generated NN configurations. The rewards and punishments will then be the latency performance and metric performance of the resulting architecture after training and evaluating it on a chosen dataset. Another key component is the term **policy**, which is the component responsible for producing an action based on the state.

Recall that in the general hyperparameter search-based NAS, the NN configuration sample acquisition is based on random sampling. In RL-based NAS approaches, the goal is not only to optimize the search process but also the acquisition process that produces NN configurations based on prior experiences. The exact methods of how the configurations can be produced, however, differ in different RL-based NAS methods.

In this section, we will dive into a few RL methods specific to NAS:

- Founding NAS based on the RL method
- **Efficient NAS (ENAS)** via parameter sharing
- **Mobile NAS (MNAS)**

Let's start with the first founding NAS based on the RL method.

Understanding founding NAS based on RL

RL can be implemented with NNs, and in the use case of NAS, a **recurrent NN (RNN)** is used to act as the missing piece needed to probabilistically generate the main NN configurations at test time. *Figure 7.4* roughly shows an architectural overview of the foundational NAS with the RL method:

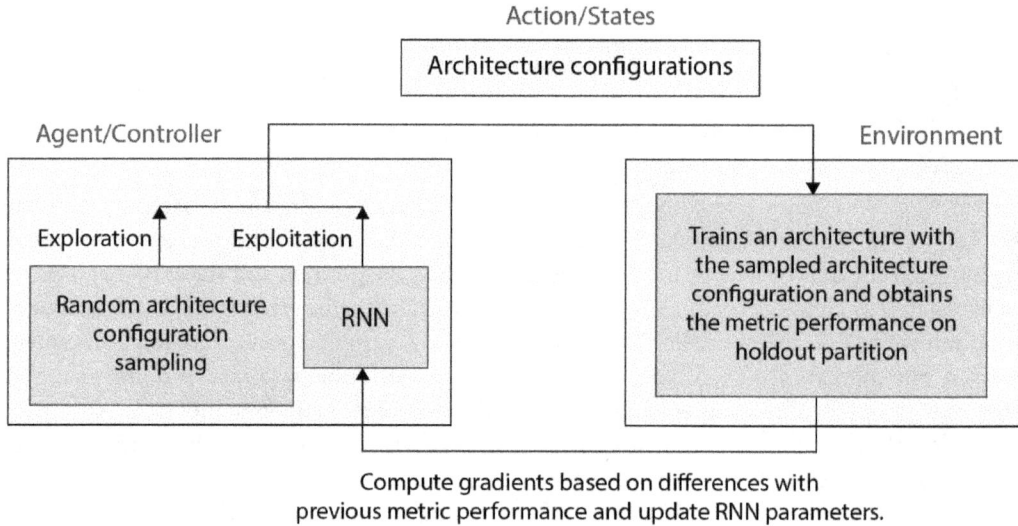

Figure 7.4 – Architectural overview of NAS with RL workflow

The RNN is the policy that determines the state after a learning step from the previous environment interaction. In the case of NAS, the action is equivalent to the state. Recall that an RNN is composed of multiple sequential recurrent-based cells where each cell is capable of producing an intermediate sequential output. In NAS, these intermediate sequential outputs are designed to predict specific configurations for the main NN. The predictions are then fed into the next RNN cell as a cell input. Consider the NAS task to search for the best CNN architecture in the image domain. *Figure 7.5* shows the structure of this task using the RNN-based NN configuration predictions:

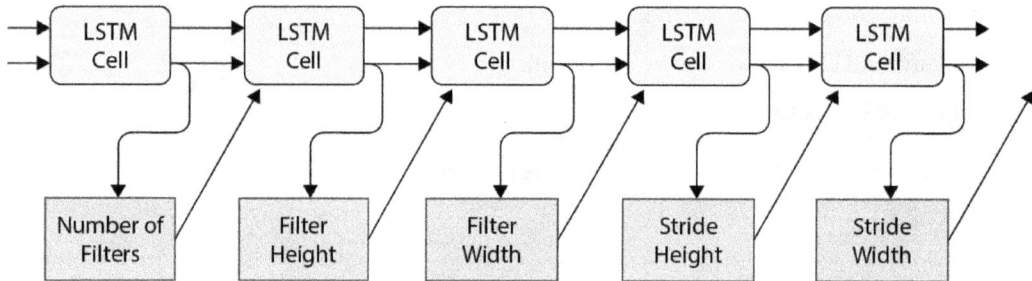

Figure 7.5 – LSTM-based CNN layer configurations prediction for NAS

A convolutional layer has a few specifications that need to be decided: namely, the number of convolutional filters, the size of the convolutional filters, and the size of the stride. *Figure 7.5* shows the predictions for a single CNN layer. For subsequent CNN layers, the same **long short-term memory (LSTM)** cells are repeatedly sequentially predicted with the state and cell outputs from the last LSTM cell as input. For a four-layered CNN, the LSTM would then be autoregressively executed four times to obtain all required configuration predictions.

As for how the parameters of the LSTM are updated, a process called **policy gradient** will be used. Policy gradient is a group of methods that uses gradients to update the policy. Specifically, the **reinforce** rule is used here to compute the gradients for updating the parameters. In more understandable terminology, the following formula shows how the gradients are computed:

gradients = average of (cross entropy loss x (reward − moving average of previous rewards)) for all sampled architectures

The cross-entropy loss here is specifically used to emphasize that the configuration prediction tasks are framed as a multiclass classification problem so that the number of search parameters can be constrained to a small number while making sure boundaries are set. For example, you wouldn't want a million filters for a single CNN layer or a million-neurons-sized fully connected layers in an MLP.

The RL process here is guided by the concept of exploration versus exploitation. If we continue to use only the predicted states of the RNN and use that as the labels for computing the cross-entropy loss, the policy will just become more and more biased toward its own parameters. Using the RNN predictions as labels is known as the *exploitation process*, where the idea is to just allow the RNN to be more confident about its own predictions. This process grows the model deeper toward its current intelligence instead of toward intelligence that can be gained from external data exploration. *Exploration* here is when network configurations are randomly sampled to act as the label for the cross-entropy loss at each RNN cell. The idea here is to start with lots of exploration and slowly reduce exploration going into later stages of policy learning and depending more on exploitation.

Until now, the steps only allow for a relatively simple form of CNN, but modifications can be added to the RNN agent to account for more complex CNN builds, such as parallel connections or skip connections from ResNet or DenseNet. In the original method, the addition of skip connections for complexity is attempted where an additional cell is added at the end of the five sequential RNN cells shown in *Figure 7.5* to act as something called the **anchor point**. An anchor point for a convolutional layer is combined individually with each anchor point of the convolutional layers prior to the referenced convolutional layer, applied with a `tanh` activation function, multiplied by a learnable weight, and finally applied with a sigmoid activation function that bounds the output values in between 0 to 1. The key information here is that the sigmoid function provides a probabilistic value that allows a binary classification task of "to add a skip connection or not" to be executed. A `0.5` value can be used to determine whether the output is `1` or `0`. One problem, however, is that the size of the outputs between different layers might not be compatible. A trick is automatically applied to solve the incompatibility by padding smaller output feature maps with zeros so that both feature maps have the same size.

This method allows you to dynamically add skip connections to a CNN in NAS. *Figure 7.6* shows the final architecture obtained from this NAS method using the CIFAR-10 image dataset:

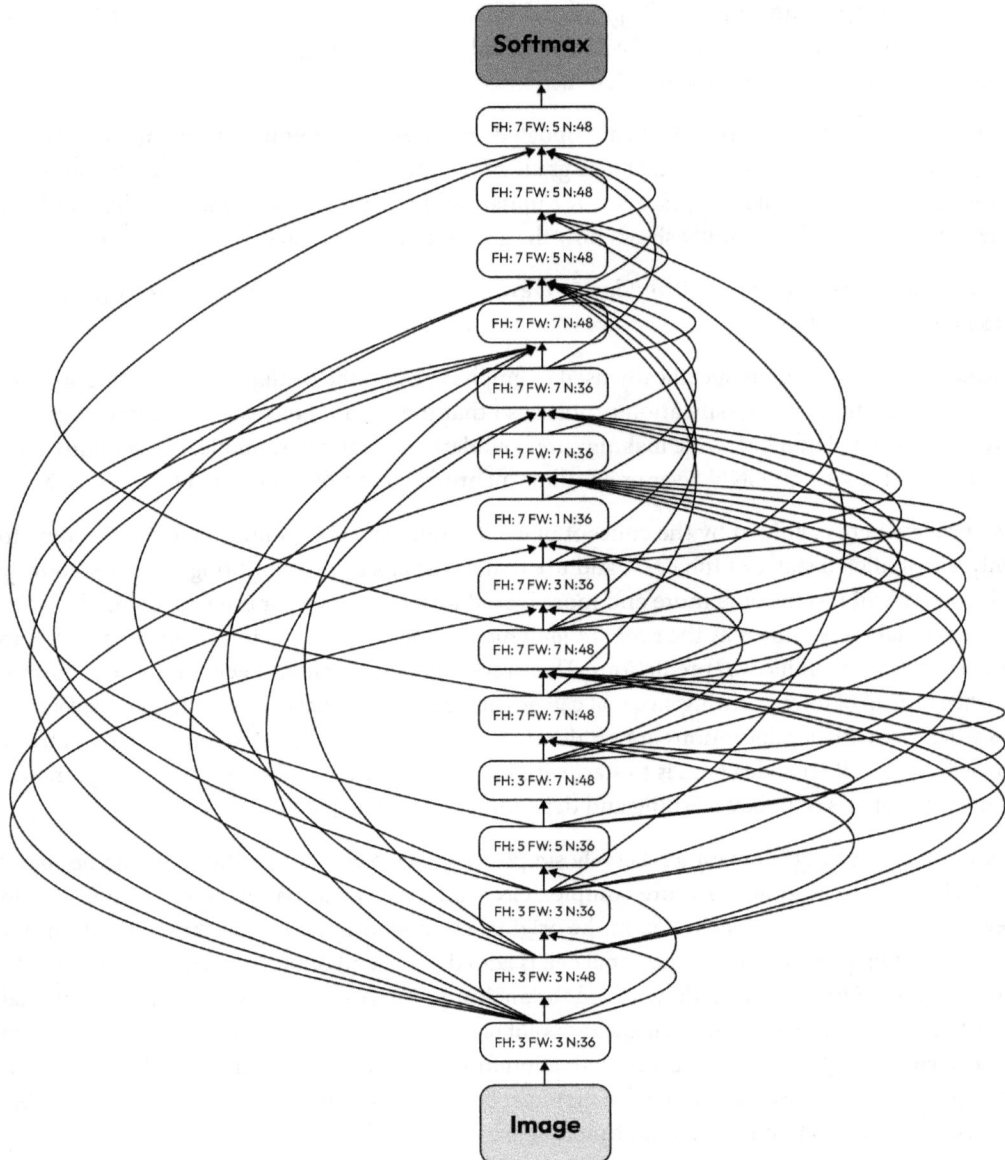

Figure 7.6 – CNN obtained through NAS with RL from the https://arxiv.org/abs/1611.01578v2 paper

The architecture, although simple, is capable of deciding the best skip connections needed to achieve a good result. This resulting architecture shows how complex an architecture can be and shows how hard it would be for a human to design this outcome manually without proper searching algorithms. Note again that any complexity and modifications can be added to the RNN policy to account for additional components such as the learning rate, pooling method, normalization method, or activation methods, which emphasizes the flexibility of the idea. Additionally, the NAS method can also be applied to search MLP or RNN main NNs. These additional adaptations and complexities, however, won't be covered here.

Note that this technique fixed the microarchitecture structure in the sense that a standard convolutional layer is used. The technique, however, enabled some form of macroarchitecture designing by allowing skip connections. One of the main problems of this foundational technique is the time needed to evaluate the randomly generated or predicted architecture configurations. Next, we will explore a method that attempts to minimize this problem.

Understanding ENAS

ENAS is a method that extends foundational NAS with the RL method by making the evaluation of generated architectures more efficient. Additionally, ENAS provides two different methods that allow either the macroarchitecture or microarchitecture to be searched. Parameter sharing is a concept that relates to **transfer learning** (**TL**), where what is learned from one task can be transferred to another task and fine-tuned for that subsequent task to get better results. Training and evaluating the main child architectures in this way provides an obvious way to speed up the process. However, it does rely heavily on weights pre-trained from the previous architectures and doesn't provide an unbiased evaluation of the final searched architecture even if it performs well. Regardless, the method still proves to be valuable when combined with the novel search space.

ENAS applies RL using an RNN as well but does so in an entirely different searching direction and predicting different components with its RNN. The search space used by ENAS is through a single **directed acyclic graph** (**DAG**) where the number of nodes determines the number of layers of the child architecture. *Figure 7.7* shows an example of a four-node DAG:

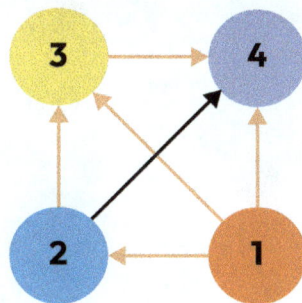

Figure 7.7 – A four-node DAG representing the search space of ENAS

The RNN will then act as the controller that predicts two components for any architecture type: namely, which previous nodes to connect to and which computation operation to use. The RNN in this case will autoregressively predict the two components four times to account for four nodes. The red lines in *Figure 7.7* show the predicted previous nodes to connect to. There will be a fixed number of computation operations that can be chosen for every node. Since there will be a random sampling of the computation operation procedure to ensure an unbiased trajectory, the procedure will be based on the same search space. The parameter-sharing method is applied in the computation operation component for these nodes. After each training iteration, the weights for each computation operation at each layer will be saved for future parameter-sharing use. Parameter sharing works in a way that each computation operation at each node number will be used as an identifier to save and reload weights whenever it is used again at the same layer with the same computation operation.

ENAS can be applied to search for RNN architectures, CNN architectures, and MLP architectures and is generally extensible to any other architecture types. Let's take the case of searching for CNN architectures for ENAS. For CNN, ENAS introduced two methods for searching; the first is to perform a macroarchitecture search, and the second is to perform a microarchitecture search. For the macroarchitecture search, six operations were proposed, which consisted of convolutional filters with filter sizes of 3 x 3 and 5 x 5, depthwise-separable convolutions with filter sizes 3 x 3 and 5 x 5, and max pooling and average pooling of kernel size 3 x 3. This set of operations allows for more diversity instead of just the plain convolutional layer, but instead of allowing more dynamic values of convolutional layer configuration, the same configurations are set to fixed values. Another implementation detail here is that when more than one previous node is selected for connection, the outputs from the layers of the previous nodes are concatenated along their depth dimension before being sent to the layer of the current node. *Figure 7.8* shows the result of a macroarchitecture search using the CIFAR-10 dataset:

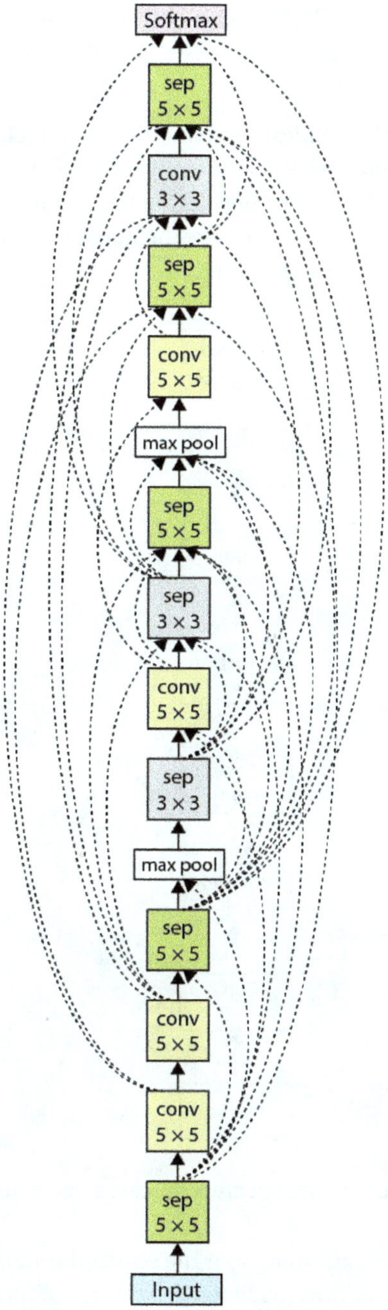

Figure 7.8 – Result of ENAS using the macroarchitecture search strategy

The result was achieved using only 0.32 days using an outdated NVIDIA GTX 1080 Ti GPU, albeit on the `CIFAR-10` dataset instead of ImageNet, and achieved only a 3.87 error rate on the `CIFAR-10` validation dataset.

As for the microarchitecture search, the idea is to build low-level logical blocks and repeat the same logical blocks so that the architecture can be scaled easily. Two different logical blocks are searched in ENAS: a logical block consisting of the main convolutional operations, and a reduction logical block intended to reduce dimensionality. *Figure 7.9* shows the macroarchitecture of the final architecture used to scale microarchitecture decisions in ENAS:

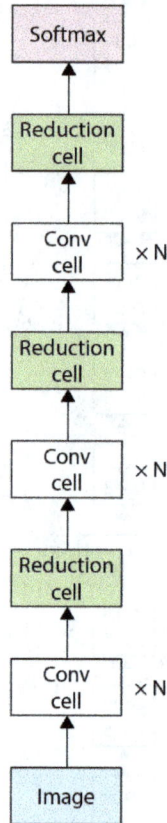

Figure 7.9 – Final macroarchitecture structure to scale a microarchitecture logical block

N is a fixed number that stays constant throughout the search duration. As this is microarchitecture, the architecture construction process was made to allow more complex interactions between layers, specifically the addition of operations to the output of previous nodes in skip connections. Due to this, for the convolutional logical block, the RNN was adapted by fixing the first two RNN cells for a node to specify which two previous node indexes to connect to and the subsequent RNN cells to predict the computation operation to apply individually for the two chosen previous node indexes. As for the reduction logical block, the main idea is to choose any operation and use a stride of two, which effectively reduces the spatial dimension of its input by two. The reduction logical block can be predicted along with the convolutional logical block using the same number of nodes and the same RNN. The parameter-sharing method in the microarchitecture case is adapted from the general case equivalently by similarly using the layer number and computation type as an identifier to save and load trained child architecture weights. *Figure 7.10* shows the result of using the microarchitecture searching strategy in ENAS:

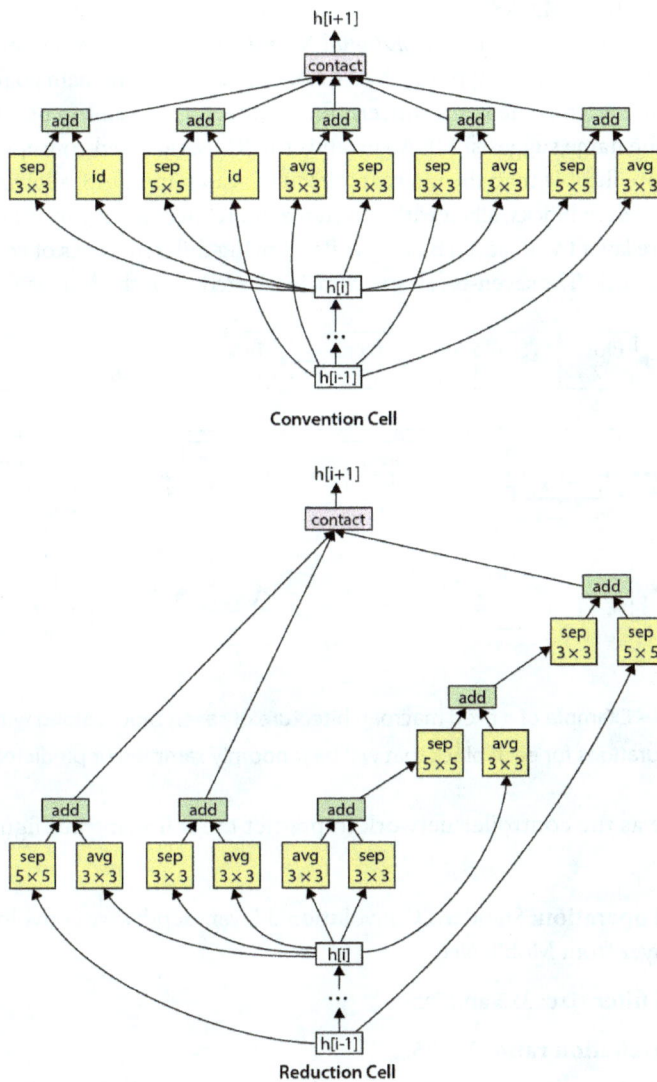

Figure 7.10 – Result of ENAS microarchitecture search strategy

The result was achieved in only 0.45 days using an outdated NVIDIA GTX 1080Ti GPU, albeit on the CIFAR-10 dataset instead of ImageNet, and achieved only a 2.89 error rate on the CIFAR-10 validation dataset.

Next, let's go through the final RL-based NAS method, called MNAS.

Understanding MNAS

MNAS is the searching method that was used to create the CNN architecture called **MnasNet**, which is a CNN-based architecture. MNAS was later utilized to build the EfficientNet architecture family introduced in *Chapter 3, Understanding Convolutional Neural Networks*. However, the method can still be used to generate other architecture types such as RNN or MLP. MNAS's main goal is to account for the latency component, which is the main concern for architectures meant to be run at the edge, or in mobile devices, as the name suggests. MNAS extends the RL NAS-based concept and introduces a search space that is more flexible than the microarchitecture search in ENAS, allowing the creation of more varied layers at different blocks, albeit with a fixed macroarchitecture. *Figure 7.11* shows an MNAS fixed macroarchitecture layout with seven blocks while allowing different types of configurations with different layers in each block. The seven-block structure is adapted from the MobileNetV2 architecture:

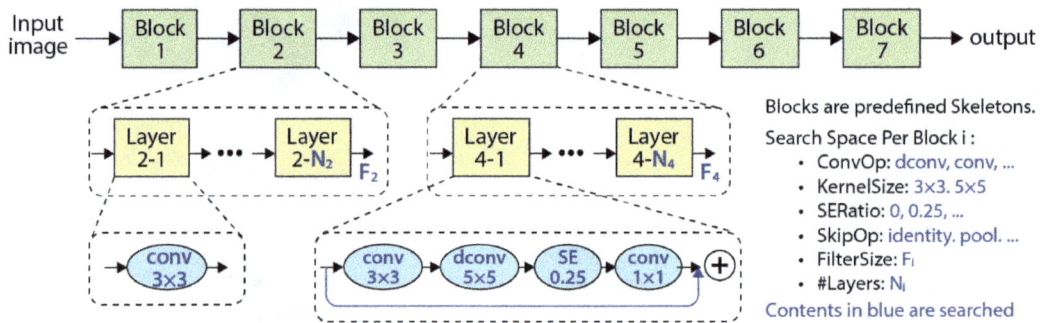

Figure 7.11 – Example of a fixed macroarchitecture of seven blocks along with the configurations for each block that will be randomly sampled or predicted

An RNN is used here as the controller network to predict the following configurations for each CNN block:

- **Convolutional operation**: Standard Convolutional layer, depthwise convolutional layer, and the MBConv layer from MobileNet.

- **Convolutional filter size**: 3x3 and 5x5.

- **Squeeze-and-excitation ratio**: 0, 0.25.

- **Skip connection operation**: Pooling, identity residual, or no skip connection.

- **The number of layers per block**: 0, +1, -1. This is structured to be in reference to the number of layers in the same block numbers in MobileNetV2.

- **Output filter size per layer**: 0.75, 1.0, 1.25. This is also structured to be in reference to the filter size of the convolution layer at the same positions in MobileNetV2.

The search space introduced is crucial to allow for a more efficient network and higher capacity to achieve better metric performance. In CNN, for example, a lot of the computation is dominated at the earlier layers as the feature sizes are much larger and require more efficient layers compared to later layers.

A big issue about latency is that it is a component that is dependent on both the software and hardware environment. For example, let's say architecture *A* is faster than architecture *B* in hardware and software *C*. When tested on another hardware and software *D*, it is possible for architecture *B* to be faster than architecture *A*. Additionally, the number of parameters and **floating-point operations per second (FLOPs)** specification of the architecture is also a proxy to the actual latency that depends also on the degree of parallelism of the architecture and the computational cores of the hardware. Based on these reasons, MobileNet adds the latency component to the reward computation by evaluating it objectively in the software and hardware environment of a mobile phone, combining both the metric computation and latency. *Figure 7.12* shows an overview of the entire MNAS process:

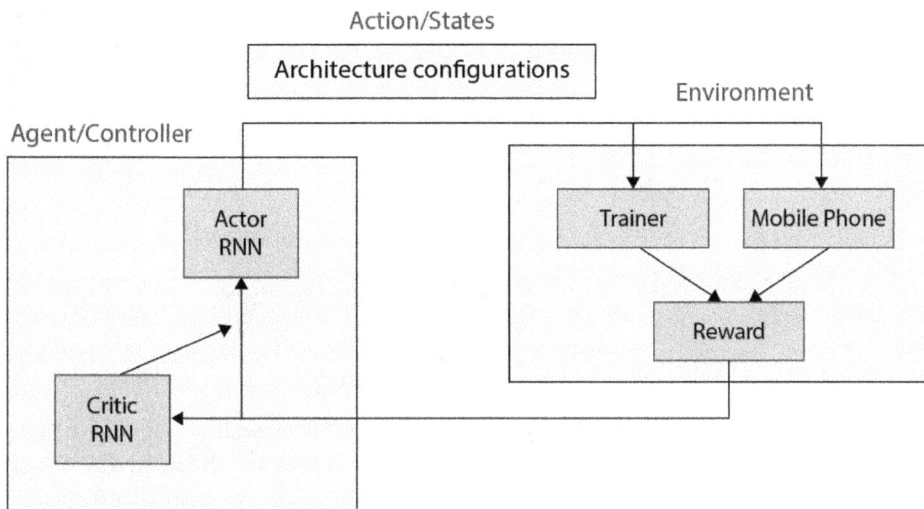

Figure 7.12 – Overview of MNAS, a platform-aware NAS with a latency guarantee

Latency can be computed on the actual target software and hardware environment and is not restricted to just using mobile phones. The reward is computed using the formula with the capability to input the desired target latency in seconds:

$$reward = accuracy \; x \left[\frac{latency}{target \; latency} \right]^w$$

Another key detail about MnasNet is that another policy gradient method called **proximal policy optimization (PPO)** by OpenAI was used to train the RNN policy network instead of the reinforce method. PPO is a method that accomplishes two things over the standard reinforcement policy gradient, namely:

- Make smaller gradient updates to the policy so that the policy learns in a stabler way and is thus capable of achieving more efficient convergence

- Use the generated probabilities themselves as sampling probabilities for the random sample generation that automatically balances exploration and exploitation

The first point is achieved by two means:

- Weighing the loss using the probabilities of the current actor network with the old probabilities generated from the actor-network before a parameter update

- Clipping the probabilities in an interval $[1 - \epsilon, 1 + \epsilon]$, where ϵ can be varied but the value of 0.2 was used

The method is performed through two networks instead of one, called actor-network and critic network. The critic network is structured to predict an unconstrained single value that serves as part of the evaluation logic of the generated architecture, along with the reward from measuring the metric performance. The actor-network, on the other hand, is the network we know is the main network responsible for generating ideal network architecture configurations. Both networks can be implemented with RNNs. This is depicted well in *Figure 7.12*. The two network parameters are jointly updated per batch. The loss of the actor-network can be computed with the following formula:

$$Loss = \frac{current \; predicted \; configuration \; probability}{old \; predicted \; configuration \; probability} x \; Advantage$$

A minimum of this loss and another version of the loss with clipped probabilities will then be used as the final loss. The clipped loss is defined as follows:

$$clipped \; loss = \left(\frac{current \; predicted \; configuration \; probability}{old \; predicted \; configuration \; probability}, 1 - \epsilon, 1 + \epsilon \right) x \; Advantage$$

The advantage here is a custom loss logic that provides a quantified evaluation number of the sampled child architecture that uses both the reward (using metric performance) and the single value predicted from the critic network. In the reinforce method, the **exponential moving average (EMA)** of previous rewards was used. Similarly, here, a form of the EMA is used to reduce the advantage at different timesteps. The logic is slightly more scientific, but for those who would like to know more, it can be computed using the following formula:

$$Advantage = discount \; t^{DISTANCE \; FROM \; FIRST \; TIMESTEP} x \; critic \; and \; reward \; evaluation \; discount = \lambda y$$

$$critic \; and \; reward \; evaluation = (\; reward + y(critic \; prediction \; value \; at \; t + 1) - critic \; prediction \; value \; at \; t)$$

The lambda, λ, and gamma, γ, are constants with values between 0 and 1. They each control the level of weight decay of the discount of advantages at each timestep moving forward. Additionally, the gamma also controls the contribution of predicted critic values at future timesteps. As for the loss of the critic network, it can be defined using the following formula:

$$critic\ loss = (advantage + future\ critic\ values - current\ critic\ value)^2$$

The final total loss will be the summation of the critic loss and the actor loss. PPO generally performs better than the vanilla reinforcement policy gradient in efficiency and convergence. This sums up the PPO logic at a more intuitive level.

The RL search space here is not efficient and takes approximately 4.5 days to train on ImageNet directly with a whopping 64 TPUv2 devices. However, this resulted in a child architecture called MnasNet that is more efficient than MobileNetV2 at the same accuracy, or more accurate than MobileNetV2 at the same latency when benchmarked on ImageNet. The same MNAS methods eventually got adopted in EfficientNet, which has become one of the most efficient CNN model families today.

Summarizing NAS with RL methods

RL allows a way for us to smartly learn the most performant NN architectures through sampling, training, and evaluation of neural architectures and apply the experience learned by predictively generating the most efficient neural architecture configurations. Simply said, NAS with RL trains an NN to generate the best NN architecture! The biggest problem with NAS with RL is still the expensive compute time needed. A few tricks carried out by different methods to try to circumvent this issue are listed here:

- Training NAS with RL on a smaller but still representative dataset as a proxy task and training and evaluating the final obtained neural architecture on the main larger dataset

- Parameter sharing by the unique layer number and computation type can, fortunately, be generically adapted for other methods from ENAS

- Balancing the macroarchitecture and microarchitecture search flexibility to reduce the search space while making sure it is flexible enough to take advantage of key differences needed at different stages of a network to achieve efficiency and good metric performance

- Directly embedding target and achieved latency as part of the reward structure and as a result searching only mostly architectures around the specified latency

Do note that the methods that were introduced here do not provide an exhaustive overview of RL and its potential.

Although RL provides a concrete way to accomplish NAS, it is not strictly necessary. In the next section, we will go through examples of a category of NAS-specific methods that do not use RL.

Understanding non-RL-based NAS

The core of NAS is about intelligently searching through different child architecture configurations by making decisions based on prior search experience to find the best child architecture in a non-random and non-brute-force way. The core of RL, on the other hand, involves utilizing a controller-based system to achieve that intelligence. Intelligent NAS can be achieved without using RL, and in this section, we will go through a simplified version of the progressive growing-from-scratch style of NAS without a controller and another competitive version of elimination from a complex fully defined NN macroarchitecture and microarchitecture.

Understanding path elimination-based NAS

First and foremost, **differentiable architecture search** (**DARTS**) is a method that extends the DAG search space defined in ENAS by removing the RL controller component. Instead of choosing previous nodes to connect to and choosing which operation to use for a node, all operations are included in a big overparameterized architecture through the same DAG system during training. A learnable weight vector with a size of the total number of operations is used to perform a weighted addition of all operations during training between nodes. This weight vector is applied with a `softmax` activation before the weighted addition process. During testing, the top k paths or operations between nodes are chosen to act as the actual network, whereas the other paths are pruned away. When the weight vector gets updated, however, the child architecture essentially changes. Instead of training and evaluating this new child architecture to obtain the new metric performance on the holdout or validation partition of the dataset, only a single training epoch is used for the entire architecture to obtain an estimate of the best validation performance using the training loss. This estimate will be used to update the parameters of the overparameterized architecture through gradient descent.

Proxyless NAS, in turn, extends upon the DARTS algorithm with a few additions. The first addition is adding a layer that uses binarized weights to represent the weight vector called `BinaryConnect`. These binarized weights act as gates that allow data to travel through only when it is enabled. This addition helps to alleviate the biggest issue with any overparameterized architectures: the GPU memory size needed to hold the parameters of the defined architecture. The second addition is the latency component to the overall loss component, which is crucial to make sure the search takes latency into consideration and doesn't attempt to utilize more paths just to get better metric performance. Let's uncover the details step by step by first describing the overall training method used in proxyless NAS:

1. Train only the BinaryConnect weight vector based on a single randomly sampled path for each node based on the probabilities specified by the `softmax` conditioned weight vector using the training dataset loss. This is achieved by freezing the parameters of the rest of the architecture and using standard cross-entropy loss.

2. Train only the architecture parameters based on two randomly sampled paths for each based on the probabilities specified by the `softmax` conditioned weight vector using the validation

dataset loss. This is achieved by freezing the parameters of the weight vector and using an approximate gradient formula for the architecture parameters:

gradient of parameters through path 1 = *gradient of binary weight* 2 *x path* 2 *probability x*
(− *path* 1 *probability*) + *gradient of binary weight* 1 *x path* 1 *probability x*
(1 − *path* 2 *probability*)

gradient of parameters through path 2 = *gradient of binary weight* 1 *x path* 1 *probability x*
(− *path* 2 *probability*) + *gradient of binary weight* 2 *x path* 2 *probability x*
(1 − *path* 1 *probability*)

The formula computes gradients for path 1 and path 2. The loss used is cross-entropy loss summed with a predicted latency after pruning the paths similar to DARTS. The latency is predicted with an external ML model trained to predict latency based on the parameters of the architecture due to the reason that latency evaluations take up too much time and usually require an average of multiple runs to get a reliable estimate. Any ML model can be used to build the latency predictor model and is a one-off process before starting the NAS process.

3. Repeat *steps 1-2* until convergence, a predefined number of epochs, or any early stopping without improvements on the validation loss for a predefined number of epochs.

    Recall that BinaryConnect is used to achieve binary weights that act as gates. One detail is that the standard unconstrained non-binary weight vector itself is still present but a binarization operation is applied. The binarization process is executed by the following steps:

    I.   Set all binary weights to 0.

    II.   Sample the desired number of chosen paths by using the `softmax` conditioned weight vector as probabilities.

    III. Set the chosen path's binary weights to 1.

4. BinaryConnect saves memory by only loading nonzero paths to memory.

PNAS manages to achieve an 85.1 top-1 accuracy on ImageNet directly without using a proxy dataset such as `CIFAR-10`, using only 8.3 days of search using the NVIDIA GTX 1080 Ti GPU.

Next, we will go through a simple progressive growth-based NAS method as an introduction.

Understanding progressive growth-based NAS

The progressive growth-based NAS method's key differentiator is that the method can be structured to be unbounded in both macroarchitecture and microarchitecture. Most of the techniques introduced in this chapter have placed a lot of domain knowledge in terms of general structures found to be useful. Growth-based NAS is naturally non-finite in terms of search space and can potentially help to discover novel macroarchitectural structures that work well. This line of NAS will continue to evolve to a stage where it can be competitive with other NAS methods, but, in this book, we will only go into a method to search the microarchitectural structure of the child architecture called **progressive NAS** (**PNAS**) to act as an introduction.

PNAS adopts a progressive growth-based approach in NAS by simply using concepts defined in Bayesian optimization introduced earlier in this chapter and searching at the microarchitecture level while fixing the macroarchitecture structure similar to the ENAS microarchitecture search method. The macroarchitecture structure is adapted to the size of the dataset, CIFAR-10, with a smaller structure, and ImageNet with a deeper structure. *Figure 7.13* shows these structures:

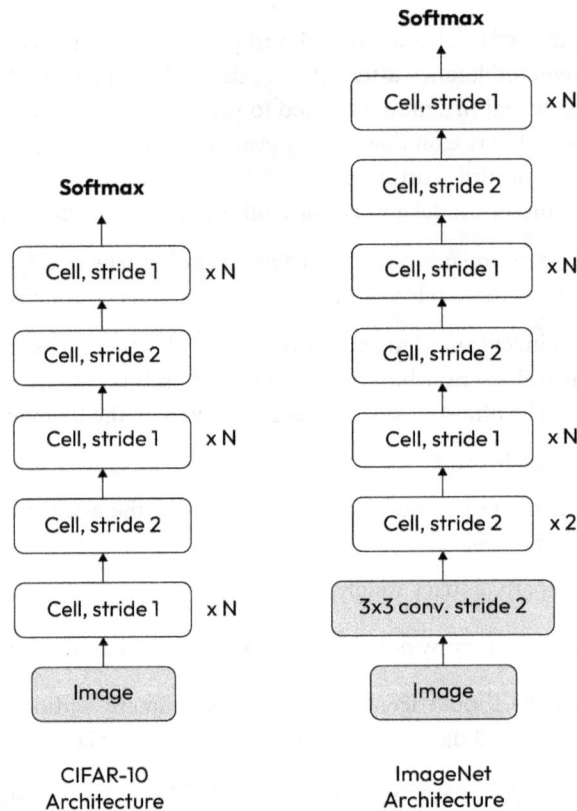

Figure 7.13 – PNAS macroarchitecture structure from the https://arxiv.org/abs/1712.00559v3 paper

The method can be accomplished with the steps defined next, along with an initial predefined max number of blocks in the cell and starting with zero blocks in the cell:

1. Start with the first block. Construct the full CNN with all cell options iteratively and evaluate all of the CNN.

2. Train an RNN surrogate model to predict the metric performance using the cell configurations.

3. Expand to the next block and predict the metric performance of the possible cell option combinations for the next block using all available chosen and evaluated previous block variations.

4. Take the top two best metric performance cell options for the next block and train and evaluate the fully constructed CNN using the two cell options.

5. Fine-tune the RNN surrogate model using the extra two data points obtained in *step 4*.

6. Repeat *step 3* to *step 5* until the total number of blocks reaches the max number of blocks or until the metric performance does not improve anymore.

Each block in a cell will have a configuration defined with five variables similar to ENAS; namely, the first input, the second input, the operation to the first input, the operation to the second input, and the method to combine the outputs of the operation to the first input and operation to the second input. The set of possible inputs is all the previous blocks, the output of the previous cell, and the output of the cell before the previous cell. This means that the cell can interact with other cells. All the possible input combinations, operation types, and combination methods are laid out in each progressive block step and fed to the RNN model to predict the metric performance.

PNAS managed to achieve an 84.2 top-1 accuracy on ImageNet but utilized a whopping 225 GPU days using an NVIDIA GTX 1080 GPU.

The progressive growth-based line of NAS methods has since progressed to the stage where it is possible to achieve an accuracy of 84 top-1 accuracy with only 5 days of searching with a method called **efficient forward architecture search**, but this is beyond the scope of this book.

To end this chapter, *Figure 7.14* shows a summary performance comparison among the methods introduced for the image domain, excluding general hyperparameter search methods:

Method	Number of params (million)	Search time (days)	CIFAR-10 test error	ImageNet test error
Vanilla RL NAS	4.2	1680	4.47	N/A
ENAS macro	21.3	0.32	4.23	N/A
ENAS micro	4.6	0.45	2.89	N/A
DARTS	3.4	4	2.83	26.9
Proxyless NAS	7.1	8.3	N/A	24.9
PNAS	5.1	225	3.41	25.8

Figure 7.14 – Performance comparison of all the introduced NAS methods, excluding general hyperparameter search methods

This table includes the number of parameters, search time, and test error rates for CIFAR-10 and ImageNet datasets for each NAS method introduced. Each NAS method has its own strengths and weaknesses in terms of latency, complexity, and accuracy. The ENAS micro method, in particular, stands out with a relatively low number of parameters, a short search time, and a low test error rate for CIFAR-10. It could be a recommended choice for neural architecture search in the image domain. However, the specific choice depends on the requirements and constraints of the project, such as available compute resources and desired accuracy.

Summary

NAS is a method that is generalized to any NN type, allowing for the automation of creating new and advanced NNs without the need for manual neural architecture design. As you may have guessed, NAS dominates the image-based field of NNs. The EfficientNet model family exemplifies the impact NAS provides to the image-based NN field. This is due to the inherent availability of a wide variety of CNN components that make it more complicated to design when compared to a simple MLP. For sequential or time-series data handling, there are not many variations of RNN cells, and thus the bulk of work in NAS for RNNs is focused on designing a custom recurrent cell. More work could have been done to accommodate transformers as it is the current state of the art, capable of being adapted to a variety of data modalities.

NAS is mainly adopted by researchers or practitioners in larger institutions. One of the key traits practitioners want when trying to train better models for their use cases is the speed to the final result. NAS by itself is still a process that takes days to accomplish, and if applied to a large dataset, it can take up to months. This deters most practitioners' usage of NAS directly. Instead, they mostly use the existing architectures from published open source implementations. Using the existing architectures makes no difference in speed when compared to using manually defined architecture and thus gives practitioners the motivation they need to use it instead. It is also widely known that pre-training helps to improve the performance of the model, thus using NAS directly means you'd have to also pre-train the resulting architecture yourself on a large generalized dataset, which further extends the time needed to complete the NAS process. Use cases in ML often require a lot of time to explore the problem setup and figure out the potential performance that is achievable from the available dataset. Thus, quick iteration between model experimentations is crucial to the success of the project. Slow experimentations dampen the time to identify success. These reasons are why NAS is mainly adopted by practitioners in bigger institutions or researchers who are willing to spend time designing generalized custom neural architectures that can be amortized across different domains instead of building custom architectures for a specific use case.

However, NAS still undoubtedly provides a unique way to find unique custom architectures for your use cases as long as time is not of concern and the goal is either to maximize the performance you can get with a target latency or to generally get the best-performing model without latency considerations.

In the next chapter, we will go into the details of different problem types in SL, along with general tips and tricks for **supervised DL (SDL)**.

8

Exploring Supervised Deep Learning

Chapters 2 to *6* explored the core workhorse behind **deep learning** (DL) technology and included some minimal technical implementations for easy digestion. It is important to understand the intricacies of how different **neural networks** (**NNs**) work. One reason is that when things go wrong with any NN model, you can identify what the root cause is and mitigate it. Those chapters are also important to showcase how flexible DL architectures are to solve different types of real-world problems. But what are the problems exactly? Also, how should we train a DL model effectively in varying situations?

In this chapter, we will attempt to answer the preceding two points specifically for supervised deep learning, but we will leave answering the same questions for unsupervised deep learning for the next chapter. This chapter will cover the following topics:

- Exploring supervised use cases and problem types
- Implementing neural network layers for foundational problem types
- Training supervised deep learning models effectively
- Exploring general techniques to realize and improve supervised deep learning-based solutions
- Breaking down the multitask paradigm in supervised deep learning

Technical requirements

This chapter includes some practical implementations in the Python programming language. To complete it, you will need to have a computer with the following libraries installed:

- `pytorch`
- `catalyst==22.04`
- `numpy`
- `scikit-learn`

You can find the code files for this chapter on GitHub at `https://github.com/PacktPublishing/The-Deep-Learning-Architect-Handbook/tree/main/CHAPTER_8`.

Exploring supervised use cases and problem types

Supervised learning requireslabeled data. Labels, targets, and ground truth all refer to the same thing. The provided labels essentially supervise the learning process of the **machine learning (ML)** model and provide the feedback needed for a DL model to generate gradients and update itself. Labels can exist in many different forms. They are **continuous numerical format, categorical format, text format, multiple categorical formats, image format, video format, audio format,** and **multiple target formats**. All of these are then categorized as either of the following supervised problem types:

- **Binary classification**: This is when the target has categorical data with only two unique values.

- **Multiclassification**: This is when the target has categorical data with more than two unique values.

- **Regression**: This is when the target has continuous numerical data.

- **Multi-target/problem**:

  - **Multilabel**: This is when the target has more than one binary associated with a single data row.

  - **Multi-regression**: This is when the target has more than one regression target associated with a single data row.

  - **Multiple problems**: Either multiple targets associated with a single data row in a single dataset or a chain of problems that is sequential in nature and multiple models are learned through different datasets.

- **Supervised representation learning**: This can be in many forms, and the main goal is to learn meaningful data representations given input data. The results of learned representation can subsequently be utilized for many purposes, including **transfer learning (TL)**, and to realize recommendation systems.

The definition of these problems may be hard to understand by itself. To create a better understanding of the different problems, we will check out an extensive set of use cases that can take advantage of DL technologies. Given the problems specified previously, the following table lists their use cases:

Problem Types	Supervised Deep Learning Model Use Cases
Binary classification	• Gender prediction of babies with ultrasound imagery • Semiconductor chip-quality rejection prediction in manufacturing with image data • Using email data that can contain text, images, documents, audio, or any data to predict spam
Multiclassification	• Document data topic classification • Hate/toxic speech or text classification • General image object classification • Sentiment prediction of text or speech data
Regression	• **Click-through rate (CTR)** prediction of advertisements with image or text or both • Human age prediction with a facial input image • Predicting GPS location from an image
Multi-target	• **Text topic classification**: Multilabel, multiple topics can exist in a single text data row. • **Image object detection**: A multiple-target problem consisting of multiple regression targets and multiclass classification. First, with an image bounding box as multiple regression targets, a single x and y numerical coordinate, and its width and height as the two extra targets forms a rectangular-shaped bounding box. Next, the bounding box will be used to extract a cropped image for multiclassification purposes to predict the type of object. • **Image segmentation**: A kind of multilabel problem, where every pixel will serve as binary targets.
Supervised representation learning	• Face feature representation for face recognition. • Audio representation for speaker recognition using **K-Nearest Neighbors (KNN)**. • Representing categories with their own representative feature vectors. This can be achieved with a method called **categorical embeddings,** which is an NN layer type that holds a feature vector for each category in a categorical feature column. It is learnable and serves as a lookup table. The method can reduce the feature dimensions of high-cardinality categorical data when compared to basic one-hot-encoding but still maintain around the same performance.

Table 8.1 – A table of DL problem use cases

Binary classification, multiclass classification, and regression problems are rather straightforward to approach. Multi-target types, however, pose complicated setups and require more architecting to be done depending on the nature of the problem. Multi-target tasks also can be straightforward, such as multilabel or multi-regression problems. This task falls into the bigger envelope of multitask solutions and will be discussed further in the *Exploring general techniques to realize and improve DL-based solutions* section of this chapter.

Next, we will implement the basic NN layers of realizing the foundational problems, which include binary classification, multiclass classification, and regression problems.

Implementing neural network layers for foundational problem types

In *Chapters 2* to *7*, although many types of NN layers were introduced, the core layers for the problem types were either not used or not explained. Here, we will go through each of them for clarity and intuition.

Implementing the binary classification layer

Binary means two options for categorical data. Note that this does not necessarily mean a strict rule for the categories to be true or false nor positive or negative in the raw data. The two options can be in any format possible in terms of raw data, in strings, numbers, or symbols. However, note that NNs can always only produce numerical outputs. This means that the target itself has to be represented numerically, for which the optimal numbers are the binary values of zero and one. This means that the data column to be used as a target for training with only two unique values must go through preprocessing to map itself into zero or one.

Generally, there are two ways to define binary outputs in NNs. The first is to use a linear layer with a size of one. The second method is to use a linear layer with a size of two. There is no significant difference between the two in terms of task-quality metrics, but method one takes slightly less space for storage and memory, so feel free to always use that version. The outputs from method 1 will be constrained to values between 0 and 1 by using a sigmoid layer. For method 2, the outputs need to be passed into a softmax layer so that the probabilities for the two outputs will add up to one. Both methods usually can be optimized using **cross-entropy**. Cross-entropy is also known as **log loss**. Log loss measures the difference between predicted probabilities and true labels using a logarithmic scale. This scale penalizes incorrect predictions more heavily, emphasizing the importance of a model's ability to assign high probabilities to the correct class and low probabilities to the incorrect class.

Translating the layer from method one into actual `pytorch` code will look like the following using the nn module from `torch`:

```
from torch import nn
final_fc_layer = nn.Sequential(
  nn.Linear(10, 1),
```

```
    nn.Softmax(),
)
```

Next, we will implement the multiclass classification layer.

Implementing the multiclass classification layer

Multiclass means the data contains multiple categories with more than two categories, differentiating it from binary classification. Likewise, in multiclass classification, the raw data can be in any format possible and must go through a preprocessing task to convert the raw categories into ordered numbers that start from 0. The ordered numbers, however, do not signify an actual ordering, nor do they encode any order information that the NN model can leverage. The output linear layer for multiclass problems needs to be configured with the same number of neurons as the number of unique classes. A softmax layer is similarly applied after the output linear layer to make sure the probabilities for the final outputs will add up to one. Cross-entropy is also used here as the standard loss. During the inference stage, the class index with the highest probability will then be used as the predicted class. Translating the multiclass prediction layer into `pytorch` code for a 100-class multiclass classification problem with 10 logits will look like this:

```
final_fc_layer = nn.Sequential(
    nn.Linear(10, 100),
    nn.Softmax(),
)
```

Another sub-problem of multiclass classification is when the classes are ordinal. This sub-problem and task is called ordinal classification. This means that the classes have an incremental relationship with each other. A plain multiclass classification layer strategy represents ordinal classes sub-optimally as the classes in the multiclass are considered to have an equal relationship with each other. A good strategy here to add the information of ordinal classes is to utilize a technique based on the multilabel classification task, which is a multiple-binary classification task.

Let's say that we have five ordinal classes represented as numerical numbers from 1 to 5 for simplicity. In reality, this could be represented by any categorical data. In an NN, five binary classification heads would be used for this case where the classes will be assigned to the respective head in an ascending ordered manner. The raw predictions from this NN will be consumed in a way where the final predicted ordinal class will be derived from the position of the furthest consecutive positive binary prediction. Once there is a negative prediction, the rest of the prediction heads on the right will then be ignored. *Figure 8.1* depicts this strategy by simulating the output predictions of the five binary classification heads:

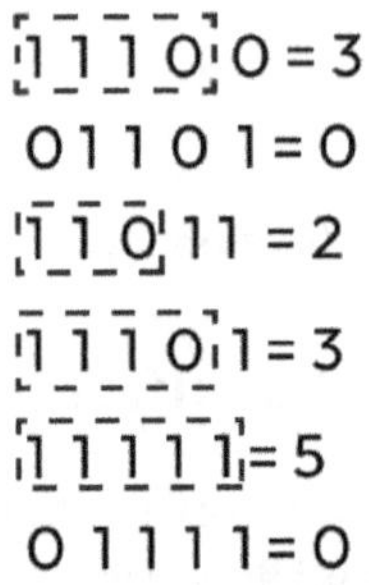

$$1\ 1\ 1\ 0\ 0 = 3$$
$$0\ 1\ 1\ 0\ 1 = 0$$
$$1\ 1\ 0\ 1\ 1 = 2$$
$$1\ 1\ 1\ 0\ 1 = 3$$
$$1\ 1\ 1\ 1\ 1 = 5$$
$$0\ 1\ 1\ 1\ 1 = 0$$

Figure 8.1 – Ordinal classification processing strategy using the output predictions of the five binary classification heads

The learning process will be, as usual, using the cross-entropy loss for multiple binary targets. Additionally, the performance at every epoch can be monitored using robust metrics that don't depend on probabilities such as recall or precision. The ordinal encoding method allows the model to learn that the targets have an ordinal relationship.

Next, we will dive into the implementation of a regression layer.

Implementing a regression layer

Regression means a single numerical target and prediction. Regression can be realized in an NN by simply having a linear layer with a single neuron so that an unbounded single numerical output can be obtained without any activation layers. Translating the layer into actual `pytorch` code will look like the following using the nn module from `torch`:

```
final_fc_layer = nn.Linear(10, 1)
```

Mean squared error (**MSE**) is the standard loss here to act as the error for optimization. However, one of the risks of unbounded numerical predictions is that values can skew out of acceptable bounds. If a bounded range is known, one way this can be enforced in an NN model is by using a scaling method such as min-max scaling to map target values into the known boundaries. Once min-max scaling is used, the target values will then be in the range of 0 and 1. Coupled with the scaling of target values, the bounds can then be enforced in the NN by using the sigmoid layer, which similarly scales activation values between 0 and 1. During the inference stage, the predicted values can then be mapped into actual values by descaling values between 0 and 1 into the known minimum and maximum boundaries specified earlier.

The unbounded method allows for some form of generalization by allowing extrapolation, and the bounded method allows the addition of informed bias to the NN. Both methods have their own benefits and disadvantages, and thus the choice of approach needs to be evaluated on a case-by-case basis.

Next, we will dive into the implementation of representation layers.

Implementing representation layers

Most methods focus on the interactions between architectures for different data modalities or the training methods that optimize the represented features. These are topics we will dive into further in the next topic after this. One key layer type that truly represents the representation layer is the embedding layer. Embeddings are a type of layer structure that maps categorical data types into learnable vectors. Through this layer, each category will be able to learn a representation that is able to perform well against the specified target. The method can be used for converting text word tokens into more representative features or plainly as a replacement for one-hot encoding. Categorical embeddings make it possible to automate the feature engineering process for categorical data types. One-hot encoding produces an encoding that enforces the same distance between all the categories to every other category. Categorical embedding, however, allows for the possibility of obtaining an appropriate distance based on its interactions with the target variable and with other data if any extra data exists.

However, categorical embeddings are also not a silver bullet for all ML use cases, even if they are decoupled from an actual NN model after training and just act as a featurizer. They can sometimes perform better against one-hot-encoding in general and, vice versa, can happen other times to perform worse against one-hot-encoding. The method still remains a key method to experiment with for any dataset with categorical data as input.

Training supervised deep learning models effectively

In *Chapter 1, Deep Learning Life Cycle*, it is emphasized that ML projects have a cyclical life cycle. In other words, a lot of iterative processes are carried out in the course of the project's lifetime. To train supervised deep learning models effectively, there are a lot of general directions that should be taken based on different conditions, but the one that absolutely stands out across every problem is proper tooling. The tooling is more commonly known as **ML operations** (**MLOps**). Good MLOps systems for DL are easy to use and provide versioning methods for datasets and model experiments, visualization methods, easy ways to use DL libraries such as `pytorch` or `keras` with `tensorflow`, ease of deployment, ease of model comparisons using different metrics, ease of model tuning, good visualization of model training monitoring, and, finally, good feedback about the progress (this can be sent through messages and notifications for alerts). If no advanced tools that truly simplify the entire process are at your disposal, you can focus on the important bits of making a model work well instead of dealing with infrastructure issues such as coordinating the saving of models into different folders like **DataRobot**, a paid-for tool, then open sourced tools such as MLflow, Kubeflow, or Metaflow will be the next-best alternative. Once the tool of choice is picked, carrying out training in DL models will be a breeze. We will be using MLflow as an example tool to demonstrate some effective methods for DL model training in the following steps:

1. Data preparation
2. Configuring and tuning DL hyperparameters
3. Executing, visualizing, tracking, and comparing experiments

Additionally, we will explore some extra tips when building a model before ending this topic.

Let's explore each in detail.

Preparing the data for DL training

Data is the core of any ML model. Data ultimately determines the achievable model performance, the quality of the final trained model, and the validity of the final trained model. In *Chapter 1, Deep Learning Life Cycle*, we explored what it takes for a dataset setup to be DL-worthy, along with the qualities needed when acquiring data, coupled with **exploratory data analysis** (EDA) to verify causality and validity. The general idea there was to identify and add extra features and data modalities that have causal effects toward the desired target. In this section, we will cover more in-depth essential steps that convert the data into a DL trainable state listed here:

1. Data partitioning
2. Data representation
3. Data augmentation

Partitioning the data for DL training

The first step we will cover is data partitioning. Having a good data-partitioning strategy for training, validating, and testing your model is essential for a good-performing model. The training partition will be the partition that will strictly be used for training. The validation partition will be the partition that will strictly be used for validating a model during training. In DL, a **validation partition** is often used as a guide to signal when to stop training or extract the best-performing weights using external data out of the training data. Since the validation data will affect the learning process of the model and add some bias that will cause overfitting toward the nature of the out-of-training validation data, a testing partition will be the final partition that will be used to verify the generalizability of the trained model. The testing partition is also known as the holdout partition. To be extra safe in preventing overfitting and to ensure the generalizability of the model, the validation partition can also be used exclusively to be validated only once after the model is trained instead of being used for validation at every epoch. This strategy, however, requires that a smaller internal validation partition is created from the original training partition. The following figure depicts the two different strategies:

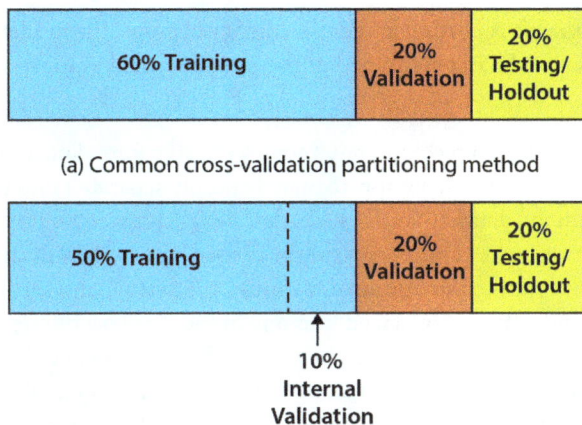

(a) Common cross-validation partitioning method

Figure 8.2 – Two different cross-validation data-partitioning strategies

This process of partitioning the data is called **cross-validation**. The preceding figure shows simple cross-validation strategies with only a single partitioning setting. This might create issues where the model's performance metrics reported are biased toward a specific resulting partitioning setting. The resulting partitioning may have some inherent distribution or nature that allowed results to perform particularly well or badly toward it. When that happens, having mismatched expectations of performance during the deployment stage will create more operational issues. To safely remove the possibility of such bias, **k-fold cross-validation** is typically used to report a more comprehensive validation score that could better reflect the performance of the model in the wild. To perform this partitioning method, a single testing set is removed from the original dataset, and validation scores are averaged across different ordered k cross-validation training and validation partitions. This is better visualized in *Figure 8.3*:

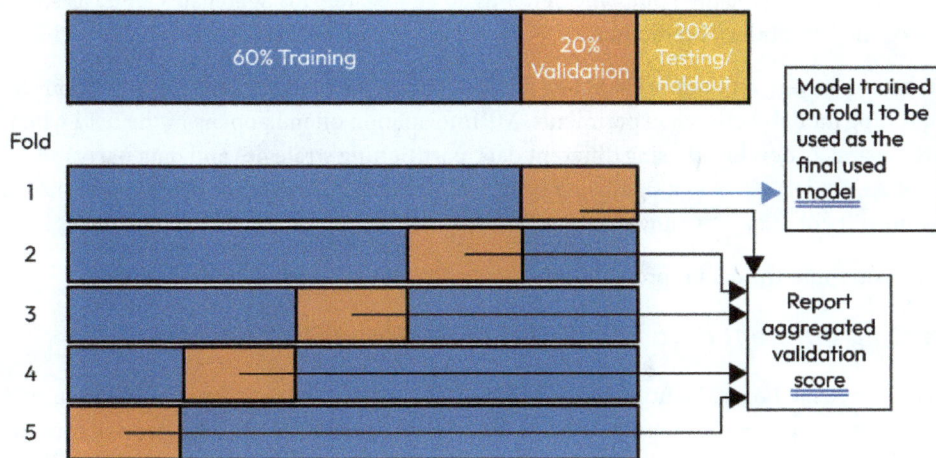

Figure 8.3 – K-fold cross-validation as a strategy to eliminate metric reporting bias

Finally, for testing performance reporting and deployment purposes, the model either gets retrained on the training and validation dataset combined or the model trained in the first fold is extracted for deployment purposes.

Recall that stratified partitioning is a recommended strategy to split your data into the three mentioned partitions. This means that the data will be approximately evenly separated into three partitions based on the label associated with the dataset. This ensures that no labels get left out in any partition, which could potentially cause misinformation. Take a simple case of binary classification where the dataset is randomly partitioned into three partitions of training, validation, and testing with prespecified sizes. Since each data row was randomly placed into one of the three partitions, there is a probability the partitions will only contain one label from the two binary labels. Since the validation or testing partition is usually assigned with smaller data sizes, they have more potential to face this issue. Let's say the model is mistakenly trained to predict only a single label. If the label is exactly the label that exclusively exists in the validation and testing of ML learning practitioners, we will mistakenly think the model is doing extremely well, but in fact, it is useless. You never know when you will be unlucky when doing full random partitioning, so use stratified random partitioning whenever you can!

The data-partitioning strategy described here builds only a single model that will be utilized during inference mode in model deployment. This strategy is the standard option when the inference runtime of the final model setup is a concern and having a faster model is more important than having small improvements in the accuracy metrics. When it is okay to trade runtime for some accuracy performance, an alternative strategy called **k-fold cross-validation ensemble** can be used. This is a method that is widely advocated in many ML competitions, especially in the ones hosted on Kaggle. The method uses the k-fold cross-validation described previously but actually uses k models trained during cross-validation and performs an ensemble of the k model's predictions. An ensembling method called blending aggregates predictions of models and almost always improves the accuracy metrics from a single model. This process can be thought of as a method that leverages the best ideas and expertise of each k model, making the final outcome better as an aggregate. This aggregate can be as simple as an average or median of the k predictions.

A final tip before moving on to the next method is to always remember to make sure partitions match when comparing models between experiments. Misinformation often happens in the field when two models are separately developed using different data-partitioning strategies and data partitions. Even when one of the models achieves a significant performance advantage over the other, it does not mean anything and will not amount to any meaningful comparisons.

Next, we will dive into the data representation component for different data modalities.

Representing different data modalities for training DL models

So far, we have brushed over the utilization of numerical, categorical, text, audio, image, and video modalities. These are the most common modalities utilized across multiple industries. Representing different data modalities is a complicated topic as, in addition to the common modalities, there are actually a lot of rare data modalities out there. Examples of rare modalities are chemical formulas (a special structured form of textual data), document data (another special form of textual data with complex positional information), and graph data. In this section, we will only discuss the representation of the common unstructured modalities here to ensure the relevancy of content to our readers. Both numerical and categorical data are considered structured data and are have been covered properly in previous sections. Let's now start with the text data modality.

Representing text data for supervised deep learning

Text data representation in general has improved tremendously over the years. The following list shows a few relevant methods:

- **Term frequency-inverse document frequency (TF-IDF) with N-grams**: The term here is implemented with N-grams. An N-gram is an adjacent sequence of n textual characters. N-grams are produced by a method called tokenization. The tokenization can be a representation as low level as single characters, or it can be a higher-level representation such as words. Once represented as N-grams, TF-IDF is computed using the following formula.

$$TF\text{-}IDF = term\ frequency \times inverse\ document\ frequency$$

Term frequency is simply the count array of a single row. **Inverse document frequency** is computed through the following formula:

$$IDF = \log\left(\frac{number\ of\ text\ samples}{number\ of\ documents\ containing\ the\ term\ for\ each\ term}\right)$$

The representation is an efficient and lightweight way to extract useful information from text, where words that are rare have higher values and more frequent words such as "the" and "and" will be suppressed. Outputs of TF-IDF can be directly fed into a simple **multilayer perceptron** (**MLP**) or any ML model to produce a predictive model. In simpler use cases, this representation will be enough to achieve a good metric performance. However, in more complex use cases that require the decoding of complex interactions that can happen with the different compositions of text and labels, it will underperform.

- **Word/token embeddings**: Word embeddings can be trained from scratch or pre-trained from a bigger dataset. Pre-trained embeddings can be pre-trained in either a supervised fashion or an unsupervised fashion, usually on a larger dataset. However, the embedding method suffers from the issue of token mismatch during the training, evaluation, and testing stages. This means that it is required to perform a lot of tinkering with the way the specific text token is preprocessed before looking up to the embeddings table. This occurrence is known as **out of vocabulary** (**OOV**) during the evaluation and testing stages. In the training stage, different variations of the same word will have their own meaning, which is inefficient in terms of learning and resource-space utilization. In practice, methods such as stemming, lemmatization, lowercasing, and known word-to-word replacements are applied to mitigate OOV, but the problem won't be mitigated completely. These word embeddings can be paired with either **recurrent NNs** (**RNNs**) or transformers.

- **Subword-based tokenization**: This family of methods attempts to solve the token mismatch issue and the large vocabulary size of tokens. *Subword* might sound unintuitive, as we as humans use full words to perceive the meaning of text. This family of algorithms only performs subword tokenization when the word can't be identified or is considered to be rare. For common words, they will remain full word tokens. Examples of such methods include **byte-pair encoding** (**BPE**), **WordPiece**, and **SentencePiece**. We will go through these methods briefly as a simple guide:

  - **BPE tokenization**: BPE treats the text as characters and groups up the most common consecutive characters iteratively during training. The number of iterations determines when the training iterations to group up the most common characters should be stopped. This formulation allows for rare words to remain as subword tokens and common words to be grouped up into a single token. This representation is notably used by **generative pre-trained transformer** (**GPT**) models. However, this representation faces an issue where there will be multiple ways to encode a particular word.

  - **WordPiece**: WordPiece improves upon BPE by utilizing a language model to choose the most likely pair of tokens to group up. This enforces a kind of intelligent choice when deciding the way to encode a particular word. This algorithm is utilized by **Bidirectional Encoder Representations from Transformers** (**BERT**) and **Efficiently Learning an Encoder that Classifies Token Replacements Accurately** (**ELECTRA**).

  - **SentencePiece**: SentencePiece is a method that optimizes the tokens generated by base tokenizers such as BPE. It uses a couple of components, such as using a form of Unicode text conversion to ensure no language-dependent logic exists and using a method called subword regularization that performs a form of subword token group augmentation (probabilistically and randomly choose a single sample from the top-k predicted subword token to group up using language models) to solve multiple representation issues. This algorithm is used by XLNet and **A Lite BERT** (**ALBERT**) most notably.

Text data are represented as tokens for DL. This also means that there will be strict limits to the number of tokens so that the NN model can be initialized with the right parameters. Pick a token size limit that is reasonable for your use case based on what's needed to get a good model performance. Since the size limit will affect model size, be sure to make sure the model size doesn't get so big that it overshoots your inference runtime requirements.

In terms of missing text data, which can happen in real-world use cases with multimodal data, using an empty string is the most natural way to work as an imputation method. Under the hood, these models would usually use all zeros to represent the text array.

Now that we've briefly covered supervised text data representations, let's discover supervised audio data representations next.

Representing audio data for DL

Audio data is time-series data with one or two arrays of values where each value in the array represents a single piece of audio data at a specific timestamp. Audio data can be either represented as a simple normalized form from the original raw data, represented as something called a spectrogram, which is a two-dimensional data, or as **Mel-frequency cepstral coefficients** (MFCCs).

A spectrogram is the resulting output of a process called **short-time Fourier transform** (STFT) that can decompose raw audio data into its respective signal across a range of frequencies. Raw audio data contains combined audio signals that can be from any audio frequency. Breaking down audio data into its signal-by-frequency representation allows ML algorithms to have an additional degree of freedom when attempting to identify key patterns. In most audio use cases, **spectrogram**-based models perform much better than using raw audio data directly. Spectrograms exist in two-dimensional format. This means that it can be treated as an image and can be fed into image-based models such as **Residual Networks** (ResNets) and **Vision Transformers** (ViTs). However, there are some models that are capable of taking advantage of the raw unprocessed audio data, such as `wav2vec 2.0`, which is a type of transformer.

The STFT process has hyperparameters that can affect the resulting representation. The following list summarizes these hyperparameters and tips on how to set it up properly:

- **Sampling rate**: This specifies the samples per second (Hertz/Hz) parameter that will be used before applying STFT. Audio data might be recorded in different sampling rates, and to build a model, this data will be required to be unified to a single sampling rate through resampling algorithms. The most typically used value is 16,000 Hz. As this value will affect the size and runtime of the model given a fixed time window a model is built to handle, be sure to only increase it if it's necessary in terms of metric performance.

- **STFT window length**: Each window size will be responsible for the data for a fixed duration and specific position. Each window will produce a single value at the same time window for a range of frequencies. The typical value of this parameter is 4096 or 2048. Configure this based on the prediction resolution you require for your use case if there is a strict requirement there. This parameter will also affect the model size.

- **Window stride**: This is similar to a convolutional layer filter stride and does not have many significant tuning methods. Using a small percentage of the window length, such as 10%, should be a good enough setting.

- **Whether to use Mel scaling**: A **Mel scale** is a logarithmic transformation of the audio signal's frequency. Fundamentally, it is a transformation to mimic how humans perceive audio. It makes higher-frequency changes matter less and lower-frequency changes matter more. Use this when it involves some form of human judgment to improve the metric performance.

As for empty audio rows, imputing them with a single pre-generated random noise audio or using an array of zeros with the same length should work well in multimodal datasets.

Now that we've briefly covered supervised audio data representations, let's discover supervised image and video data representations next.

Representing image and video data for DL

Image data doesn't require a lot of introduction here as we have gone through a few tutorials using them directly. The key is to perform some sort of normalization before feeding it to NN models such as CNNs, and the NN will extract great representations. Transformers have also been making tight competition with CNNs in image-based tasks and can be used to both extract representative features and predict directly on the task. One thing to note is that unless the resolution of the image is crucial in identifying certain patterns, it is usually much more effective as a model to utilize a smaller image resolution. One might not be able to visually certain patterns when the resolution of the image is smaller, but a computer would still be able to. The resolution of the image affects the runtime of the training, the runtime of the model in production, and sometimes the model size, as for transformers, so make sure this is done conservatively.

Video data, however, is an extended form of image data where a number of images are aligned sequentially to form a video. This means that video data is a form of sequential data just like text without absolute timestamp information. Each sequential image is known as a **frame**. Video can have a variety of frame rates. Commonly, this would be in rates of 24, 30, or 48 **frames per second (FPS)** but can generally be any number. For **computer vision (CV)** use cases, make sure to set a low FPS so that the processing load can be reduced depending on the use case. For example, the use case of lip reading has lower FPS requirements than for asking a model to identify whether a person is running or not. For the frame resolution, the same guide for image resolution applies here. Once the video properties have been decided, representative features have to be learned and extracted. The current SoTA features are extracted through models similar to image-based use cases. Examples of such models are 3D CNNs and 3D transformers.

There is, however, an intersection of these two data types, which are images extracted through video data. For this type of image data, it is possible to reduce the probability of predictive models making wrong predictions. ML models are not perfect predictors, so whenever there is a chance to reduce incorrect predictions such as false positives or false negatives without compromising the true predictions, do consider taking it. Consider using manual image processing techniques from the OpenCV library to perform any preliminary steps before a model takes the image as input. For example, the motion detection technique in OpenCV can be used as a preliminary condition checker before feeding the video array to the DL model. Since motion is required to identify most use cases of video data, it doesn't make sense to predict anything if nothing is moving. This also reduces any false predictions that can happen from predicting on multiple unchanged video frames. The motion detector in OpenCV utilizes a simple change in pixel value without using a probabilistic model and thus is a far more reliable indicator of motion.

Now that we've covered representing different data modalities, let's move on to the topic of data augmentation.

Augmenting the data for training better DL models

Augmentation is widely used in DL to increase the generalization of the resulting trained model and increase the metric performance of the model. By accounting for the additional unique variations brought in by augmentation, the model would be able to attend to these unique variations on external data during the validation, testing, and inference stage. Naturally, this will also reduce any over-dependence on a specific pattern and any benefits that come with a sufficiently sized dataset. Augmentation increases the amount of training data and thus the variations of patterns in the training data. The process is usually done randomly and individually in every training iteration in memory. This makes sure there are no limitations to the additional data variations available for training and also removes the need for additional storage. However, you can't just randomly use all the available types of augmentation known for a specific modality. The following list shows the different types of augmentation you can perform on your data:

- **Image**: Image sharpening through **Contrast Limited Adaptive Histogram Equalization** (**CLAHE**), hue and saturation variation, color channel shuffling, contrast variation, brightness variation, horizontal/vertical flip, grayscale conversion, blurring, image masking, mixup (weighted combination of images and their labels), cutmix (mixup, but only by random patches from the original image), and more. Look into `https://github.com/albumentations-team/albumentations` for more than 70 augmentations!

- **Text**: Synonym replacement, back translation (a process that translates a text into another language and then translates it back into the original language), and more.

- **Video**: Video mixup (same as image mixup but for videos), all the same augmentation for images.

- **Audio**: Random white noise, random pink noise, reverberation, stretching the time, back resampling (similar to back translation, but changing the sampling rate instead), random STFT window types, random masking, random bandpass noise, additional noise from randomly chosen real audio clips (you can either get this from YouTube or from permissive public audio datasets). These can be accessed through the `librosa` library.

Choosing the type of augmentation to use requires some understanding of the expected environment that you will face when you deploy a model. Bad choices add noise to the model and might confuse the model during the training process, resulting in a degraded metric performance. Good choices revolve around estimating the variations that can realistically happen in the wild. Let's take an example of a manufacturing use case where the goal is to deploy an image-based model that will predict product characteristics on a conveyor belt for sorting purposes using a camera sensor. If you can assume the camera will be fixed almost perfectly straight on the machine in all the setups, using image rotation augmentation likely wouldn't be smart. Even if you want to use the augmentation, the rotation variation you should use should only be in the low end, such as below 10 degrees variation. Grayscale augmentation would also be unintuitive if the cameras do not provide grayscale images.

This concludes the data preparation stage of training effective models. Next, we will dive into the model training stage of the workflow.

Configuring and tuning DL hyperparameters

Hyperparameter configuration and tuning play a crucial role in training DL models effectively. They control the learning process of the model and can significantly impact the model's performance, generalization, and convergence. In this section, we will discuss some essential hyperparameters and their impact on training DL models.

The most general set of impactful hyperparameters that need to be configured for training each NN model are its epochs, early stopping epochs, and learning rate. These three parameters are considered a set of parameters that together form a **learning schedule**. There have been a few notable learning schedules that focused on obtaining the best model with the least amount of time spent. However, they depend a lot on the initial estimation of the total number of epochs a model can converge with the method. Methods that depend on an estimation of the total number of epochs needed are fragile, and their configuration strategies are not easily transferable from one problem to another. Here, we will focus on using a validation dataset to track the number of epochs needed to achieve the best possible metric performance.

Early stopping epochs is a parameter that controls how many epochs you want to keep training before you stop. This strategy means that the epochs' hyperparameter can either be set to an infinite number or a very large number so that the best-performing model on the validation dataset can be found. Early stopping reduces the number of training epochs you need to spend dynamically based on the validation dataset. By saving the best-performing model weights on the validation dataset, when the model is stopped early, you will then be able to load the best-performing weights. The typical early stopping epoch is 10.

As for the learning rate, there are two general directions that work consistently well in practice. One is to immediately start with a large learning rate such as 0.1 and to gradually decay the learning rate. The gradual decay of the learning rate can be through percentage reductions from validation score monitoring when it doesn't improve after 3 to 5 epochs. The second method is to use a smaller learning rate as a warmup method in initializing a base weight for the NN before using method 1. In the initial stage of learning, as models are initially in a randomized state, the learning process can be very unstable where the loss will seem to not follow a proper improvement trend. Using a warmup helps to initialize the foundation needed to make stable loss progressions. Note that if pre-trained weights are used to initialize the model, a warmup is usually not needed as the model will already be in a stable state, especially if the pre-trained weights are obtained from a similar dataset.

Batch size is another crucial hyperparameter in the training of DL models, as it determines the number of training samples used in a single update of the model's weights during the optimization process. The choice of batch size can significantly impact the model's training speed, memory requirements, and convergence. Smaller batch sizes, such as 16 or 32, provide a more accurate estimate of the gradient, leading to more stable convergence, but may require more training iterations and can be slower due to less parallelism in computation. On the other hand, larger batch sizes, such as 128 or 256, increase the level of parallelism, speeding up the training process and reducing memory requirements, but may lead to a less accurate gradient estimate and potentially less stable convergence. In practice, it's essential to experiment with different batch sizes to find the one that provides the best balance between training speed and convergence stability for your specific problem. Additionally, modern DL frameworks often support adaptive batch size techniques, which can automatically adjust the batch size during training to optimize the learning process.

The foundational strategy discussed here is robust and can easily obtain the best-performing model most of the time while sacrificing some additional training time. It is worth noting that regularization methods, optimizers, and different activation functions have been covered in *Chapter 2, Designing Deep Learning Architectures*, and I encourage you to refer to that chapter for more information on those topics.

As much as there can be a manual strategy for the configuration of these hyperparameters, there will always be space to tune the hyperparameters further to optimize the metric performance of the model. Commonly, tuning can be executed through either grid search, random search, or tuning through more intelligent searching mechanisms. Grid search, otherwise known as brute-force searching, explores and validates all possible combinations of specified hyperparameter values to identify the optimal configuration for a given problem through cross-validation. For more intelligent tuning methods, refer back to *Chapter 7, Deep Neural Architecture Search*, for more insights on it. Additionally, as model evaluation metrics contribute to this hyperparameter-tuning process, we will explore more on this in *Chapter 10, Exploring Model Evaluation Methods*.

The core of hyperparameter tuning depends on the process and workflow to iteratively execute, visualize, track, and compare modeling experiments, with each configuration being part of a modeling experiment. This brings us to the next topic, diving into the actual workflow of training DL models effectively.

Executing, visualizing, tracking, and comparing experiments

The key to executing ML projects effectively is to iterate quickly between experiments. A lot of exploration is needed in any ML project, both in the initial stage of the project to gauge the viability of the use case, and in the later stage to improve the model's performance. When this exploration process can be optimized, things meant to fail can fail quickly, and things that are viable can succeed quickly. Failure in ML projects is very common in practice. Once we acknowledge that and fail quickly, we can utilize the recovered time to tackle more valuable use cases. A good MLOps platform will help us execute, visualize, track, and compare experiments effectively and efficiently.

Let's go through an example practically with the Iris dataset and an MLP using the MLflow MLOps platform. We will also be using the `catalyst` library, which is also considered to be an MLOps platform, albeit partially and mostly focused on providing common `pytorch` DL model training tools. Since `catalyst` provides most of the model versioning and model storing mechanisms, we will only utilize the tracking feature in MLflow. The steps for this example are as follows:

1. First, let's import all the necessary libraries:

```python
import json
import osimport numpy as np
import torch
import torch.nn as nn
import torch.nn.functional as F
from catalyst import dl, utils
from catalyst.contrib.datasets import MNIST
from sklearn import datasets
from sklearn.metrics import log_loss
from sklearn.model_selection import train_test_split
from sklearn.preprocessing import MinMaxScaler
from torch import nn as nn
from torch import optim
from torch.utils.data import DataLoader
from torch.utils.data import TensorDataset
from catalyst.loggers.mlflow import MLflowLogger
```

2. Next, as we will be using the pytorch-based MLP, we will again set the random seed in pytorch:

```python
torch.manual_seed(0)
```

3. The dataset we will be using for the practical implementation here is the Iris dataset again. The dataset consists of the petal and sepal lengths of various flowers with three different iris types. We will now load this dataset:

```python
iris = datasets.load_iris()
iris_input_dataset = iris['data']
target = torch.from_numpy(iris['target'])
```

4. Scaling is a type of regularization method that can reduce memorization and reduce bias. Let's perform a straightforward minimum and maximum scaling here:

```python
scaler = MinMaxScaler()
scaler.fit(iris_input_dataset)
iris_input_dataset = torch.from_numpy(scaler.transform(iris_
input_dataset)).float()
```

5. To train a model, we need a proper cross-validation strategy to verify the validity and performance of the model. We will use 77% of the data for training and 33% for validation. Let's prepare the data for cross-validation:

```python
X_train, X_test, y_train, y_test = train_test_split(iris_input_
dataset, target, test_size=0.33, random_state=42)
```

6. Next, we will need to prepare the data loaders using the prepared data in numpy format for cross-validation:

```
training_dataset = TensorDataset(X_train, y_train)
validation_dataset =  TensorDataset(X_test, y_test)
train_loader = DataLoader(training_dataset, batch_size=10, num_
workers=1)
valid_loader = DataLoader(validation_dataset, batch_size=10,
num_workers=1)
loaders = {"train": train_loader, "valid": valid_loader}
```

7. Since this is a multiclass problem of three classes, we will use the cross-entropy loss in pytorch:

```
criterion = nn.CrossEntropyLoss()
```

8. We will be using the pytorch high-level wrapper library called catalyst here. To train a model in catalyst, we have to define a model trainer class instance called a runner:

```
runner = dl.SupervisedRunner(
    input_key="features", output_key="logits", target_
key="targets", loss_key="loss"
)
```

9. We will be using an MLP in this project with an MLP constructor class that allows us to specify the input data size, the hidden layer configuration, and the output data size. The hidden layer configuration is a list of layer sizes that simultaneously specifies the number of layers and the layer size at each layer. Let's say that we want to randomly obtain 20 different hidden layer configurations and find out which performs the best on the validation partition. Let's first define a method that generates the configuration randomly:

```
def get_random_configurations(
  number_of_configurations, rng
):
  layer_configurations = []
  for _ in range(number_of_configurations):
    layer_configuration = []
    number_of_hidden_layers = rng.randint(low=1, high=6)
    for _ in range(number_of_hidden_layers):
      layer_configuration.append(rng.randint(low=2, high=100))
      layer_configurations.append(
        layer_configuration
        )
      layer_configurations = np.array(
        layer_configurations
        )
  return layer_configurations
```

```
rng = np.random.RandomState(1234)
number_of_configurations = 20
layer_configurations = get_random_configurations(
  number_of_configurations, rng
)
```

10. Now, let's define a method that will allow us to train and evaluate the different MLP configurations:

```
def train_and_evaluate_mlp(
    trial_number, layer_configuration, epochs,
):
```

The trial number here plainly differentiates the different experiments. Apart from layer configuration, we can also configure the epochs that we want to run. In this method, we will create an MLP model instance based on the layer configuration passed in:

```
model = MLP(
    input_layer_size=iris_input_dataset.shape[1],
    layer_configuration=layer_configuration,
    output_layer_size=len(np.unique(target)),
)
```

11. We will use the Adam optimizer for gradient descent and set the checkpoint directory:

```
optimizer = optim.Adam(model.parameters(), lr=0.02)
checkpoint_logdir = "experiments"
```

12. Next, we will define the MLflow logger helper class available in catalyst to log experiments in MLflow format. In this setup, we log the mean and standard deviation of the training and validation log loss:

```
loggers = {
  "mlflow": MLflowLogger(
    experiment="test_exp", run="test_run"
  )
}
```

13. Finally, we will start the training process that trains for the specified number of epochs:

```
runner.train(
    model=model,
    hparams=hparams,
    criterion=criterion,
    optimizer=optimizer,
    loaders=loaders,
    num_epochs=epochs,
```

```
        callbacks=[
            dl.CheckpointCallback(
                logdir=checkpoint_logdir,
                #save_n_best=0,
                loader_key="valid",
                metric_key="loss",
                mode="model",
            )
        ],
        logdir="./logs",
        valid_loader="valid",
        valid_metric="loss",
        minimize_valid_metric=True,
        verbose=verbose,
        loggers=loggers
)
```

14. As the last code step, we will loop through each randomly generated layer configuration and perform the training and evaluation process:

```
for layer_config in layer_configurations:
    train_and_evaluate_mlp(
        trial_number,
        layer_config,
        epochs=10,
        load_on_stage_start=False
    )
```

15. Now, we need to start up the MLflow server service. We can do this by running the following command in the command line in the same directory as the directory that contains the introduced code:

```
mlflow server –backend-store-uri mlruns
```

After running this command, the same directory should contain the following file named .catalyst, which instructs catalyst to enable MLflow support. This file should have the following content:

```
[catalyst]
cv_required = false
mlflow_required = true
ml_required = true
neptune_required = false
optuna_required = false
```

Once the command is executed, and by opening the HTTP website link, you should see the screen of MLflow, as shown in *Figure 8.4*:

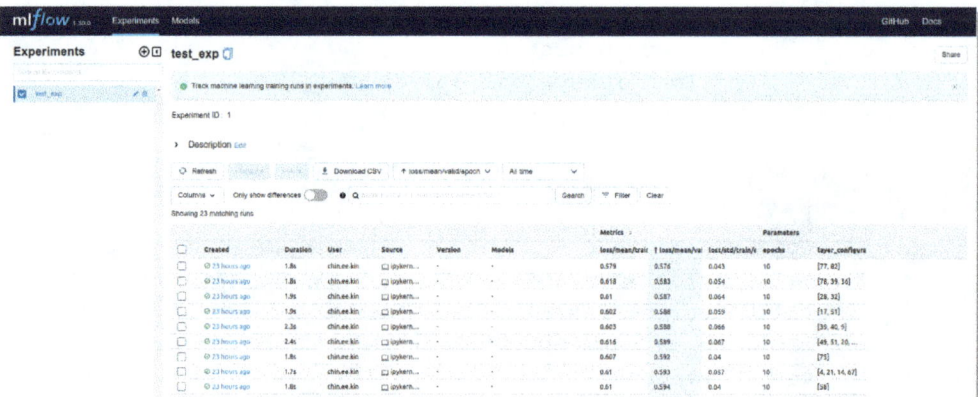

Figure 8.4 – MLflow interface

The interface shows a convenient way to visualize performance differences between different experiments while showing the utilized parameters. The numerical metric values can be sorted to obtain the best-performing model on the validation partition, as shown in the figure. The process of preparing data and training a model requires iterative comparisons to be made between different setups or experiments. Experiments can be compared more objectively with quantitative metrics with their experimentation parameters. When displayed visually automatically through code in an interface instead of plugging it manually into an Excel or Google sheet, this makes the process much more dependable and organized. Additionally, if you click into any of the experiments, you'll be able to check out the loss curves at each epoch interactively, as shown in *Figure 8.5*:

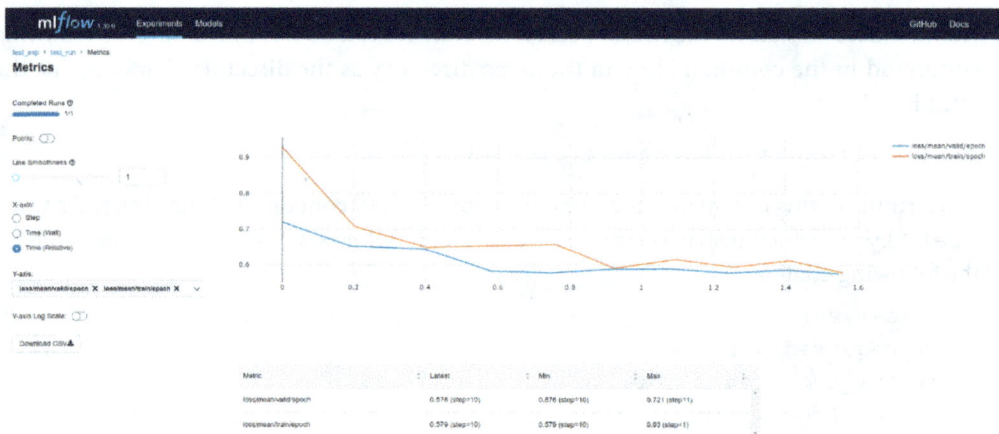

Figure 8.5 – MLflow interface showing an example loss curve of the best model

While ensuring speed of iteration, it is also required to organize and track all artifacts you generated for your model properly. This means that you need to version your model, your dataset, and any key components that affect the resulting model output. Artifacts can be model weights, metric performance reports, performance plots, embedding visualization, and loss visualization plots. This task can obviously be done manually through manual coding. However, organizing the artifacts of models' built-in experiments gets messy when the number of experiments goes up. Any custom files, graphs, and metrics can be tied into each of these experiment records and viewed in the MLflow interface. Experiments here can differ by using different models, different datasets, different featurization methods, different hyperparameters of a model, or a different sample size of the same dataset. Additionally, models can be stored directly in MLflow's model registry, which allows MLflow to deploy the model directly.

Exploring model-building tips

This practical content serves as an example of how an MLOps platform such as MLflow can ease the process of building and choosing the right model programmatically and visually. As much as MLOps is great and helps in training models efficiently, there are a few things that an MLOps platform does not handle for you but are considered key components before a model can be properly utilized and have its predictions consumed. These components are listed next:

- **Prediction consistency validation test**: This is a test that ensures the predictions made by the same trained model are consistent on the same data. A model's predictions can't be utilized if its logic is not deterministic. This will be discussed further in *Chapter 10, Exploring Model Evaluation Methods.*

- **Reproducibility of model training experiments**: Reproducibility is a responsibility that has to be upheld by the ML engineer. This means that we have to ensure that a model can be deterministically reproduced using the configuration that was used to train it the second time around. As the ML life cycle is an iterative cyclical process, when there is a need to retrain the model with new data, the same performance can be approximately achieved again. The trick to ensuring reproducibility is to ensure all components that require random data generation are seeded deterministically. A seed ensures that the random number generator generates random numbers deterministically. In `pytorch`, this can be done globally through the following code:

```
import torch
import random
import numpy as np

torch.manual_seed(0)
random.seed(seed)
np.random.seed(seed)
```

In `tensorflow` and `keras`, this can be done globally through the following code:

```
import tensorflow as tf
tf.keras.utils.set_random_seed(seed)
```

This method automatically seeds both the `random` and `numpy` libraries. These global settings can help to set the random seed for layers that did not explicitly set random number generator seeds locally.

One last piece of advice in experimentation is to make sure a baseline is created at the start of the project. A baseline is the simplest version of a solution possible. The solution can even be a non-DL model with simple features. Having a baseline can help ensure that any improvements or complications you add are justified by metric performance monitoring. Refrain from adding complications for the sake of them. Remember that the value of an ML project is not how complicated the process is but the results that can be extracted from it.

Next, we will dive into actual techniques that can be used to realize and improve a solution that utilizes DL.

Exploring general techniques to realize and improve supervised deep learning based solutions

Notice that earlier in the chapter we focused on use cases based on problem types and not the problems themselves. Solutions in turn solve and take care of the problem. DL and ML in general are great solvers of issues related to staffing difficulties and for the automation of mundane tasks. Furthermore, ML models in computers can process data much quicker than an average human can, allowing a much quicker response time and much more efficient scaling of any process. In many cases, ML models can help to increase the accuracy and efficiency of processes. Sometimes, they improve current processes, and other times, they make previously unachievable processes possible. However, a single DL model may or may not be enough to solve the problem. Let's take an example of a solution that *can* be solved sufficiently with a single DL model.

Consider the use case of using a DL model to predict the genders of babies with ultrasound imagery. Traditionally, a doctor would perform a visual-based gender analysis of the resulting ultrasound imagery of a baby in the mother's womb in real time and offline before finally providing their prediction of the gender. Based on the amount of prior experience and knowledge, the doctor would have different levels of competency and accuracy in decoding the gender. Things might get more complicated when there are abnormalities in the baby. The probable underlying problem would be that experienced and capable doctors are scarce and expensive to hire. If we had a system that could decode the gender from the ultrasound imagery automatically, it would either be of good assistance to the judgment of real doctors or a replacement as a cheaper alternative. The same analogies can be applied to identifying diseases or symptoms in any advanced imaging results such as X-ray images.

This example depicts a DL model as a component of a solution and a solution where a single DL model is enough to obtain the desired output. This is an example of staffing issues but not so much on the efficiency side. Note that for some use cases, it is required to explain in some form the reasons that drove the predictions that were made. In other words, you'd have to explain the decisions that were made by the model. To provide assistance to a doctor, pinpointing where and which types of patterns contributed to the decision would be more helpful than the decision itself, as doctors would be able to utilize the extra information to make their own decisions. This will be thoroughly introduced in *Chapter 11, Explaining Neural Network Predictions*.

Explanations aside, not all solutions to problems can be accomplished by a single ML or DL model alone. At times, DL models have to be coupled with general ML methods, and at others, multiple DL models have to be coupled together. In some special cases, intermediate data needs to be specially processed and prepared before feeding it to the next task in a pipeline. Creating and architecting logical pipelines are essential when dealing with such problems. Let's take an example of a problem and solution that requires multiple datasets, multiple DL models, and constructing a task pipeline.

Consider the problem of finding criminals who have just robbed a bank. By using CCTV cameras deployed in the city, you can use a face detection and recognition solution if you have identified the face of the criminals that did the robbery. The following figure shows an example solution task pipeline for this problem:

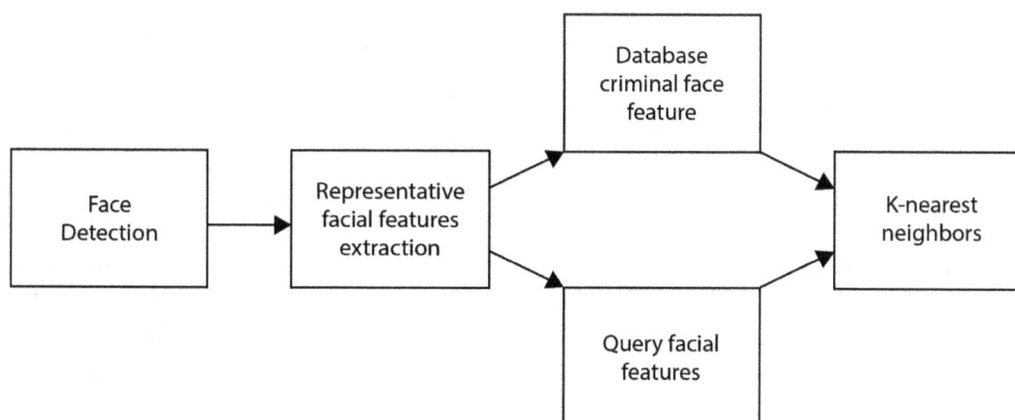

Figure 8.6 – Task pipeline of the solution for finding criminals through CCTV cameras

Face detection is an image-object-detection process where there is an image bounding box regressor and a binary classifier that predicts whether the bounding box is a face or not. The representative facial features extraction utilizes a DL model that can be trained using supervised representation learning methods that are trained against the goal of optimizing the discriminative effects of facial features against the facial features of different persons. Next, a separate task is needed to build the database of criminal facial features that will be passed into the KNN ML algorithm to find the matched facial ID based on queried facial features obtained from CCTV cameras deployed in the city. This solution shows the need to break a solution into multiple components in order to obtain the final result of finding the criminals.

The preceding example is part of a larger paradigm called multitask learning and multitask problems. The multitask paradigm is a set of topics that allows for greater advancement in the ML space, not only for DL but definitely much more achievable through DL, due to its inherent flexibility. In the next topic, we will dive into the multitask paradigm.

Breaking down the multitask paradigm in supervised deep learning

Multitask is a paradigm that covers a wide spectrum of tasks that involves the execution of ML models on multiple problems coupled with their respective datasets to achieve a goal. This paradigm is usually built based on two reasons:

- To achieve better predictive performance and generalization.

- To break down complicated goals into smaller tasks that are directly solvable using separate ML models. This reiterates the point made in the previous topic.

Let's dive into four multitask techniques, starting with multitask pipelines.

Multitask pipelines

This variation of multitask systems revolves around realizing solutions that can't be directly solved by using a single ML model. Breaking down highly complicated tasks into smaller tasks can allow solutions to be made with multiple ML models handling different smaller tasks. These tasks can be sequential or parallel in their paths and generally form a **directed acyclic graph** (**DAG**)-like pipeline, similar to the example shown in *Figure 8.6*.

However, this does not mean that the tasks should exclusively be ML models. Problems for different industries and businesses can be in many forms, and being flexible in assigning components needed to produce a solution is key to deriving value from ML technology. For example, if human supervision is needed to accomplish a certain task after breaking down the larger task, do not hesitate to utilize it along with ML models to achieve value. Let's go through another use case that utilizes multitask pipelines to create a solution, which is **recommendation systems**. First, we need to perform either supervised or unsupervised representation learning for feature extraction. Second, using the features extracted, create a database used to match extracted query features. Third, obtain the top-k closest data from the database and apply a regression model to predict the rank of the top-k data for fine-tuned e-tuned high-performance ranking.

Next, we will discover another paradigm of multitasking, called **TL**.

TL

TL is a technique that involves using what was learned from one task in another task. The core reasons can be one of the following:

- Increasing the metric performance.

- Decreasing the required number of epochs needed for the network to reach a state of convergence.

- Allowing more stable learning trajectories. In other cases, networks just take longer to converge when the initial learning process is unstable. However, in some other cases, networks cannot converge at all when networks don't have a stable foundation to start learning. TL can help models that originally fail to learn anything reach convergence in the learning process.

- Increasing generalization and reducing the probability of overfitting. When the second task involves only a small subset of variations from the actual data population, knowledge learned from a first task that covers a wider range of variations helps to prevent narrow oversights.

Concretely, TL in DL is achieved by using the network parameters that were learned from the first task in the second task. The parameters involve all the weights and biases that are associated with a network. The parameters can be used as an initialization step for the same network instead of the usual randomly initialized parameters. These are known as **pre-trained weights**. The process of network learning with pre-trained weights is called **fine-tuning**. Additionally, the network parameters can also opt to be completely frozen and plainly act as a **featurizer** component that provides features for another SL algorithm.

There are a couple of automated strategies that focus on improving the results you can get with fine-tuning. However, these methods are not silver bullets and can take a lot of time to carry out. The practical strategy to achieve a better performance using TL is to choose the number of layers you want to train by gauging the transferability component of the two tasks. *Table 8.2* shows an easy way to decide on a TL strategy based on task similarity and dataset size of the second task:

		Dataset Size	
		Small	**Big**
Task similarity	Low	Train the entire network as usual.	Train the entire network as usual.
	High	Freeze all base network parameters, add an extra linear prediction layer, and only train this linear layer on the dataset.	Train the entire network as usual.

Table 8.2 – Deep TL (DTL) strategy guide

For clarity purposes, let's say "big" is when the dataset has at least 10,000 examples. For task transferability/similarity, human intuition is required to obtain an evaluation on a case-by-case basis. Here in the guide, we assume that a big dataset size means a dataset with large variations that represent the population adequately. A hidden component not presented in the preceding figure, however, is the size of the dataset of the first task. TL has the best impact when the task similarity is high, the second task dataset size is small, and, additionally, when the dataset size of the first task is big. The size of the first dataset usually also limits the range of NN sizes that can be used. Let's say that the first dataset size is small; the best-performing models in this case are usually smaller-sized models. When TL is highly beneficial, even when the dataset size of the second dataset size is medium or big, the smaller models can still outperform bigger-sized models. An act of balancing is required in complex cases such as this to obtain the ideal model.

One prominent issue in TL is called **catastrophic forgetting**. This is a phenomenon where the network performance regresses to earlier tasks as the network trains on new tasks. If the performance of the previous task is not of concern to you, this issue can be ignored. Practically, if it is required to maintain the performance of the previous task, you can follow these steps:

1. Use a unified metric that takes care of the performance of the first task and second task by additionally validating on the validation dataset of the first task.

2. Combine the dataset from the first task and second task and train it as a single model. If the targets are not relevant to each other, use different fully connected layer prediction heads.

Lastly, there is an additional popular technique for TL known as **knowledge distillation**. This method involves two models, where one pre-trained teacher model is used to distill its knowledge to a student model. Typically, the teacher model is a bigger model that has the capacity to learn more accurate information but is slower in runtime, and the student model is a smaller model that can be run at reasonable speeds during runtime. The method distills knowledge by using an additional similarity-based loss of a chosen layer output between the teacher and student model, which is typically the logit layer, on top of the base cross-entropy loss. This method encourages the student model to produce similar features to the teacher model. The technique is typically used to obtain a smaller model with better accuracy than if trained without knowledge distillation, so the deployment infrastructure can be cheaper. This technique will be practically introduced in *Chapter 13*, *Exploring Bias and Fairness*, as a key technique to also mitigate bias.

Next, we will dive into another type of multitask execution, called multiple objective learning.

Multiple objective learning

Multiple objective learning is a type of multitasking process that involves training with simultaneously different goals. Different goals direct the learning trajectory of a network toward different paths. Multiple objective learning can be further broken down into the following options:

- Multiple losses on the same outputs.

- Multiple targets, which are taken care of by separate NN prediction heads, each with their respective losses. This can be further broken down into the following categories:

- Multiple targets with real impact and usage.

- A single or multiple main targets and a single or multiple auxiliary targets. Auxiliary targets are paired with their own losses called auxiliary losses.

Aside from option *2(A)*, the other options (that is, *1* and *2(B)*) for multiple objective learning are mainly used to improve metric performance. Metrics can be as simple as accuracy or log loss, or more intricate, such as the degree of bias toward a minority class. A simple more straightforward example of multiple objective learning is the multilabel target type. Multilabel is where multiple labels can be associated with a single data row. This means that the setup will be a multiple binary classification target, which is the case with *2(A)*.

Multiple targets and their associated losses mean there might be issues of conflicting gradients during the learning process. This phenomenon is more commonly known as **negative transfer**. A more extreme case of negative transfer is when gradients from the two losses cancel each other out when they have the same magnitude in exactly opposite directions. This will block the learning process of the model where the model will never converge. In reality, this issue can be at a lower scale and dampen the speed of convergence or, worse, introduce huge fluctuations that make it hard to learn anything. Unfortunately, there are no silver-bullet mitigation methods here other than to understand the background behind why a model learns poorly. Iterative experiments are usually required to figure out how to balance these losses properly to encourage a stable learning process.

Next, we will dive into multimodal NN training.

Multimodal NN training

Multimodal NNs are a type of multitask system in the sense that networks responsible for different modalities learn in the same task in completely different paths. A common method of handling multimodality in NNs is to assign different neural blocks at the initial stage for different data modalities. Neural blocks contain the networks specific to each modality. The neural blocks for different modalities will then be merged using a series of intermediate fully connected layers and an output fully connected layer. This is depicted in *Figure 8.7*:

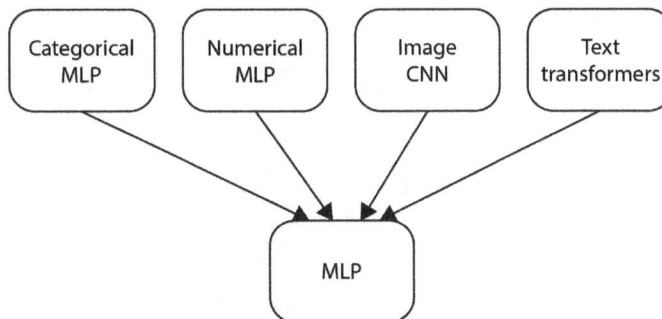

Figure 8.7 – Typical multimodal NN structure

The idea of leveraging multimodality is that the additional data input can allow for more comprehensive patterns to be identified and thus should improve the overall metric performance. In reality, this commonly will not be the case without careful handling of the training process. Different modalities exist in entirely different distributions and learn at different rates with different paths. A single global optimization strategy applied to all the data modalities will likely produce suboptimal results. A common and effective strategy is to do the following:

- Pretrain the individual modality neural block (unimodal) with a temporary prediction output layer until a certain degree of convergence

- Remove the temporary prediction output layer and train the multimodal NN as usual with the pre-trained weights from the unimodal training process

Other than that, freezing the weights of the unimodal trained NN and only training the multimodal aggregation fully connected layer prediction head is also a sound strategy. Many more complex strategies exist to tackle this issue but are out of scope in this book.

Summary

In this chapter, we explored supervised deep learning, including the types of problems it can be used to solve and the techniques for implementing and training DL models. Supervised deep learning involves training a model on labeled data to make predictions on new data. We also covered a variety of supervised learning use cases on different problem types, including binary classification, multiclassification, regression, and multitask and representation learning. The chapter also covered techniques for training DL models effectively, including regularization and hyperparameter tuning, and provided practical implementations in the Python programming language using popular DL frameworks.

Supervised deep learning can be used for a wide range of real-world applications in tasks such as image classification, **natural language processing** (**NLP**), and speech recognition. With the knowledge provided in this chapter, you should be able to identify supervised learning applications and train DL models effectively.

In the next chapter, we will explore **unsupervised learning** for DL.

9

Exploring Unsupervised Deep Learning

Unsupervised learning works with data that does not have any labels. More broadly, unsupervised learning aims to uncover the intrinsic patterns hidden within the data. The most rigorous and expensive part of a supervised machine learning project is the labels required for a given data. In the real world, there is tons of unlabeled data available with tons of information that could be learned from. Frankly, it's impossible to obtain labels for all of the data that exist in the world. Unsupervised learning is the key to unlocking the potential of the abundant unlabeled digital data we have today. Let's explore a hypothetical situation below to understand this better.

Imagine that it costs 1 USD and 1 minute to obtain a label for a row of data for whatever use case it could be, and a single unit of information can be obtained through supervised learning. To get 10,000 units of information, 10,000 USD would need to be spent, and 10,000 minutes need to be contributed to obtain 10,000 pieces of labeled data. Both time and money are painful things to burn. However, for unsupervised learning, it costs 0 USD and 0 minutes to obtain 0.01 units of information through the same data without a label. Since the amount of data is not impeded by time or money, we can easily get 100 times more data than the 10,000 samples and get the same information that can be learned through a model. When money and time aren't an issue, the amount of information your model can learn is endless, assuming that your unsupervised learning model has the capacity and ability to do so.

Deep learning provides a competitive edge to the unsupervised learning field, given the huge capacity and ability to learn complex information. If we can crack the code of unsupervised deep learning, it will set the stage for models that can come close to general intelligence one day! In this chapter, we will explore the notable components of unsupervised deep learning by taking a look at the following topics:

- Exploring unsupervised deep learning applications
- Creating pretrained network weights for downstream tasks
- Creating general representations through unsupervised deep learning
- Exploring zero-shot learning

- Exploring the dimensionality reduction component of unsupervised deep learning
- Detecting anomalies in external data

Technical requirements

This chapter includes some practical implementations in the **Python** programming language. To complete it, you will need to have a computer with the following libraries installed:

- `pandas`
- `CLIP`
- `numpy`
- `pytorch`
- `pillow`

You can find the code files for this chapter on GitHub at `https://github.com/PacktPublishing/The-Deep-Learning-Architect-Handbook/tree/main/CHAPTER_9`.

Exploring unsupervised deep learning applications

Today, practitioners have been able to leverage unsupervised deep learning to tap into their unlabeled data to achieve either one of the following use cases. These have been put in descending order in terms of their impact and usefulness:

- Creating pretrained network weights for downstream tasks
- Creating general representations that can be used as-is in downstream supervised tasks by predictive supervised models
- Achieving one-shot and zero-shot learning
- Performing dimensionality reduction
- Detect anomalies in external data
- Clustering the provided training data into groups

To start, note that pure clustering is still a core application of unsupervised learning in general, but not for deep learning. **Clustering** is where unlabeled data is grouped into multiple arbitrary clusters or classes. This will be useful in use cases such as customer segmentation for targeted responses, or topic modeling to figure out trendy topics people are discussing on social media. In clustering, the relationship between the unlabeled data samples is leveraged to find groups of data that are close together. Some clustering techniques group this data by assuming a spherical distribution in each cluster, such as **K-means**. Some other clustering techniques are more adaptive and can find clusters of multiple distributions and sizes, such as **HDBSCAN**.

Deep learning methods have not been able to produce any significant improvement from non-deep learning clustering methods such as the simple k-means algorithm or the HDBSCAN algorithm. However, efforts have been made to utilize clustering itself to aid the unsupervised pretraining of neural network models. To realize it in a neural network model as a component, the clustering model has to be differentiable so that gradients can still be propagated to the entire network. These methods are not superior to non-deep learning techniques such as k-means or HDBSCAN and are simply a variation so that the concept of clustering can be realized in a neural network. An example application of clustering in unsupervised pretraining is **SwaV**, which will be introduced in the next section. However, for completeness, one example of a neural network-based clustering algorithm that is used traditionally is self-organizing maps, but the network itself is not considered a deep network.

In the next few sections, we will discover the other five applications more comprehensively, ordered by their impact and usefulness, as shown previously.

Creating pretrained network weights for downstream tasks

Also known as unsupervised transfer learning, this method is analogous to supervised transfer learning and naturally reaps the same benefits as described in the *Transfer learning section in Chapter 8, Exploring Supervised Deep Learning*. But as a recap, let's go through an analogy. Imagine you're a chef who has spent years learning how to cook a variety of dishes, from pasta and steak to desserts. One day, you're asked to cook a new dish you've never tried before; let's call it "Dish X." Instead of starting from scratch, you use your prior knowledge and experience to simplify the process. You know how to chop vegetables, how to use the oven, and how to adjust the heat, so you don't have to relearn all of these steps. You can focus your energy on learning the specific ingredients and techniques required for Dish X This is similar to how transfer learning works in machine learning, which applies to both unsupervised learning and supervised learning. A model that has already been trained on a related task can be used as a starting point, allowing the model to learn new tasks more quickly and effectively.

Besides being able to use the unlimited amount of unlabeled data available in the world, unsupervised transfer learning also bears another benefit. Labels in supervised learning often hold biases that the model will adopt. The biases acquired through learning can obstruct the acquisition of more generalized knowledge that would be more useful for downstream tasks, to varying extents. Other than biases in the labels, there are also situations where labels are wrong. Being unsupervised means that the model is stripped of any possibility of learning biases or errors from any labels. However, note that biases are more prominent in some datasets. A dataset with a complex task that has quality-related labels derived qualitatively from human judgment tends to have more biases compared to a simple task such as classifying whether a picture has a face or not.

Now, let's dive into the techniques. Unsupervised transfer learning has been rated as the most impactful and useful application due to contributions that were made in the NLP field. Transformers, introduced in *Chapter 6, Understanding Neural Network Transformers*, paved the way for the paradigm to pre-train your model with an unsupervised learning technique more commonly known as **self-supervised**

learning. How this method is categorized here is a matter of perspective and not everybody would agree with it. Self-supervised learning leverages only the relationship of co-occurring data to pre-train a neural network with relational knowledge without labels. Seeing it this way, check out the following question, and decide your own answer.

Try it yourself

Traditional unsupervised learning data preprocessing techniques such as **Principal Component Analysis (PCA)** leverage the relationship of co-occurring data to build new features that can better represent impactful patterns. Do you see PCA as its own category with self-supervising, under the umbrella of supervised learning, or under the umbrella of unsupervised learning?

The author's answer: unsupervised!

The unsupervised learning techniques that are used in transformers are masked language modeling and next-sentence prediction. These tasks help the model learn the relationships between words and sentences, allowing it to better understand the meaning and context of language data. A model that has been trained on masked language modeling and next-sentence prediction tasks can use its understanding of language to perform better on a variety of NLP downstream tasks. These are proven by the SoTA predictive performances on various datasets from leading transformers today, such as DeBERTa, as introduced in *Chapter 6, Understanding Neural Network Transformers*.

Let's briefly go through other examples of unsupervised pre-training:

- **A simple framework for contrastive learning of visual representations (SimCLR)**: SimCLR utilizes a method called **contrastive learning** to pretrain convolutional neural networks. Contrastive learning is a key technique in unsupervised deep learning that helps neural networks learn representations of data by optimizing the distance between related features. The core idea behind contrastive learning is to bring features of similar data points closer together and push features of dissimilar data points further apart in the feature space, using a contrastive loss function. While there are various forms of contrastive loss today, the general idea is to minimize the distance between similar examples and maximize the distance between dissimilar examples. This distance can be measured in various ways, such as Euclidean distance or cosine distance. Although this method requires labels for learning and is technically a supervised learning loss, the features learned, along with the selection of similar and dissimilar samples for label-free samples, make this loss function a crucial technique in unsupervised deep learning. The simplest representation of such a contrastive loss is as follows:

  - For dissimilar samples:

$$loss = -distance$$

  - For similar samples:

$$loss = distance$$

SimCLR focuses on image data and uses crafted image augmentation techniques to generate image pairs that could optimize the network to produce closer features. Random cropping and random color distortions are the most general augmentations that can be useful setups for most image datasets to perform unsupervised pre-training with SimCLR.

- **Swapping assignments between multiple views of the same image (SwaV)**: SwaV adopts a similar concept to SimCLR in utilizing image augmentations with convolutional neural networks. It also uses the concept of clustering and embeddings to optimize the model to produce features that make sure the two images are mapped to the same feature space.

The learning technique is executed as follows:

I.   A pre-set amount of cluster number is determined, called K.

II.   Featurize the two images with the same convolutional neural network.

III. The two sets of features will then be assigned independently to specific clusters by using an external technique called Optimal Transport Solver using an embedding layer that represents the representative features of the K clusters. Two features will always be assigned to different clusters.

IV.   Dot products between the convolutional features and all the cluster embeddings are computed and a softmax operation is applied.

V.   The assigned clusters for both image features will then be swapped, where cross-entropy between the swapped cluster assignments and the resulting values from the softmax operation will be used to optimize the weights of both the CNN and the embeddings layer.

The idea is to jointly learn the embedding weights and convolutional network weights that consistently categorize together multiple augmentations of the same image. The technique can be described as contrastive clustering. Both SwaV and SimCLR are competitively close in multiple downstream task performances.

- **SEER**: SEER is the combination of SwaV, an extremely high amount of unlabeled data for images at the billion scales instead of the more common million scale, and using high-capacity models to pre-train using random, uncurated, and unlabeled images. This allowed SEER to achieve SoTA downstream supervised task performance and outperformed both SimCLR and SwaV alone.

- **UP-DETR, by the researchers from SCTU and Tencent Wechat AI**: This method pre-trains the transformer that has an encoder-decoder architecture with CNN features for image object detection tasks in an unsupervised way. UP-DETR managed to improve the performance of transformers on downstream supervised image object detection datasets. The interesting thing to remember here is that it structured the network in a way that allowed random image patches to be fed separately to the decoder to predict the bounding box of these patches on the original image. The original image is fed into the encoder part of the transformer and combined with the random image patches at the decoder part.

- **Wav2vec 2.0**: Wav2vec 2.0 showed how feasible it is to train a reliable speech recognition model with limited amounts of labeled data by leveraging self-supervised pretraining as a pretext task. It also uses contrastive, loss that is simply the cosine similarity between samples. The method uses CNNs as an audio feature extractor and quantizes the representations into a discrete array of values that can be trained before passing it into a transformer. The unsupervised tasks of masked speech modeling and contrastive loss are applied here. Let's look at how these two methods can be combined:

    i.   A random location of the quantized latent speech representations is masked.

    ii. The transformers output at the same location of the masked quantized latent speech representation will be used as the prediction of the missing masked quantized latent speech representation.

    iii. A few parts of the non-masked quantized latent representations of the same sample will be used to compute the contrastive loss against the predicted missing masked quantized latent speech representation, which effectively enforces parts of the same audio sample to be in a similar latent domain.

This process is demonstrated in *Figure 9.1*:

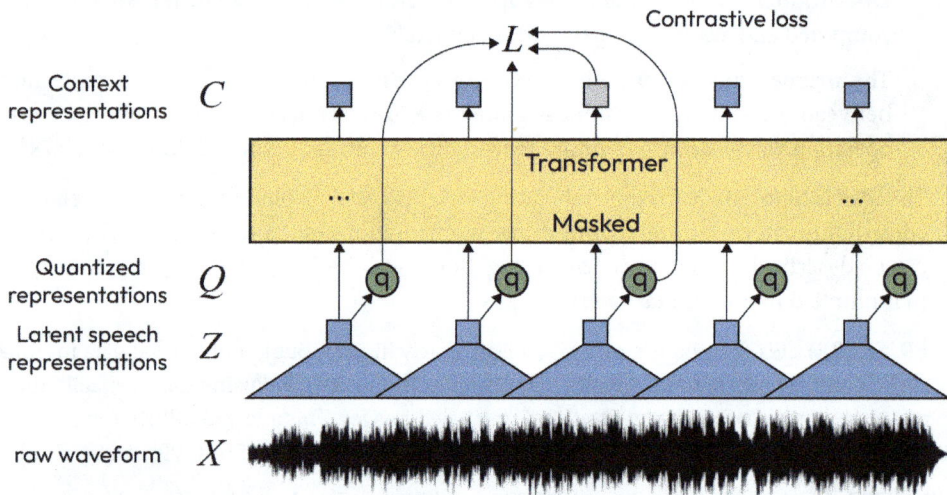

Figure 9.1 – Wav2vec 2.0 model structure

The methods we've introduced here were meant for a specific modality. However, with some effort methods, they can be adapted to other modalities to replicate performance. For example, augmentations were used for image-based methods along with contrastive learning. To adapt this method to text-based modality, using text augmentations that preserve the meaning, such as word replacement or back translation, should work well. The modalities involved in the unsupervised methods introduced here were images, text data, and audio data. These modalities were chosen due to their generalizability factor for downstream tasks. Other forms of modality, such as graph data or numerical data, are highly customizable to individual use cases where there is fundamentally no information that can be transferred to downstream tasks. Before you attempt to run an unsupervised deep learning method to create pre-trained weights, consider listing the information that can be transferred and evaluate qualitatively whether it makes sense to proceed.

But what if the pre-trained network weights are already capable of producing very generalizable features across different domains? Just like how CNNs trained on the ImageNet dataset in a supervised way can be used as feature extractors, there is no limiting the immediate usage of networks trained by unsupervised methods such as SwaV or Wav2vec 2.0 as feature extractors. Feel free to try it out yourself! However, a few unsupervised learning techniques use neural networks that are made to use their generated features instead of their weights directly. In the next section, we will discover exactly that.

Creating general representations through unsupervised deep learning

The representations that are learned through unsupervised deep learning can be directly used as-is in downstream supervised tasks by predictive supervised models or consumed directly by end users. There are a handful of generally impactful unsupervised methods that utilize neural networks that are meant to be used primarily as feature extractors. Let's take a look at a couple of unsupervised feature extractors:

- **Unsupervised pre-trained word tokenizers**: These are used heavily by variants of the transformers architecture and were introduced in *Chapter 8*, *Exploring Supervised Deep Learning*, in the *Representing text data for supervised deep learning* section.

- **Unsupervised pre-trained word embeddings**: These methods leverage unsupervised learning and attempt to perform language modeling, similar to masked language modeling in transformers. However, word embeddings-based methods have been overtaken by transformer-based pretraining with sub-word-based text tokenization in terms of metric performance. The word embeddings method still stays relevant today due to the runtime efficiency it has over transformer-based methods. Note that not every project has the GPU available that is needed to run a big transformer in a reasonable runtime. Some projects only have access to CPU processing, and word embeddings provide the perfect inference runtime versus metric performance tradeoff. Additionally, word embeddings are a natural solution for some use cases that require words as a result, such as word-to-word translation from one language into another language, or even finding synonyms or antonyms. Examples of methods that produce pre-trained word embeddings are **fastText** and **word2vec**. *Figure 9.2* exemplifies the architecture of word embedding methods:

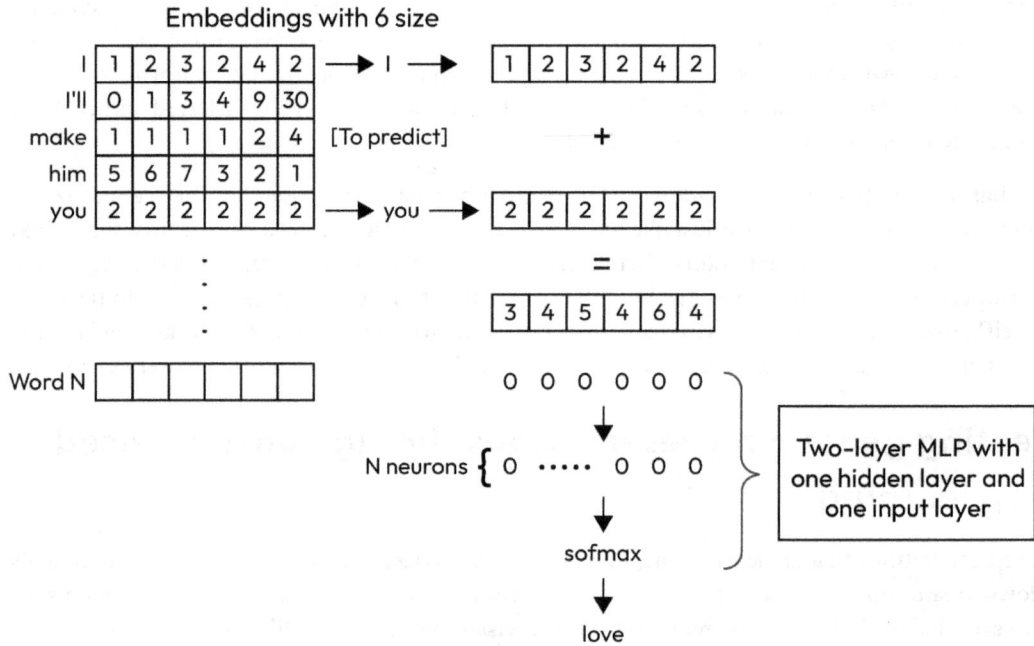

Embeddings with 6 size

I	1	2	3	2	4	2
I'll	0	1	3	4	9	30
make	1	1	1	1	2	4
him	5	6	7	3	2	1
you	2	2	2	2	2	2

Figure 9.2 – The word embeddings architecture with a two-layer MLP and trainable embeddings

The task is either to predict the middle word based on the summed embeddings of the surrounding words or to predict the surrounding words based on the embeddings of the current word. *Figure 9.2* shows the former case. After training the word embeddings with the MLP with a dictionary of N words, the MLP is then dumped, and the word embeddings are saved as a dictionary for simple look-up utilization during inference. FastText differs from word2vec by using not words but subwords to generate embeddings and thus can better handle missing words. A word embedding for FastText is produced by summing up embeddings of subwords that form the full word. Go to `https://github.com/facebookresearch/fastText` to learn how to use word embeddings that have pre-trained for 157 languages, or how to pre-train FastText embeddings on your custom dataset!

- **Autoencoders**: Autoencoders are encoder-decoder architectures that can be trained to denoise data and reduce the dimensionality of the data while optimizing it to be reconstructible. They are generally trained to extract useful and core information by limiting the number of features in the bottleneck section of the architecture right after the encoder and right before the decoder. Go back to *Chapter 5, Understanding Autoencoders*, to find out more!

- **Contrastive Language-Image Pretraining (CLIP):** CLIP is a method that trains a text language transformer encoder and a CNN image encoder with contrastive learning. It provides a dataset consisting of around 400 million image text pairs constructed by combining multiple publicly available datasets. This method produces a powerful image and text feature encoder that can be used independently. This method became a main part of the current SoTA of text-to-image methods by assisting during training to optimize to generate an image that has CLIP-encoded image embeddings close to the CLIP-encoded text embeddings. *Figure 9.3* shows the CLIP architecture:

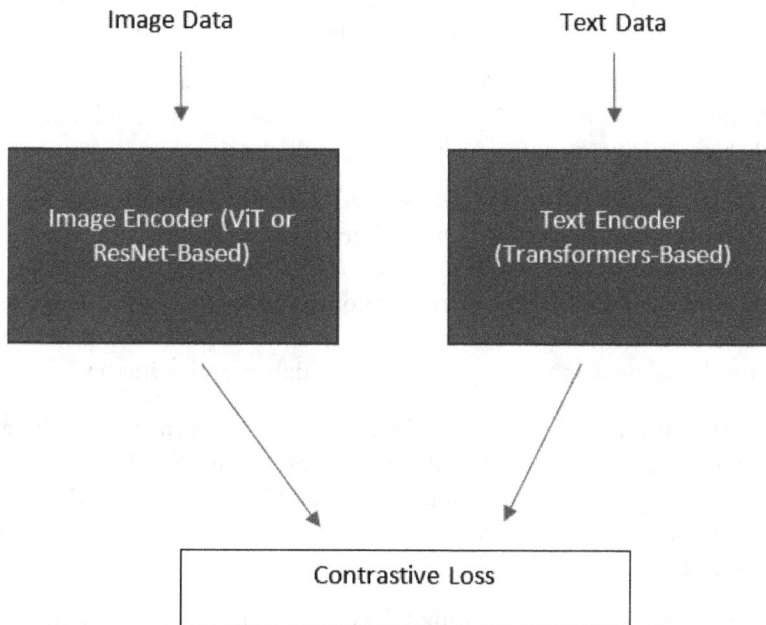

Figure 9.3 – CLIP architecture

A small implementation detail here is that the outputs of the image encoder and text encoder are fed separately to a linear layer so that the number of features matches up for contrastive learning loss to be computed. Specifically, for CLIP, the contrastive loss is applied in the following way:

```
cosine_similarity = np.dot(Image_embeddings, Text_embeddings.T)
* np.exp(learned_variable)
labels = np.arange(number_of_samples)
loss_image = cross_entropy_loss(cosine_similarity, labels,
axis=0)
loss_text = cross_entropy_loss(cosine_similarity, labels,
axis=1)
loss = (loss_image + loss_text)/2
```

Notice that cross-entropy is still applied after applying pairwise cosine similarity between the image and text embeddings, which exemplifies the many variations of contrastive loss, where the core workhorse still comes down to using the distance metric. Since CLIP leverages co-occurring data, it does not belong to the self-supervised learning sub-category.

- **Generative models**: Let's look at some examples of these models:

  - **Transformer models**. Examples include GPT-3 and ChatGPT. Both are transformer models that are trained with masked language modeling and next-sentence prediction tasks. The difference with ChatGPT is that it is fine-tuned using reinforcement learning through human feedback and training. Both generate new text data by predicting in an autoregressive manner.

  - **Text to image generators**. Examples include DALL.E 2 and Stable Diffusion. Both methods utilize the diffusion model. At a high level, the Stable Diffusion method slowly generates an extremely high-quality image from a base image full of noise.

  - **Generative adversarial networks** (**GANs**): GANs utilize two neural network components during training, called **discriminators** and **generators**. The discriminator is a classifier that is meant to discern fake from real images. The generator and discriminator are trained iteratively until a point where the discriminator can't discern that the image generated by the generator is fake. GANs can generate good-quality images but are generally surpassed by diffusion-based models as they produce much higher-quality images.

It's important to note that the methods to learn and create feature representations with unsupervised learning are not limited to those based on neural networks. PCA and TF-IDF, which are considered data pre-processing techniques, also belong in this category. However, the key difference between them and deep learning-based methods is that the latter require more training time but offer better generalization capabilities.

The key in unsupervised representation learning is leveraging relationships between co-occurring data. The techniques that were introduced in the last two sections have public repositories that can be utilized immediately. For most of them, there are already pre-trained weights that you can leverage out of the box to either fine-tune further on downstream supervised tasks or use as plain feature extractors. In the next section, we will explore a special type of utilization of CLIP called zero-shot learning.

Exploring zero-shot learning

Zero-shot learning is a paradigm that involves utilizing a trained machine learning model to tackle new tasks without training and learning directly on the new task. The method implements transfer learning at its core but instead of requiring additional learning in the downstream task, no learning is done. The method that we will be using to realize zero-shot learning here is CLIP as a base and thus is an extension of an unsupervised learning method.

CLIP can be used to perform zero-shot learning on a wide variety of downstream tasks. To recap, CLIP is pre-trained with the task of image-text retrieval. So long as CLIP is applied to downstream tasks without any additional learning process, it can be considered as zero-shot learning. The tested use cases include tasks such as object character recognition, action recognition in videos, geo-localization based on images, and many types of fine-grained image object classification. Additionally, there are basic ways people have been testing and giving demos on zero-shot learning for object detection.

In this chapter, we will implement a non-documented zero-shot application of CLIP, which is image object counting. Counting means the model will be performing regression. Let's start the implementation:

1. First, let's import all the necessary libraries. We will be using the open source `clip` library from `https://github.com/openai/CLIP` to utilize a pretrained version of CLIP using a visual transformer model:

    ```
    import os
    from tqdm import tqdm
    import numpy as np
    import clip
    import torch
    from PIL import Image
    import pandas as pd
    ```

2. Next, we will load the pretrained CLIP model in either CPU mode or GPU mode if the CUDA toolkit is installed:

    ```
    device = "cuda" if torch.cuda.is_available() else "cpu"
    model, preprocess = clip.load('ViT-B/32', device)
    ```

    The model variable we loaded here is a class that contains the individual methods to encode images and text with the encoder model loaded with pretrained weights. Additionally, the preprocess variable is a method that needs to be executed on the input before it's fed to the image encoder. It performs normalization on the data that was used during training.

3. To predict with this model more conveniently, we will create a helper method with the following structure.

    ```
    def predict_with_pil_image(image_input, clip_labels, top_k=5):
        return similarity_score_probability, indices
    ```

    Remember that CLIP is trained to generate the image encoder and text encoder outputs in a way that they're mapped into the same feature space. Matching text descriptions to an image will produce text and image features that are close together in distance. Since the metric we used is cosine similarity, the higher the similarity value, the lower the distance. The main technique to realize zero-shot learning from CLIP is to think of multiple descriptions that represent a certain label and choose the label that has the highest similarity score against an image. The similarity score will then be normalized against all the other labels to obtain a probability score.

Additionally, `top_k` is used to control how many top highest similarity score text description indices and scores to return. We will come back to how to design the descriptions objectively for zero-shot learning later, once we've defined the code that will belong in the predict method we defined previously.

4. The first part of this method will be to preprocess the provided single image input and multiple text descriptions, called `clip_labels`. The image will be preprocessed according to the provided preprocessor, while the multiple text descriptions will be tokenized according to the sub-word tokenizer used in the text transformer provided by the `clip` library:

```
image_input = preprocess(cars_image).unsqueeze(0).to(
device)
text_inputs = torch.cat([clip.tokenize(cl) for cl in clip_
labels]).to(device)
```

5. Next, we will encode the preprocessed image and text inputs using the image encoder and text encoder, respectively:

```
with torch.no_grad():
    image_features = model.encode_image(image_input)
    text_features = model.encode_text(text_inputs)
```

In PyTorch, remember that we need to make sure no gradients are computed during inference mode as we are not training the model and don't want that extra computation or RAM wastage.

6. Now that the image and text features with the same column dimensions have been extracted, we will compute the cosine similarity score of the provided image input features against all the text description label features:

```
image_features /= image_features.norm(dim=-1, keepdim=True)
text_features /= text_features.norm(dim=-1, keepdim=True)
similarity_score_probability = (100.0 * image_features @ text_
features.T).softmax(dim=-1)
```

In addition to the similarity score, the similarity among all the text descriptions is normalized using Softmax so that the scores will all add up to one. View this as a way to compare the similarity scores against other samples. This will essentially convert distance scores into a multiclass prediction setting where the provided text descriptions as labels are all the possible classes.

7. Next, we will extract the `top_k` highest similarity score and return their similarity score probability and indices to indicate which label the score belongs to:

```
percentages, indices = similarity[0].topk(5)
similarity_score_probability = percentages.numpy() * 100.0
indices = indices.numpy()
```

8.   Now that the method is complete, we can load an image using Pillow, create some descriptions, and feed it into the method to perform zero-shot learning! In this tutorial, we will work on object counting. This can range from counting how many cars to properly account for parking availability to counting how many people to account for personnel that are required to service the people. In this tutorial, we will count the number of paper clips as a pet project. Note that this can easily be extended to other counting datasets and projects. We will use the dataset at `https://www.kaggle.com/datasets/jeffheaton/count-the-paperclips?resource=download` to achieve this. Be sure to download the dataset in the same folder where the code exists. Let's load up a simple version with fewer paper clips than 2:

```
image = Image.open('clips-data-2020/clips/clips-25001.png')
```

We will get the following output:

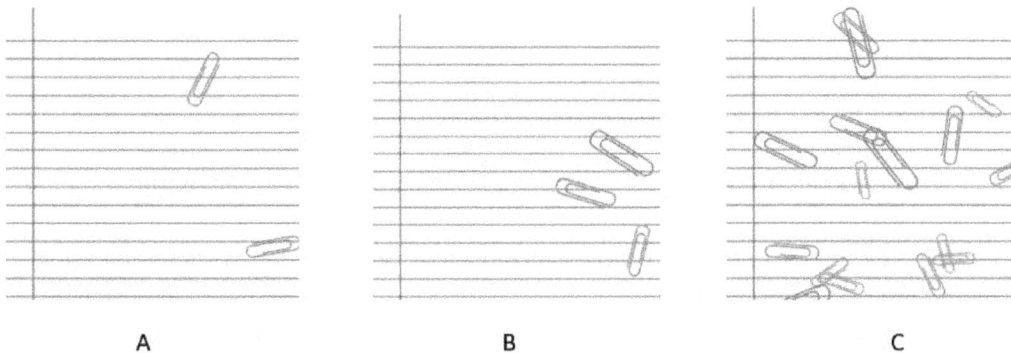

Figure 9.4 – Example paper clip counting from easy (A), medium (B), to hard (C)

Figure 9.4 B and *Figure 9.4 C* are harder examples that we will explore after we go through predictions for *Figure 9.4 A*.

9.   Now, we need multiple sets of text descriptions to account for all the possible counts of paper clips. A simple description that just uses numbers as text is not descriptive enough to even get close to a good similarity score. Instead, let's add some additional text along with the number, as shown in the following code:

```
raw_labels = list(range(100))
clip_labels = ['{} number of paper clips where some paper clips
are partially occluded'.format(label) for label in raw_labels]
```

In this code, we use all numbers from 0 to 99 with the same surrounding text. More effort can be put into designing more variations of the pretext needed. It is also possible to utilize multiple pretexts with the same raw labels. The more descriptive the text is of the image, the more likely there will be a description that produces the closest features to the image.

10. Let's run a prediction on this example and see how it performs:

```
percentages, indices = predict_with_pil_image(image, clip_
labels)
print("\nTop 5 predictions:\n")
for percent, index in zip(percentages, indices):
    print(f"{raw_labels[index]}: {percent:.2f}%")
```

This produces the following results:

```
Top 5 predictions: 2: 4.64%, 4: 4.13% ,3: 4.03%, 0: 3.83%, 1:
3.67%
```

It accurately predicted two paper clips!

11. Now, let's go through two more harder examples from *Figure 9.4 B* and *Figure 9.4 C*:

```
medium_image = Image.open('clips-data-2020/clips/clips-25086.
png'))
hard_image = Image.open('clips-data-2020/clips/clips-25485.png')
```

Predicting on *Figure 9.4 B* produces the following results:

```
Top 5 predictions: 4: 3.98%, 3: 3.78%, 2: 3.71%, 6: 3.23%, 5:
3.16%
```

This produces an error of 1 since the image contains three paper clips. Predicting on *Figure 9.4 C* produces the following results:

```
Top 5 predictions: 18: 1.32%, 9: 1.29%, 0: 1.26%, 19: 1.22% ,16:
1.20%
```

This produces an error of two since the image contains 16 paper clips. Not bad at all!

12. Now, let's evaluate the mean error on a partitioned validation dataset of 1,000 examples:

```
testing_data = pd.read_csv('train.csv')
errors = []
for idx, row in tqdm(testing_data.iterrows(), total=1000):
    image = Image.open(
      'clips-data-2020/clips/clips-      {}.png'.
format(row['id'])
    )
    percentages, indices = predict_with_pil_image(image, clip_
labels, 1)
    errors.append(abs(row['clip_count'] - raw_
labels[indices[0]]))
    if idx == 1000:
        break
print('{} average count error'.format(np.mean(errors)))
```

This results in a 23.8122 average count error. All of a sudden, this doesn't look that usable. In some of the examples, the model couldn't properly count clips that were partially occluded, even when in the description it was specified that there would be some partially occluded. It might make sense to add a description stating that the clips are in different sizes, or even that it is on top of a piece of lined paper. Try it for yourself and do some experiments!

Training a ridge regressor model with a pretrained SqueezeNet featurizer on this dataset separately and validated on the same validation partition in *step 12* results in a root mean squared error of 1.9102. This shows that zero-shot learning by itself is not as reliable as a supervised model for this use case, but it might be able to discern and predict nicely on simpler images. In the paper, the authors emphasized that CLIP-based zero-shot learning can work well and achieve performance close to supervised learning techniques in some specific use cases and datasets. However, in most use cases and datasets, CLIP-based zero-shot learning still falls way behind proper supervised learning methods. This tutorial is an example of where it could work, but it doesn't work so reliably that it can be utilized in the real world. However, it does show promise in the base unsupervised method that CLIP was trained in. A few more years and some more research down the line, and we'll be sure to see an even better performance that will likely be generally the same supervised learning or better!

Next, let's explore the dimensionality reduction component of unsupervised learning.

Exploring the dimensionality reduction component of unsupervised deep learning

Dimensionality reduction is a technique that can be useful in cases where a faster runtime is needed to train and perform inference on your model or when the model has a hard time learning from too much data. The most well-known unsupervised deep learning method for dimensionality reduction is based on autoencoders, which we discussed in *Chapter 5, Understanding Autoencoders*. A typical autoencoder network is trained to reproduce the input data as an unsupervised learning method. This is done through the encoder-decoder structure. At inference time, using only the encoder will allow you to perform dimensionality reduction as the outputs of the encoder will contain the most compact representation, which can fully reconstruct the original input data. Autoencoders can support different modalities, with one modality at any one time, which makes it a very versatile unsupervised dimensionality reduction method.

Other examples include unsupervised methods to create word embeddings that use shallow neural networks, such as FastText or Word2vec. The number of unique words that exist in a language is huge. Even if this gets scaled down to the total amount of unique words in the training data, this can balloon up to 100,000 words easily. One simple way to encode the words is by using one-hot-encoding or TF-IDF. Both methods produce 100,000 column features when the dataset contains 100,000 unique words. This will easily blow up the RAM requirements and can make or break the potential of a solution. However, word embeddings can be tuned to have the desired amount of feature columns as you can choose the embedding size during pretraining with the language model you're using.

Finally, let's learn how to detect anomalies in external data.

Detecting anomalies in external data

Anomaly detection is also considered to be an important application of unsupervised learning in general. Anomaly detection can be used in cases where you want to perform any kind of filtering of your existing data, called **outlier detection**, and also act as a real-time detector during the inference stage given new external data, known as **novelty detection**. Here are some examples of end user use cases of anomaly detection:

- Removing noise in your dataset that will be fed into a supervised feature learning process to enable more stable learning.

- Removing defective products in the production line. This can range from the manufacturing production of semiconductor wafers to egg production.

- Fraud prevention by detecting anomalous transactions.

- Scam detection through SMS, email, or direct messenger platforms.

Anomaly detection is a two-class or binary problem. This means that an alternative way people approach these example use cases listed is to use supervised learning and collect a bunch of negative samples and positive samples. The supervised learning approach can work well when a good-quality dataset containing negative/anomalous data has been collected and labeled because usually, there will already be a good amount of positive/normal data. Go back to the *Preparing data* section of *Chapter 1, Deep Learning Life Cycle*, to recap what it means to have quality data for machine learning! However, in reality, we can't possibly capture all the possible variations and types of negative samples that can occur. Anomaly detection-specific algorithms are built to handle anomalies or negative samples more generally and can extrapolate well to unseen examples.

One of the more widely known deep learning methods to achieve anomaly detection on external data that's not used for training is the autoencoder model. The model performs something called a **one-class classifier**, analogous to the traditional one-class support vector machines. This means that autoencoders can be trained to reconstruct the original data that's fed into it to achieve either or a combination of four tasks – dimensionality reduction, noise remover, general featurizer, and anomaly detection. Depending on the objective at hand, the autoencoder should be trained differently. For dimensionality reduction, noise remover, and general featurizer, the autoencoder should be trained normally as introduced. For anomaly detection, the following things need to be done to ensure it achieves the desired behavior:

1. The data without anomalies is to be used as the training data. If you know there might be anomalies, use a simple outlier detection algorithm such as the local outlier factor, and remove outliers in the data.

2. Use a combination of anomalous and non-anomalous data to form the validation and holdout partition. If such categorization is not known beforehand, use an outlier-based anomaly detection algorithm from *step 1* to label the data.

3. If *step 2* is somehow not possible, train and overfit the training data using the mean squared error loss. This is so that you can make sure the model will only be able to reconstruct the data that has the same characteristics as the training data. Overfitting can be done by training and making sure the training reconstruction loss becomes lower each epoch.

4. If *step 2* is possible, train, validate and evaluate with the given cross-validation partitioning strategy and follow the tips introduced in the *Training supervised deep learning models effectively* section in the previous chapter.

5. The predict function of the autoencoder for testing, validation, and inference can be set based on how well the model can reproduce the input data indicated through the mean squared error. A threshold will need to be used as a cut-off mechanism to determine which data is anomalous or not. The higher the mean squared error, the more probable it is that the provided data is anomalous. Since anomaly detection is a binary classification problem, once labels are available, perform performance analysis by using ROC curves, confusion metrics, and the log loss error using different mean squared error thresholds. This will be introduced more comprehensively in the next chapter.

In anomaly detection in general, note that anomalies are a vague description of what it means to be an anomaly. Different algorithms encode their definition of what an anomaly is given a dataset, its feature space, and its feature distribution. When an algorithm does not work well for your predefined case of anomalies, it does not mean that the algorithm is not a good model in general. The autoencoder approach won't always be the best method that captures your intuition of anomalies. When it comes to unsupervised anomaly detection learning, make sure you train a bunch of models with proper hyperparameter settings and analyze individually which models match your intuition of what makes data anomalous.

Summary

Deep learning has made significant contributions to the field of unsupervised learning, leading to the development of several innovative methods. But by far the most impactful method is unsupervised pretraining, which leverages the abundance of free data available on the internet today to improve the model performance of the downstream supervised tasks and create generalizable representations. With deeper research and time, unsupervised learning will aid in closing the gap toward general artificial intelligence. Overall, deep learning has been a valuable tool in the unsupervised learning domain, helping practitioners make the most of the large amounts of free data available on the internet today.

In the next chapter, we will dive into the first chapter of the second part of this book, which is meant to introduce methods that provide insights about a trained deep learning model.

Part 2 – Multimodal Model Insights

In this part of the book, we delve into the fascinating world of multimodal model insights, taking you on a comprehensive journey through various aspects of evaluating, interpreting, and securing deep learning models. This part offers a comprehensive understanding of various facets of model assessment and enhancement while emphasizing the importance of responsible and effective AI deployment in real-world applications. Throughout these chapters, you will explore methods for evaluating and understanding model predictions, interpreting neural networks, and addressing ethical and security concerns, such as bias, fairness, and adversarial performance.

By the end of this part, you will have a solid understanding of the importance of model evaluation, interpretation, and security, enabling you to create robust, reliable, and equitable deep learning systems and solutions that not only excel in performance but also consider ethical implications and potential vulnerabilities while standing the test of time for real-world applications.

This part contains the following chapters:

- *Chapter 10, Exploring Model Evaluation Methods*
- *Chapter 11, Explaining Neural Network Predictions*
- *Chapter 12, Interpreting Neural Networks*
- *Chapter 13, Exploring Bias and Fairness*
- *Chapter 14, Analyzing Adversarial Performance*

10
Exploring Model Evaluation Methods

A trained deep learning model without any form of validation cannot be deployed to production. Production, in the context of the machine learning software domain, refers to the deployment and operation of a machine learning model in a live environment for actual consumption of its predictions. More broadly, model evaluation serves as a critical component in any deep learning project. Typically, a deep learning project will result in many models being built, and a final model will be chosen to serve in a production environment. A good model evaluation process for any project leads to the following:

- A better-performing final model through model comparisons and metrics
- Fewer production prediction mishaps by understanding common model pitfalls
- More closely aligned practitioner and final model behaviors through model insights
- A higher probability of project success through success metric evaluation
- A final model that is less biased and fairer and produces more trusted predictions

Generally, the model evaluation process leads to more informed decisions across the entire machine learning life cycle. In this first chapter of *Part 2* of the book, we will discuss all the categories of model evaluation that will help to achieve these benefits. Additionally, we will dive deep into some of these categories in the next four chapters, which all belong to *Part 2* of the book. Specifically, we will cover the following topics in this chapter:

- Exploring the different model evaluation methods
- Engineering the base model evaluation metric
- Exploring custom metrics and their applications
- Exploring statistical tests for comparing model metrics
- Relating the evaluation metric to success
- Directly optimizing the metric

Technical requirements

For this chapter, we will have a practical implementation using the Python programming language. To complete it, you will only need to install the `matplotlib` library in Python.

The code files are available on GitHub at `https://github.com/PacktPublishing/The-Deep-Learning-Architect-Handbook/tree/main/CHAPTER_10`.

Exploring the different model evaluation methods

Most practitioners are familiar with accuracy-related metrics. This is the most basic evaluation method. Typically, for supervised problems, a practitioner will treat an accuracy-related metric as the golden source of truth. In the context of model evaluation, the term "accuracy metrics" is often used to collectively refer to various performance metrics such as accuracy, F1 score, recall, precision, and mean squared error. When coupled with a suitable cross-validation partitioning strategy, using metrics as a standalone evaluation strategy can go a long way in most projects. In deep learning, accuracy-related metrics are typically used to monitor the progress of the model at each epoch. The monitoring process can subsequently be extended to perform early stopping to stop training the model when it doesn't improve anymore and to determine when to reduce the learning rate. Additionally, the best model weights can be loaded at the end of the training process defined by the weights that achieved the best metric score on the validation dataset.

Accuracy-related metrics alone do not provide a complete picture of a machine learning model's capabilities and behavior. This is particularly true in unsupervised projects where accuracy metrics are superficial and only relevant to specific distributions. Gaining a more complete understanding of a model allows you to make more informed decisions across the machine's life cycle. Some examples of the ways you can gain more understanding of the model includethe following:

- Model insights can be a proxy to assess the accuracy of the data being used. If the data is deemed to be inaccurate or has some slight flaws, you can transition back into the data preparation stage of the machine learning life cycle.

- In cases where bias is detected, the model may need to be retrained to remove the bias.

- It is also important to evaluate whether the model can recognize patterns in the way that domain experts do. If it cannot, a different model or data preparation method may be required.

- It is necessary to consider whether the model can exhibit common sense in its predictions. If not, it may be necessary to apply special post-processing techniques to enforce common sense.

- Exposing other performance metrics such as inference speed and model size can be critical when choosing a model. You don't want a model that is too slow or too big to fit into your targeted production environment.

In addition to helping a project progress toward success, gathering insights from a model can also help identify potential issues early on. In the machine learning life cycle, which was introduced in *Chapter 1, Deep Learning Life Cycle*, it is important to remember that projects may fail during planning or when delivering model insights. Failing is a natural part of the machine learning process and, in fact, many projects are not meant to succeed. This could be due to a variety of factors, such as data engineering not being suited for machine learning or the task being too complex. However, it is important to fail fast and dump the project in order to be able to redirect resources to other use cases that have a higher chance of success. In cases where the project is critical and cannot be dumped, identifying the root cause of the failure quickly can improve the execution efficiency of the project by diverting resources to fix it and cyclically transitioning between the stages of the machine learning life cycle. In order for a project to fail fast, you have to have a responsible and confident way to be able to determine whether the model is not working. In summary, this ability to fail quickly can be very beneficial, as it saves time and resources that might have otherwise been wasted.

The following list shows a sufficient range of methods that can be utilized to evaluate deep learning models, with some of them being general methods that can work for non-deep learning models too:

- **Evaluation metric engineering**: While evaluation metrics are commonly used in many projects, the practice of evaluation metric engineering is often overlooked. In this chapter, we will take a closer look at metric engineering and explore the process of selecting appropriate evaluation metrics. We'll start by discussing foundational baseline evaluation metrics that are suitable for various types of problems. Then, we'll move on to explore how to upgrade the baseline evaluation metric to one that is specific to the domain and use case of the project. So, in short, this chapter will help you understand the importance of metric engineering and guide you through the process of selecting the right metrics for your project.

- **Learning curves**: Learning curves determine the level of fit for a deep learning model.

- **Lift charts**: A lift chart offers a visual representation of the performance of a predictive model. It shows how much better the model is at predicting positive outcomes compared to random chance.

- **Receiver operating characteristic (ROC) curves**: A graphical representation of the performance of a binary classification model. They plot the **true positive rate (TPR)** against the **false positive rate (FPR)** at various classification thresholds.

- **Confusion matrix**: A confusion matrix is a performance evaluation tool that measures the classification accuracy of a machine learning model. It compares the predicted and actual outcomes of a model's predictions and presents them in a matrix format.

- **Feature importances**: This is the process of determining which features in a dataset have the most influence on the output of a machine learning model. Additionally, it is useful for identifying the most important factors in a given problem and can help improve the model's overall performance.

- **A/B testing**: A/B testing for machine learning involves comparing the performance of two different models or any algorithms on a specific task, in order to determine which model performs better in practice. This can help practitioners make more informed decisions about which models to use or how to improve existing models.

- **Cohort analysis**: Cohort analysis is a technique for evaluating the performance of a model on different subgroups or cohorts of users. It can help identify whether the model is performing differently for different groups and can be useful for understanding how to improve the model for specific segments.

- **Residual analysis**: Residual analysis is a technique used to check the goodness of fit of a regression model by examining the difference between the observed values and the predicted values (**residuals**). It helps identify patterns or outliers in the residuals that may indicate areas for improvement in the model.

- **Confidence intervals**: Confidence intervals are a measure of the uncertainty in an estimate. They can be used to determine the range of values in which the true performance of a model is likely to fall with a certain level of confidence. Confidence intervals can be useful for comparing the performance of different models or for determining whether the performance of a model is statistically significant.

- **Gathering insights from predictions**: This will be covered in *Chapter 11, Explaining Neural Network Predictions*.

- **Interpreting neural networks**: This will be covered in *Chapter 12, Interpreting Neural Networks*.

- **Bias and fairness analysis**: This will be covered in *Chapter 13, Exploring Bias and Fairness*.

- **Adversarial analysis**: This will be covered in *Chapter 14, Analyzing Adversarial Performance*.

In this book, we will only be covering methods that relate to neural networks in some way. In the next section, we will start with the engineering baseline evaluation method, which is the model evaluation metric.

Engineering the base model evaluation metric

Engineering a metric for your use case is a skill that is often overlooked. This is most likely because most projects work on a publicly available dataset, which almost always already has a metric proposed. This includes projects on Kaggle and many public datasets people use to benchmark against. However, this does not happen in real life and a metric doesn't just get served to you. Let's explore this topic further here and gain this skillset.

The model evaluation metric is the first evaluation method that is essential in supervised projects, excluding unsupervised-based projects. There are a few baseline metrics that exist to be the *de facto* metrics depending on the problem and target type. Additionally, there are also more customized versions of these baseline metrics that are catered to special objectives. For example, generative-based tasks can be evaluated through a special human-based opinion score called the mean opinion score. The recommended go-to strategy here is to always start with a baseline metric and work your way up to metrics that reflect how errors should be distributed properly in different conditions in your use case, similar to how it is recommended to build models.

Here are baseline metrics for different conditions:

- **Binary classification problems**:

    - **Accuracy**: This is the percentage of correctly classified examples (true positives and true negatives) out of all examples. It is the most widely known evaluation metric across any domain. However, in reality, this is a skewed metric that can cloud the actual positive prediction performance of the model due to the natural oversupply of negatives in most datasets. If there are 99 negative examples and 1 positive example, predicting negative all the time without a model can get you 99% accuracy! Accuracy still remains the main method for model evaluation, but it is not practically used. When somebody says the model is *accurate*, they probably aren't using the accuracy metric.

    - **Precision**: This is the proportion of true positives among all predicted positive examples. It is a robust alternative that focuses on false positives. A prediction threshold is needed here for binary classification projects.

    - **Recall**: This is the proportion of true positives among all actual positive examples. It is a robust alternative that focuses on false negatives. A prediction threshold is needed here for binary classification projects.

    - **F1 score**: The F1 score is the harmonic mean of precision and recall. It provides a balanced measure of a model's performance. The formula of harmonic mean is

        $$\frac{n}{\frac{1}{x1}+\frac{1}{x2}\dots\frac{1}{xn}}$$

        where is the total number of samples and is the individual sample value. There is also the F2 score that weighs recall more than precision. Use the F1 score if you care about both false positives and false negatives equally, and use the F2 score if you care about false negatives more than false positives. For example, in an intrusion detection use case, if you want to capture any intruders, you can't afford to have false negatives, but you can afford to have false positives, so using the F2 score would be better. Note that the harmonic mean is utilized instead of the arithmetic mean to ensure that extreme values are penalized. For example, a recall of 1.0 and a precision of 0.01 will result in 0.012 instead of something close to 0.5. A prediction threshold is needed here.

    - **Area under the receiver operating characteristic curve (AUC ROC)**: This is the area under the ROC curve, which provides a measure of a model's ability to distinguish between positive and negative examples. No threshold is needed here. It is recommended for use when the positive and negative classes are balanced so you don't need to tune the prediction threshold like for F1 and F2.

- **Mean average precision (mAP)**: This is an extension on top of the precision metric where instead of a single threshold, multiple thresholds are used to compute precision and averaged up to obtain the more robust precision value at different thresholds. For multiple classes, average precision is computed independently and averaged up to obtain mAP.

- **Multiclass classification problems**:

  - **Macro**: Calculate any binary classification metric for each class individually and then average them

  - **Micro**: Determine the overall true positive and false positive rates by considering the highest predicted class, and then use these rates to calculate the overall precision

- **Regression problems**:

  - **Mean squared error (MSE)**: This is the average of the squared differences between predicted and actual values.

  - **Root mean squared error (RMSE)**: This is the square root of the MSE. It provides values at the same scale as the target data and is recommended over MSE.

  - **Mean absolute error (MAE)**: This is the average of the absolute differences between predicted and actual values. Use this over RMSE when you care about differences between predicted and actual labels without caring about its sign.

  - **R-squared**: A measure of how well the model fits the data, which ranges from 0 (poor fit) to 1 (perfect fit).

- **Multilabel**:

  - **Label ranking average precision (LRAP)**: This is the average precision of each ground truth label assigned to a particular sample. It takes into consideration the ranking of labels predicted against the ground truth labels and assigns scores appropriately according to how far or close a ground truth label is in the ranks. LRAP is beneficial for use cases such as movie recommendation systems, where predicting multiple labels and their rankings is important. It evaluates the model's ability to predict the correct labels and their order of relevance, making it an ideal metric for tasks that require accurate and meaningful rankings, such as genre recommendations.

- **Image and video labels**: Raw image and video frames, when used directly as labels, require their own set of custom metrics that can provide more meaningful evaluations compared to standard regression metrics. These are as follows:

  - **Peak signal-to-noise ratio (PSNR)**: This is a measure of the quality of a reconstructed image or video based on the difference between the original and the reconstructed image.

- **Structural similarity index (SSIM)**: This is a measure of the structural similarity between two images or video frames.

- **Text labels**: Although standard classification loss and metrics can be used for text prediction tasks, some metrics can measure much higher-level concepts that are more in tune with human intuition. These are as follows:

  - **Bleu score**: This is a measure of the similarity between machine-generated text and human-generated text based on n-gram overlap.

  - **Word error rate (WER)**: This is a measure of the error rate in automatic speech recognition systems based on the number of errors in word sequences.

- **Human quality-based metrics**: These are non-programmatically computable metrics that can only be evaluated by humans manually:

  - **Mean opinion score (MOS)**: This is a subjective quality rating given by human observers, which can be used to validate and calibrate objective quality metrics

  - **User engagement**: Metrics such as time spent on a website or app, click-through rate, or bounce rate can be used to measure user engagement and satisfaction

  - **Task completion rate**: This is the proportion of users who successfully complete a given task or goal, which can be used to evaluate the usability and effectiveness of a product or service

The baseline metrics are a set of common metrics that should be your first choices depending on your problem types and conditions. With that said, the choice of which base metric to use still depends on the specific problem and the trade-offs between the different aspects of the model's performance that are important for the task at hand. Here are step-by-step recommendations on how to actually choose and utilize an appropriate model evaluation metric:

1. Understand the problem. Consider the nature of the problem, the data, and the desired outcomes. This is a key step that will help you identify the key criteria that are most important for your task.

    A. Be mindful of the quality of the data. The pillars of data quality (representativeness, consistency, comprehensiveness, uniqueness, fairness, and validity) introduced in *Chapter 1, Deep Learning Life Cycle*, will all affect what the metric actually represents. If you are evaluating a model's performance on a bad dataset, then the chosen metric may not reflect the model's true performance on real-world data.

    B. Consider the perspective of the users who will be interacting with the model or the output of the model. What are their expectations and requirements? What are the relevant quality factors that need to be taken into account?

    C. Define clear objectives for what the predictions need to accomplish and the opposite of what they need to do.

2. Choose a metric that aligns with your defined objective. The metric you choose should align with your overall objective. For example, in a binary classification medical use case for detecting cancer, making a false positive diagnosis of cancer can ruin many years of a patient's life, so choose a metric such as precision to allow you to quantitatively reduce the number of false positives.

    A. Consider base metrics that are commonly used in similar problems and how they might need to be adapted or modified to suit the current problem. Are there any unique aspects of the problem that require a different type of metric or a modification of an existing metric?

    B. A single metric may not always capture the full performance of a model. Consider using multiple metrics that evaluate different aspects of the model's performance. A clear example of this is the creation of an F1 score that combines two metrics: precision and recall.

3. Consider the trade-offs between metrics. For example, in a binary classification project, increasing recall performance by modifying the model's prediction threshold can adversely affect precision. Evaluate the trade-offs (if any) between the metrics and choose the one that best aligns with your objectives.

4. Cross-validate the metric. Making sure the metric is computed on a validation and holdout set instead of just the training set is essential to estimating the model's real-world performance. Computing the metric on both the training data and validation data in every epoch will also allow you to visualize the learning curve using the chosen metric directly instead of using the utilized loss.

5. Consider optimizing your model to the metric directly. Some metrics can be approximately reproduced in the deep learning libraries directly and utilized as loss. Using direct optimization can sometimes help you get a better model specifically for the metric you've chosen. However, there might also be some pitfalls that can happen. We will discuss this more extensively later in the *Directly optimizing the metric* section.

6. Build, evaluate, and compare many models against the evaluation metric by iteratively improving it or just using a diverse variety of techniques. The process of building a machine learning model might just be the shortest process in the entire ML lifecycle. But remember that the amount of effort you put into the building and experimentation process can determine whether a project fails or succeeds. Building a variety of models and iteratively improving them can be a hard and time-consuming task. Having boilerplate code that can adapt to most use cases can make this process way faster and more seamless so that you can put your effort into more pressing issues in a project. Alternatively, you can also consider using AutoML tools to consistently get a variety of models trained for each use case. Look forward to the last chapter in this book where we will get a feel for how AutoML tools can streamline the model-building process!

7. Make sure you define a success criterion that relates to the chosen metric. To translate a chosen evaluation metric into an actual success metric, it is important to establish a clear understanding of what constitutes success for the given problem. This will typically involve defining a threshold or target level of performance that the model needs to achieve in order to be considered successful. Sometimes, the success criteria can have a more fine-grained definition based on specific types of error or specific groups of data. We will explore this in depth later in this chapter.

By carefully considering these recommendations, it is possible to select a unique metric that accurately measures the relevant aspects of the problem. Ultimately, this can help to develop more accurate and effective machine learning solutions that will lead to better performance and more successful outcomes.

Base metrics are a group of metrics that are commonly used by many practitioners to evaluate the performance of a model. However, in some cases, a particular problem may have additional criteria and unique behaviors that need to be considered when assessing model performance. In such situations, it may be necessary to adopt alternative or customized metrics that better suit the specific needs of the problem at hand. Base metrics can be further adapted to the additional ideals you want to use to judge your models. The next topic will explore custom metrics and their applications, including when it is appropriate to use them as the most suitable metric for a specific use case.

Exploring custom metrics and their applications

Base metrics are generally sufficient to meet the requirements of most use cases. However, custom metrics build upon base metrics and incorporate additional goals that are specific to a given scenario. It's helpful to think of base metrics as a bachelor's degree and custom metrics as a master's or PhD degree. It's perfectly fine to use only base metrics if they meet your needs and you don't have any additional requirements.

Custom ideals often arise naturally early on in a project and are highly dependent on the specific use case. Most real use cases don't expose their chosen metrics to the public, even when the prediction of the model is meant to be utilized publicly, such as **Open AI's ChatGPT**. However, in machine learning competitions, companies with real use cases accompanied by data publish their chosen metric publicly to find the best model that can be built. In such a setting for a project, the company that hosts the competition is incentivized to perform good-quality metric engineering work that reflects its ideals for its use case. A metric can affect the resulting best model and will ultimately cost the company money when it doesn't engineer a good metric that matches their ideals. Some competitions provide prizes of up to 100,000 USD!

In this section, we will present some common and publicly shared custom ideals along with associated metrics and use cases from machine learning competitions that could be useful for you to consider, regardless of your particular use case:

Ideals	Use case	Custom metric
For time-series regression point-based forecasting, the targets are seasonal and can fluctuate widely based on the season the data is in. We want a metric that can make sure errors aren't weighted heavily to any one season.	**M5 forecasting—accuracy (Kaggle):** This involves predicting the number of Walmart retail goods units sold. The competition provided sales time-series data from Walmart that followed a hierarchical structure, beginning at the item level and progressing to department, product category, and store levels. The data was generously provided and covered three regions in the United States: California, Texas, and Wisconsin.	**Weighted root mean squared scaled error (WRMSSE):** The main part here is the RMSSE, which is a modification of RMSE. Before applying the root of MSE, RMSSE divides the standard MSE by the MSE that uses most recent observation as ground truth. This makes sure all RMSE from any season will be scaled into the same range of values.
Some labels/classes don't matter as much in reality, either because they don't occur as much or they just don't impact post-prediction decisions as much. Don't judge the model too much on unimportant labels; put further emphasis on the errors of more important labels/classes.	**RSNA intracranial hemorrhage detection (Kaggle):** This is a multi-label classification problem to detect the location and type of any hemorrhage present in a medical image of the human brain. A special label called any was made to account for all other types of hemorrhage not accounted for with a specific label. The any label had 2-3 times more data than any other label alone.	**Weighted multilabel log loss:** The any label was weighted more than any other label even though it had more data. This shows that the significance of a label for a metric is not exclusive to the scarcity of the data associated with the label in a dataset; it really depends on the specific problem context.
	M5 Forecasting—accuracy (Kaggle): This involves predicting the number of Walmart retail goods units sold. The competition provided time-series sales data from Walmart that followed a hierarchical structure, beginning at the item level and progressing to department, product category, and store levels. The data was generously provided and covered three regions in the United States: California, Texas, and Wisconsin.	**WRMSSE:** The competition author valued more unit sales forecast from products that provided more significant sales in dollars. The weight for each product is obtained by using the sales volumes for the product in the last 28 observations of the training sample (sum of units sold multiplied by their respective prices).
	Walmart Recruiting—Store Sales Forecasting (Kaggle): This involves forecasting the sales of Walmart goods.	**Weighted mean absolute error (WMAE):** Walmart weighed holiday weeks forecast error more than non-holiday weeks by five-fold, as they have much higher sales during holiday weeks.

We don't really care too much about small errors, as they can be tolerated, but we care about big errors because they can result in the triggering of unwanted actions by consuming the predictions.	**Google Analytics customer revenue prediction (Kaggle)**: This is a regression problem about predicting the total revenue generated by a customer for an online store. The revenue values were highly skewed and contained many zero values, which made it challenging to evaluate the performance of the participating models using traditional metrics such as MSE or MAE.	**Root mean squared log error (RMSLE)**: RMSLE applies a logarithmic transformation to the predicted and actual values before computing RMSE, which helps to penalize large errors more heavily than small errors. A natural way of doing this is to simply train the model to predict the log values of the target and apply RMSE to achieve RMSLE.
	Diabetic retinopathy detection (Kaggle): This is a multiclass problem about predicting whether high-resolution retina images have diabetic retinopathy or not. The problem has five classes: one for no disease and the other four for different severities of the disease.	**Quadratic Weighted Kappa (QWK) score**: The Kappa score is a statistical measure that provides a single value to quantify the degree of agreement between predicted and actual labels in multi-class classification tasks. The quadratic weighing mechanism enables the Kappa score to become more robust to minor errors and more sensitive to larger ones. The quadratic weighting scheme can address issues where the Kappa score may be inflated by a high proportion of agreement in easy-to-classify categories while still providing an accurate representation of the level of agreement for the more difficult cases.
	Two Sigma Connect Rental listing inquiries (Kaggle): This involves predicting how popular an apartment rental listing is based on the listing content, such as text descriptions, photos, number of bedrooms, and price.	**Log loss**: Log loss is an evaluation metric that emphasizes wrong predictions by penalizing models that are confident about incorrect predictions.

In a multiclass problem, we have a higher tolerance for our multiclass model in our use case where we can consume several of the top predicted classes instead of using the single most probable class to maximize the true positive hit rate performance of the model.	**Airbnb new user bookings (Kaggle):** This is a multiclass problem for predicting the country in which a new user will make their first booking with Airbnb (including a no-booking class). The predicted class will allow Airbnb to share more personalized content with their community, decrease the average time to first booking, and better forecast demand. Airbnb had a relaxed requirement for the multiclass predictions a model produced, where they could perform personalized content based on the top five countries instead of a single country to improve the true positive hit rate.	**Normalized discounted cumulative gain (NDCG) of top-k classes:** NDCG is a measure of the ranking effectiveness of top-k classes, usually used for recommendation use cases. This metric pairs nicely with the tolerance Airbnb has for multiclass predictions. It would work well for any multiclass use case that can tolerate using several top predicted classes instead of the single predicted top class.
For multi-object video-based tracking use cases, we want to use a metric that penalizes undesired tracking behaviors, such as false object identification and failure to track objects consistently over time. Additionally, we need a metric that can be computed by the multiple trajectories created by a model through time against a fixed set of ground truth tracking trajectories.	**Multi-camera people tracking (AICity)** `https://www.aicitychallenge.org/2023-data-and-evaluation/`: The dataset includes multiple camera feeds captured in various settings, including warehouse environments within a building, as well as synthetic data generated using the NVIDIA Omniverse Platform in multiple indoor settings. To enhance the diversity and size of our dataset for `Track 1`, we have created a large-scale synthetic dataset of animated people. All camera feeds in our dataset are high-resolution (1080p) feeds with frame rates of 30 frames per second.	**Identity F1 score (IDF1):** IDF1 handles this by evaluating the overall performance of the tracking algorithm based on how well it matches predicted tracks to ground truth tracks, taking into account the identity of each object over time. The algorithm penalizes false positives, false negatives, and identity switches, which occur when the algorithm incorrectly identifies two detections as the same object or incorrectly assigns two different identities to the same object over time. Most importantly, the model can dynamically determine matching trajectories based on a version of the **intersection over union (IOU)** algorithm between predicted and ground truth trajectories.

We care about bias in our model's performance a lot more than the overall model's performance.	**Jigsaw unintended bias in toxicity classification (Kaggle)** `https://www.kaggle.com/competitions/jigsaw-unintended-bias-in-toxicity-classification/overview/evaluation`: This involves predicting the different intensities of toxicity in text data. The data contains identities that the competition author wishes to optimize against in the name of bias. The identities involved are male, female, transgender, other gender, heterosexual, homosexual (gay or lesbian), bisexual, other sexual orientation, Christian, Jewish, Muslim, Hindu, Buddhist, atheist, other religion, black, white, Asian, Latino, other race or ethnicity, physical disability, intellectual or learning disability, psychiatric or mental illness, and other disability.	**Bias-focused AUC:** The competition focuses on a weighted combination of multiple AUC metric that are computed on a different subset of the dataset based on specific identities that can be mentioned in the text data row. Overall, AUC is combined equally with the next three bias-focused metrics. **Identity-based subgroup AUC:** This involves analyzing a dataset that only includes comments mentioning a specific identity subgroup. A low score in this metric indicates that the model struggles to distinguish between toxic and non-toxic comments related to that identity.
		Background positive, subgroup negative (BPSN) AUC: This involves evaluating the model's performance on non-toxic examples mentioning the identity and toxic examples that do not. A low score in this metric suggests that the model may incorrectly label non-toxic examples related to the identity as toxic.
		Background negative, subgroup positive (BNSP) AUC: This involves assessing the model's performance on toxic examples mentioning the identity and non-toxic examples that do not. A low score in this metric indicates that the model may incorrectly label toxic examples related to the identity as non-toxic.

Table 10.1 – A table of custom ideals with example use cases and metrics used

In the second ideal, the general metric idea is to apply weights to any metric you'd like to choose based on what you deem more important. Additionally, weights can be applied more flexibly to any auxiliary data that does not act as an input to a model, nor as a target to the model. Finally, for the last ideal, look forward to *Chapter 13, Exploring Bias and Fairness*, to discover methods to optimize against bias.

While the examples of custom ideals and metrics we've discussed are useful guidelines, it's important to remember that there are many different metrics and ideals that may be relevant to your specific use case. Don't be afraid to dig deeper into your problem domain and identify metrics that are unique to your particular situation.

The examples we've given can serve as a helpful cheat sheet for developing custom metrics that are tailored to your needs. By understanding the reasoning behind the use of these special metrics in specific domains, you can gain insight into the factors that should be considered when evaluating model performance. Ultimately, the key is to choose metrics that align with your goals and capture the most important aspects of your problem domain.

Next, we will discuss a robust strategy to compare the metric performance of different models across multiple metric values computed from different cross-validation folds or dataset partitions.

Exploring statistical tests for comparing model metrics

In machine learning, metric-based model evaluation often involves using averages of aggregated metrics from different folds or partitions, such as holdout and validation sets, to compare the performance of various models. However, relying solely on these average performance metrics may not provide a comprehensive assessment of a model's performance and generalizability. A more robust approach to model evaluation is the incorporation of statistical hypothesis tests, which assess whether observed differences in performance are statistically significant or due to random chance.

Statistical hypothesis tests are procedures used to determine whether observed data provides sufficient evidence to reject a null hypothesis in favor of an alternative hypothesis, helping to quantify the likelihood that the observed differences are due to random chance or a genuine effect. In statistical tests, the null hypothesis (H0) is a default assumption that states there is no effect or relationship between variables, serving as a basis for comparison against the alternative hypothesis with the goal of determining whether the observed data provides enough evidence to reject this default assumption. For the purpose of comparing model metric performance across multiple partitions and datasets, the null hypothesis is typically that there is no difference between the performances of the models, while the alternative hypothesis is that there are differences.

Overall, statistical tests offer a formal framework to objectively determine whether differences in performance are significant or due to chance. Additionally, statistical tests offer a comprehensive understanding of model performance by accounting for variability and uncertainty in metrics. The following table shows common statistical tests that you can consider using, along with the Python code needed to execute it, how to interpret the result, and when to use it:

Statistical test	Python code	Result interpretation	Recommended use
| Paired t-test | ```
from scipy.stats
import ttest_rel

t_stat, p_value =
ttest_rel(model1_
scores, model2_
scores)
``` | If $p\_value < 0.05$, there's a significant difference between the models. | Use this when comparing two dependent samples with normally distributed differences. |
| Mann-Whitney U test | ```
from scipy.stats
import mannwhitneyu

u_stat, p_value =
mannwhitneyu(model1_
scores, model2_
scores)
``` | If $p\_value < 0.05$, there's a significant difference between the models. | Use this when comparing two independent samples with non-normally distributed data or ordinal data. |
| **Analysis of variance (ANOVA)** | ```
from scipy.stats
import f_oneway
f_stat, p_value =
f_oneway(model1_
scores, model2_
scores, model3_
scores)
``` | If $p\_value < 0.05$, there's a significant difference among the models. | Use this when comparing three or more independent samples with normally distributed data and equal variances. |
| Kruskal-Wallis H test | ```
from scipy.stats
import kruskal

h_stat, p_value
= kruskal(model1_
scores, model2_
scores, model3_
scores)
``` | If $p\_value < 0.05$, there are significant differences between the models. | Use this when comparing three or more independent samples with non-normally distributed data or ordinal data. |

Table 10.2 – Common statistical tests with details on their Python implementation, result interpretations, and recommendations on when to use them

These recommendations can help you choose the appropriate statistical test based on the conditions and assumptions of your data, such as the number of samples, the type of data (dependent or independent), and the distribution of the data (normal or non-normal).

Next, we will discuss how the outcome of metric engineering can be converted to a success criterion.

Relating the evaluation metric to success

Defining success in a machine learning project is crucial and should be done at the early stages of the project as introduced in the *Defining success* section in *Chapter 1, Deep Learning Life Cycle*. Success can be defined as achieving higher-level objectives, such as improving the efficiency of processes or increasing the accuracy of processes in comparison to manual labor. In some rare cases, machine learning can enable processes that were previously impossible due to human limitations. The ultimate success of achieving these objectives is to save costs or earn more revenue for an organization.

A model with a metric performance score of 0.80 F1 score or 0.00123 RMSE doesn't really mean anything at face value and has to be translated to something tangible in the use case. For instance, one should answer questions such as what estimated model score can allow the project to achieve the targeted cost savings or revenue improvements. Quantifying the success that can be obtained from model performance is essential, particularly as machine learning projects can be expensive to execute. Sometimes, the return on investment can be low if the model fails to perform at a certain level. After selecting the evaluation metric, it's important to establish a metric threshold for success that is realistic, achievable, and based on the business objective.

As a finale to the topic, let's go through an example workflow to relate the evaluation metric to success based on a hypothetical use case.

Let's consider a use case to identify defective products in a manufacturing process using image data. Let's assume that the cost of producing a product is $50 and it has a retail price of $200. If that product is defective and it makes it through to a customer, this will result in an additional chargeback of $1,000 and the return of the $200 paid by the customer to compensate for the defective product, which may have caused harm. On the other hand, if a good product is identified as defective and scrapped, it results in a cost of $250; $50 is used to produce it and $200 is lost opportunity cost as it is scrapped.

Let's say that we have a dataset of 10,000 product images, which is the amount of products produced a month. If we don't use a model and produce and ship all 10,000 products, we would have 95 defective products, which would cost $500,000 in production costs (10,000 x $50) and $95,000 in chargeback costs, resulting in a total cost of $595,000. After selling all the non-defective products, the company will gain $1,981,000 in sales (9,905 x $200). The total money earned will be $1,386,000.

Consider a case in which we use a trained deep learning model to classify these images as either good (negative) or defective (positive). To make sure it is worth using a deep learning model to optimize this process, let's say the company needs to gain at least $120,000 a year in cash to make this worthwhile. Let's also consider that maintaining a machine learning model costs $20,000 per month. The threshold in any metric needs to relate to the result of making at least $30,000 gains per month.

Say we want to use an F score-based metric. Since false negatives affect the total money more than false positives, we might want to use the F2 score instead of the F1 score. But will models with the same metric score exhibit different monetary returns for either of the two metrics? This is essential to understand so that a proper success threshold can be set. Let's attempt to use Python code to analyze score behavior:

1. First, we define the number of actual positives and actual negatives and the total data:

    ```
    actual_positive = 95
    total_data = 10000
    actual_negative = total_data - actual_positive
    ```

2. Next, let's define methods to compute precision, recall, F1 score, and F2 score:

    ```
    def precision(tp, fp):
        denominator = tp + fp
        if denominator == 0:
            return 0
        return tp/ denominator
    def recall(tp, fn):
        denominator = tp + fn
        if denominator == 0:
            return 0
        return tp/denominator
    def f1score(tp, fp, fn):
        prec = precision(tp, fp)
        rec = recall(tp, fn)
        denominator = prec+ rec
        if denominator == 0:
            return 0
        return 2 * (prec * rec) / denominator
    def fbeta(tp, fp, fn, beta=0.5):
        prec = precision(tp, fp)
        rec = recall(tp, fn)
        denominator = beta**2 * prec + rec
        if denominator == 0:
            return 0
        return (1+beta**2) * (prec * rec) / denominator
    ```

3. Next, let's define the method to compute the final cash we will have:

    ```
    def compute_total_cash(tp, fp, fn, tn):
        return tp * -50 + fp * -50 + fn * -1050 + tn * 150
    ```

4. Let's use the compute_total_cash method to compute the baseline cash for the current setup where no model is used. This is so that we can find out the cash success threshold required for a model to be considered valuable enough to be used:

    ```
    baseline_cash = compute_total_cash(0, 0, actual_positive,
    actual_negative)
    threshold_cash_line = baseline_cash + 30000
    ```

5. Next, we will simulate every possible combination of true positives (`tp`), false positives (`fp`), true negatives (`tn`), and false negatives (`fn`) and compute the F1 and F2 scores:

```
f1_scores = []
f2_scores = []
total_cash = []
for tp in range(0, 96):
    fn = 95 - tp
    for fp in range(0, actual_negative):
        tn = total_data - tp - fp - fn
        f1_scores.append(f1score(tp, fp, fn))
        f2_scores.append(fbeta(tp, fp, fn, beta=2.0))
        total_cash.append(compute_total_cash(tp, fp, fn, tn)
```

6. Now, let's plot both of the scores independently against the total cash return while drawing horizontal lines using `threshold_cash_line` and `baseline_cash`:

```
import matplotlib.pyplot as plt
fig, axs = plt.subplots(2, figsize=(18, 15))
axs[0].scatter(f1_scores, total_cash, alpha=0.01)
axs[1].scatter(f2_scores, total_cash, alpha=0.01)
for i in range(2):
    axs[i].axhline(y=baseline_cash, color='r',
linestyle='dotted')
    axs[i].axhline(y=threshold_cash_line, color='y',
linestyle='dashdot')
```

This will result in the following figure:

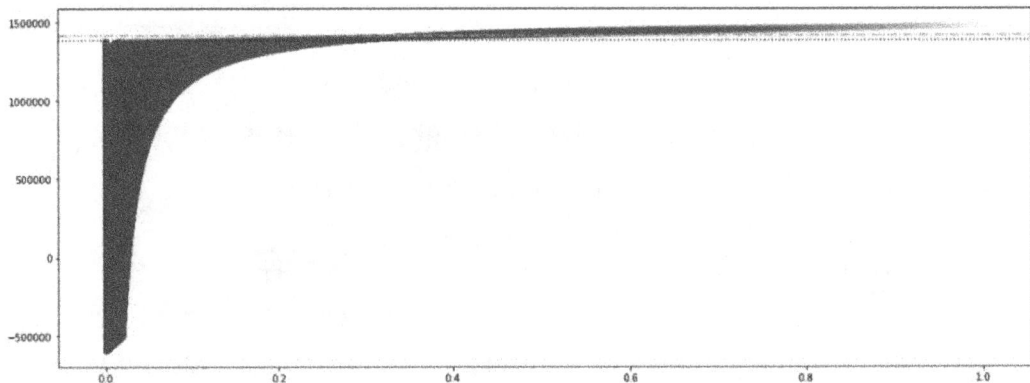

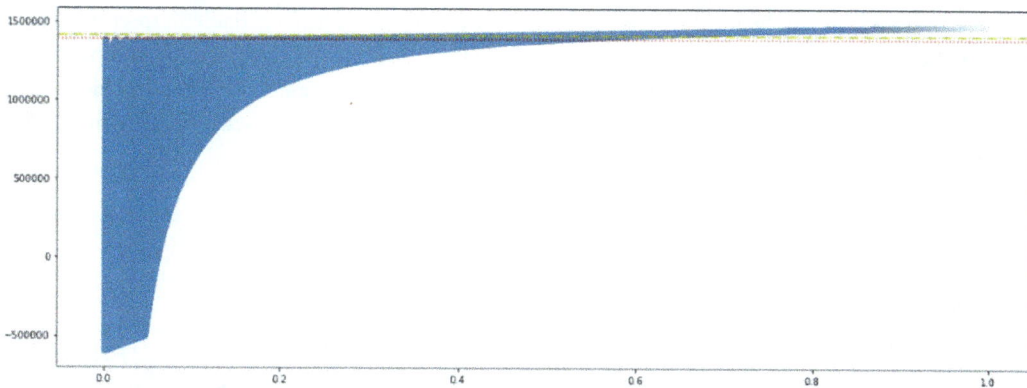

Figure 10.1 – Cash versus F1 score and cash versus F2 score

7. The figure suggests that the F2 score has huge fluctuations in the lower score range even though it weights recall higher. The goal here is to make sure a metric can be properly linked to success, so having wider cash fluctuations with the same score isn't a desirable trait. F1 score would probably be the wiser choice here. Using the topmost horizontal line (which references the minimum 30,000 cash threshold we need to hit) as a reference, we want to find a point where it is not even possible to get a lower score than the threshold in the F1 score graph. Roughly, a 0.65 F1 score should guarantee that the model can produce a score.

This example demonstrates the level of analysis required to properly choose a metric and find a threshold that can be directly linked to success while taking into consideration both monetary profit and loss. However, it is important to note that not all machine learning projects can be measured in terms of dollar cost. Some projects may not be directly related to cash, and that is perfectly acceptable. However, to be successful, it is important to quantify the value of the model in a way that stakeholders can understand. If nobody understands the value that the model provides, the project is unlikely to be successful.

Next, let's dive into the idea of directly optimizing the metric in a deep learning model.

Directly optimizing the metric

The loss and the metric used to train a deep learning model are two separate components. One of the tricks you can use to improve a model's accuracy performance against the chosen metric is to directly optimize against it instead of just monitoring performance for the purpose of choosing the best performing model weights and using early stopping. In other words, using the metric as a loss directly!

By directly optimizing for the metric of interest, the model has a chance to improve in a way that is relevant to the end goal rather than optimizing for a proxy loss function that may not be directly related to the ultimate performance of the model. This simply means that the model can result in a much better performance when using the metric as a loss directly.

However, not all metrics can be used as a loss, as not all metrics can be differentiable. Remember that backpropagation requires all functions used to be differentiable so that gradients can be computed to update the neural network weights. Note that discontinuous methods are not all differentiable. Here are some common discontinuous functions that are not differentiable, along with NumPy methods to look out for easy identification:

- **Max and min functions**: NumPy methods to look out for are `np.min`, `np.max`, `np.argmin`, and `np.argmax`
- **Clipping functions**: NumPy methods to look out for are `np.clip`
- Other functions in NumPy to look out for are `np.sign`, `np.piecewise`, `np.digitize`, `np.searchsorted`, `np.histogram`, `np.fft`, `np.count_nonzero`, `np.round`, `np.cumsum`, and `np.percentile`

Additionally, using a metric as a loss function can sometimes lead to suboptimal performance because the metric does not always capture all aspects of the problem that the model needs to learn in order to perform well. Some important aspects of a problem may be difficult to measure directly or to include in a metric. These aspects may form the foundation needed for a model to learn before it can proceed to slowly get better at the chosen metric. For example, in image recognition, the model needs to learn to recognize more abstract features, such as texture, lighting, or viewpoint before it can attempt to get better at accuracy. If these features are not captured in the metric, the model may not learn to recognize them, resulting in suboptimal performance. A good solution here is to experiment with more conventional loss functions initially and then fine-tune the model using either only the metric as a loss or a combination of the original loss and the metric as a loss.

While using a metric as a loss function can be beneficial in some cases, it's not a surefire method of improving performance. The efficacy of this approach largely depends on the specific use case and the complexity of the problem being addressed. The performance boost achieved through this method might be minimal and too nuanced for some projects to even consider. However, when used successfully, it can lead to meaningful improvements in performance.

Summary

In this chapter, we briefly explored an overview of different model evaluation methods and how they can be used to measure the performance of a deep learning model. We started with the topic of metric engineering among all the introduced methods. We introduced common base model evaluation metrics. On top of this, we discussed the limitations of using base model evaluation metrics and introduced the concept of engineering a model evaluation metric tailored to the specific problem at hand. We also explored the idea of optimizing directly against the evaluation metric by using it as a loss function. While this approach can be beneficial, it is important to consider the potential pitfalls and limitations, as well as the specific use case for which this approach may be appropriate.

The evaluation of deep learning models requires careful consideration of appropriate evaluation methods, metrics, and statistical tests. Hopefully, after reading through this chapter, I have helped ease your journey into metric engineering, encouraged you to take the first step toward deeper metric engineering by following the guidelines provided, and highlighted the importance of metric engineering as a valuable component to improve model evaluation.

However, some information can be hidden behind a single metric value no matter how good and appropriate the final chosen metric is. In the next chapter, we will cover a key method that can help uncover either wanted or unwanted hidden behaviors of your neural model based on how your neural network model makes predictions.

11

Explaining Neural Network Predictions

Have you ever wondered why a facial recognition system flagged a photo of a person with a darker skin tone as a false positive while identifying people with lighter skin tones correctly? Or why a self-driving car decided to swerve and cause an accident, instead of braking and avoiding the collision? These questions illustrate the importance of understanding why a model predicts a certain value for critical use cases. By providing explanations for a model's predictions, we can gain insights into how the model works and why it made a specific decision, which is crucial for transparency, accountability, trust, regulatory compliance, and improved performance.

In this chapter, we will explore neural network-specific methods for explaining model predictions. Additionally, we will discuss how to quantify the quality of an explanation method. We will also discuss the challenges and limitations of model explanations and how to evaluate their effectiveness.

Specifically, the following topics will be covered:

- Exploring the value of prediction explanations
- Demystifying prediction explanation techniques
- Exploring gradient-based prediction explanations
- Trusting and understanding integrated gradients
- Using integrated gradients to aid in understanding the predictions
- Explaining prediction explanations automatically
- Exploring common pitfalls in prediction explanations and how to avoid them

Technical requirements

This chapter includes some practical implementations in the Python programming language. To complete it, you will need to have a computer with the following libraries installed:

- `pandas`
- `captum`
- `transformers-interpret`
- `transformers`
- `pytorch`
- `numpy`

The code files are available on GitHub: `https://github.com/PacktPublishing/The-Deep-Learning-Architect-Handbook/tree/main/CHAPTER_11`.

Exploring the value of prediction explanations

First off, the concept of explaining a model through its predictions is referred to by many other names, including explainable AI, trustable AI, transparent AI, interpretable machine learning, responsible AI, and ethical AI. Here, we will refer to the paradigm as **prediction explanations**, which is a clear and short way to refer to it.

Prediction explanations is not a technique that is adopted by most machine learning practitioners. The value of prediction explanations highly depends on the exact use case. Even though it is stated that explanations can increase transparency, accountability, trust, regulatory compliance, and improved model performance, not everybody cares about these points. Instead of understanding the benefits, let's look at it from a different perspective and explore some of the common factors that drove practitioners to adopt prediction explanations that can be attributed to the following conditions:

The prediction explanations technique provides the following benefits regarding the utilization of your built model:

- **Transparency**: Prediction explanations allow model prediction consumers to have access to reasons, which will, in turn, enforce the confidence of the consumer in the predictions made. A transparent model allows consumers to objectively gauge their intelligence, which increases trust and increases adoption.
- **Accountability**: Prediction explanations allow consumers to perform root cause analysis, which is especially important in critical use cases or in use cases where there is a human in the loop to make the final decision with a model's prediction as a reference.

- **Trust**: Without trust, nobody will use the prediction of the model. Prediction explanations provide a small boost toward achieving higher trust.

- **Regulatory compliance**: Some governments enforce laws that require decisions made by computer systems to be explainable in certain industries such as banks and insurance.

- **Metric performance**: The capability to perform root cause analysis can lead to a better understanding of either the model's behavior or the training dataset. This will, in turn, allow machine learning practitioners to improve or fix the issues found using prediction explanations, and eventually lead to improved metric performance.

This list of benefits makes it worth it to utilize prediction explanations on any use case and model. However, the value of using prediction explanations increases exponentially with certain conditions. When the utilization is based on a specific goal, the method becomes twice as useful. Let's take a step further and explore some of the common conditions that drove practitioners to utilize the prediction explanations technique:

- **Critical and high-impact use case**: In these use cases, the model's decisions can typically result in significant consequences for human welfare, safety, or well-being. Understanding the model's behavior in every way can help mitigate the worst case from happening. This can range from billions of money lost to actual human life lost.

- **Failing to achieve the required threshold for success**: A machine learning project can't move forward to the model deployment stage in the machine learning life cycle if it can't even achieve success thresholds during the model development stage. Understanding the behavior of the model toward different inputs can help signal whether the data is of bad quality, help indicate whether the data has biased patterns that promote overfitting, and generally help debug how to improve the model's performance.

- **The model makes wrong predictions with the simplest examples**: Making wrong predictions with the most complex examples is expected, especially when humans also could have made mistakes in making the same decisions. When it comes to the simplest examples, making an error would indicate that the model is not learning the right things. Understanding what a model is focusing on could be key to figuring out why a model failed.

- **Regulatory laws mandate accountability for using machine learning models in decision-making**: This means prediction explanations will be required not typically as a method for the machine learning practitioner to understand how the model behaves or with different inputs, but instead, to be used after a model is deployed, which will allow the people that consume the predictions understand why a certain decision has been made for accountability.

- **The model fails to make proper predictions after it is deployed**: Have you ever thought about whether a good model with high accuracy metric performance in your cross-validation setup means that the model pays attention to the right things? Even with cross-validation partitioning strategies, data can still be very biased toward a certain set of conditions. This means that when the model gets deployed into the real world, the data it encounters during the production operation can be from a distribution and set of conditions that are different than what was available in the original dataset used for model development. When this happens, prediction explanations can help uncover biases and unwanted behaviors of the model either in the new data or data used for model development. In other words, you'll need to quantify wrong predictions from right predictions.

It is fine to not utilize prediction explanation techniques in your use case and development workflow. However, when these edge cases happen, know that prediction explanation techniques are your key tool to help overcome your obstacles. But what exactly do prediction explanations explain? In the next section, we will discuss a short overview of prediction explanation methods and dive into the method category we will be introducing in this chapter, which is specifically for neural networks.

Demystifying prediction explanation techniques

Prediction explanations is a technique that attempts to explain the logic behind a model's decision, given input data. Some machine learning models are built to be more transparent and explainable out of the box. One example is a decision tree model, which is built from the ground up using explicit conditioning rules to split data into multiple partitions that result in specific decisions, allowing the predictions to be explained through the explicit rules that were used to predict the data sample. However, models such as neural networks are treated like a black box without any straightforward way to retrieve the reasons the decision was made directly.

The logic of a model's decision on a data sample can be explained and presented in a variety of ways, so long it contains information on how the final decision was made. Additionally, predictions made by a machine learning model can be explained in either a model-agnostic or model-specific way. There are a few types of explanations that can be made using the model's predictions:

- **Saliency-based explanation**: This is also known as importance attribution, feature importance, or feature impact

- **Transparent model logic-based explanation**: Provide a rationale on why a decision is made

- **Exemplar-based explanation**: Using similar data to reason why a label is predicted

Neural networks, being a black box out of the box, can only be explained through saliency-based explanation or exemplar-based explanation. However, workarounds have been invented to achieve indirect transparent model logic-based explanations for neural networks using knowledge distillation methods from a neural network to more transparent and interpretable models such as linear models. Additionally, attention mechanisms implemented in transformers provide a shallow way to check for feature importance through its attention maps as a saliency method but fall short in the following components:

- Attention weights are not feature-specific as they are obtained by calculating the interactions between all input tokens and the output token of interest. This means the weight does not represent a true reliance on an isolated feature.

- The inputs attended to by the attention mechanism can still not be used in later parts of the network.

- Attention maps can be biased and cause some features to be neglected.

Additionally, neural-networks-based explanation techniques should be opted over model-agnostic explanation techniques as they are capable of providing more detailed and nuanced explanations while being more efficient in terms of both computational requirements and time. In this chapter, we will focus on saliency-based explanations for neural networks using more reliable methods.

In the next section, we will dive into the core workhorse behind neural network model-specific explanations, which are gradient-based saliency explanations.

Exploring gradient-based prediction explanations

Most up-to-date neural network-based explanation techniques today are variations of using the gradients that can be obtained through backpropagation. Gradient-based explanations for neural network models work because they rely on the fundamental principle of how the weights in a neural network are updated during the training process using backpropagation. During backpropagation, the partial derivatives of the loss function concerning the weights in the network are calculated, which gives us the gradient of the loss function concerning the weights.

This gradient provides us with a measure of how much the input data contributes to the overall loss. Remember that gradients measure the sensitivity of the input value concerning the loss function. This means it provides the degree of fluctuation of the predictions when you modify the specific input value, which represents the importance of the input data. Input data can be chosen to be the weights of the neural network or the actual input to the entire neural network. In most cases, the actual input-based explanations are enough to provide clarity on the important feature groups. However, sometimes, a more fine-grained explanation is needed to decode the underlying characteristics of the highly attributed and important actual input data.

For example, consider the use case of identifying animal breeds with image data using a CNN. For an image with a dog standing on grass, if an attribution method signals that both the dog and grass are important, why is the grass important? This can be for one of the following reasons:

- The fur of the dog and the grass are both identified by a single convolutional filter.

- Fur and grass are explicitly separately identified by different filters but shown as important similarly because the model might have overfitted to think that dogs need to be accompanied by grass. This might mean that the training dataset only contains images of dogs with grass and signals the need to add more dog images without grass.

This will require diving deeper into important filters that are attributed highly among the filters that are activated and have something to do with the grass and the dog. Even after pinpointing the filters that are both important, highly activated for the image, and have something to do with the grass and the dog, you can't know for sure what patterns the filter is identifying. Is it a high-level feature such as the dog's shape? Or is it fur? In cases like this, it is highly beneficial to visualize the patterns directly. This topic will be discussed in more detail in *Chapter 12, Interpreting Neural Networks*.

Gradients by themselves, however, are not reliable enough to be used as-is as an explanation method. More generally, the quality of pure gradients to be used as a feature importance explanation method is low. However, the quality of explanation methods is not a quantitative metric. The perceived quality of the explanations generated is subjective to the requirements of the consumer.

Consider a scenario where a financial institution wants to develop a credit scoring model to determine loan approvals. The institution is interested in using an explanation technique to understand which features are most important in the model's predictions. A bank might prioritize an explanation technique that focuses on the features that have the greatest impact on creditworthiness, such as credit history, income, and outstanding debts. This would help the bank make informed lending decisions and manage its risks effectively.

On the other hand, an individual loan applicant may prefer a more comprehensive explanation that provides insight into how the model evaluated their specific financial situation, including factors beyond the top three most important features, such as their employment history and recent financial hardships. This would help the applicant make informed decisions about how to improve their creditworthiness in the future. Selecting an appropriate explanation technique is important as it should cater to the audience and their specific needs to make the explanation clear and useful for its intended purpose.

Even so, there are axiomatic-based evaluations, which involve evaluating based on a set of principles or axioms that are considered desirable. Of all techniques, a technique called **integrated gradients** stands out for its focus on neural network models while being developed to satisfy widely accepted axioms.

Next, we'll understand how integrated gradients explain a neural network prediction and understand how the method satisfies widely accepted axioms.

Trusting and understanding integrated gradients

First off, the integrated gradient technique is available off-the-shelf in a few open source libraries such as **shap** and **captum**. We can utilize the method from the library without reimplementing the method. However, obtaining explanations without understanding the technicalities behind the technique can reduce trust in the explanations. If you don't trust the technique, the explanation results themselves hardly mean anything. Subsequently, if you don't explain your predictions, the prediction results themselves hardly mean anything! In this section, we will dive into what integrated gradients do so that you can trust what integrated gradients can explain to you.

Integrated gradients add a few extra components to the basic gradient-based feature importance-based explanations. These components are geared toward satisfying a few critical axioms that make or break the reliability of an explanation method. These axioms are as follows:

- **Sensitivity**: A model that produces two different predictions on two data samples that differ in that only a single feature should be given a non-zero importance score. Activation functions such as ReLU break this axiom with a zero input value as the gradients would also be reduced to zero, even when the predictions are different between the two predictions.

- **Implementation invariance**: Two models that have the same performance across all possible data must produce the same importance scores. Assume that any external data and data received during the entire lifetime of the model is included.

- **Completeness**: One of the best ways to make a good-quality explanation for both human and machine learning explanations is to provide counterfactual reports. To evaluate model accuracy performance, having a baseline model works the same way. For explanation methods, this involves engineering a baseline data sample that can produce a neutral prediction. Completeness means that the importance score for every feature column must add up to the difference in prediction score between the baseline sample and the targeted sample. This axiom can be more useful in some use cases, such as in regression, where the prediction is directly used as an output rather than in a multiclass setting where the predictions are only used for choosing the top class.

- **Linearity**: If you linearly combine two neural network models, which is a simple weighted addition, the explanation of the combined neural network on a data sample must be the same weighted addition of the individual explanation of the two neural network models. Linear behaviors that are implemented in the model should be respected. A straightforward axiom is specifically designed as some methods generate unaccountable importance values.

Now that we understand the core axioms that can be used to compare methods, know that integrated gradients satisfy all of these axioms. Simply put, integrated gradients integrate the gradients from samples that are sampled from a straight-line path (linear interpolation) from a chosen baseline data sample and the target data sample and multiply it by the difference between the predictions of the target sample and baseline sample. The path integral of the gradients is the area under the curve of the gradient values along the straight path. This value represents the change in the model output as the input feature's value changes along the path. The rate of change directly translates to feature importance. By multiplying this value by the difference between the actual input value and the baseline input value, we get the contribution of the feature to the model output for integrated gradients. *Figure 11.1* showcases this process with the results obtained by computing integrated gradients on the predictions of a pre-trained ResNet50 model on an image of an orange. The model is capable of predicting the orange class and predicts the image to be an orange with a 46% probability:

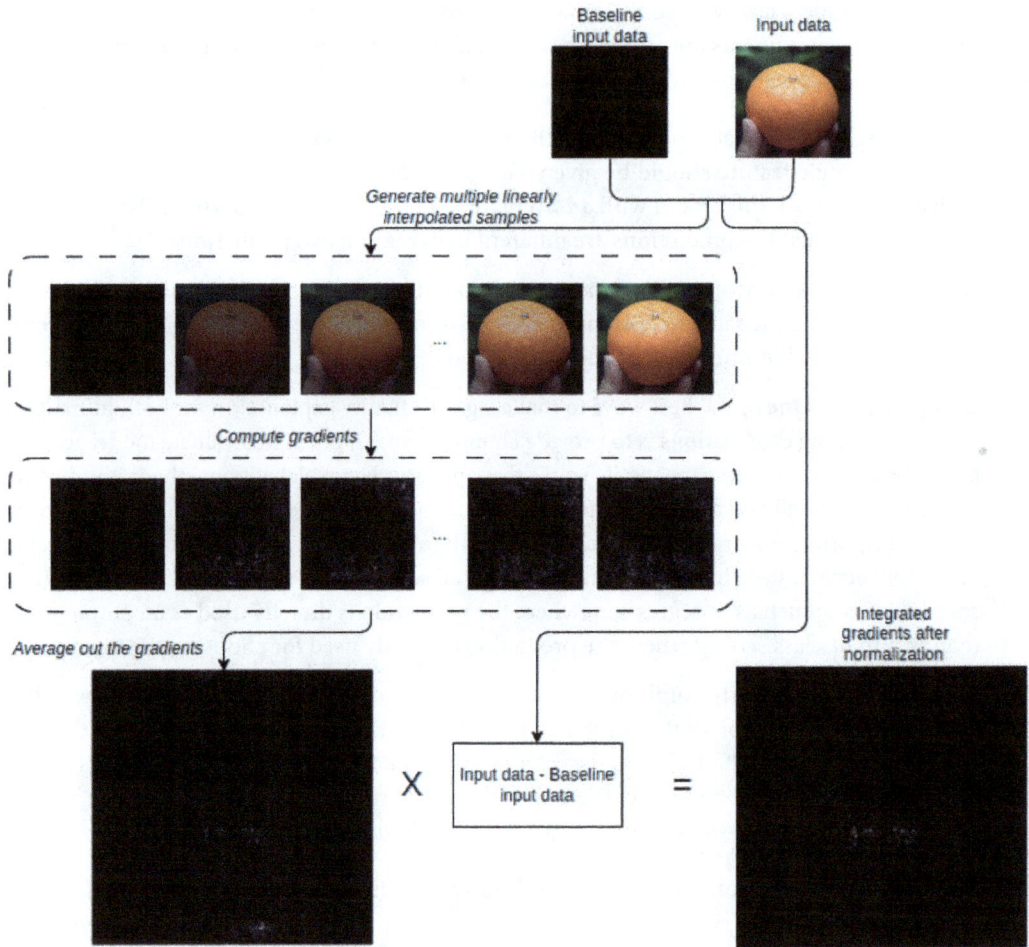

Figure 11.1 – An illustration of how integrated gradients are computed for an image of an orange with a pre-trained ResNet50 model that is capable of predicting the orange class

More intuitively, this means that integrated gradients measure the rate of change of model predictions as the target input feature's value changes toward a neutral baseline sample and provides accountable importance that is properly scaled according to the predictions:

Feature importances = prediction difference x Area under gradients along path curve

This allows it to capture the sensitivity of the model's output to the feature across its entire range of values. Integrated gradients ensure that the contribution of each feature is proportional to its effect on the model's output by aggregating the gradients from samples in the linear interpolated path between the baseline and target. Using plain gradients as the core allows integrated gradients to satisfy implementation invariance as the chain rule used to obtain partial derivatives allows gradients to be computed partially.

For integrated gradients, if a feature has twice the effect on the model's output compared to another feature, its attribution score will be twice as high because the gradients along the path will be twice as high for that feature, thus satisfying linearity. Additionally, since the importance is scaled by the difference in predictions between the baseline input and target input, integrated gradients will satisfy the completeness axiom. These are local explanations that provide reasons for predictions through individual data samples. As a bonus, by aggregating the local feature importance of all data samples through mean or median, you can obtain the global feature importance of the model.

Next, we will attempt to use integrated gradients to explain predictions from a model.

Using integrated gradients to aid in understanding predictions

At the time of writing, two packages provide easy-to-use classes and methods to compute integrated gradients, which are the `captum` and `shap` libraries. In this tutorial, we will be using the `captum` library. The `captum` library supports models from TensorFlow and PyTorch. We will be using PyTorch here. In this tutorial, we will be working on explaining a SoTA transformer model called **DeBERTA** on the task of text sentiment multiclass classification. Let's go through the use case step by step:

1. First, let's import the necessary libraries and methods:

    ```
    from transformers import (
        DebertaForSequenceClassification,
        EvalPrediction,
        DebertaConfig,
        DebertaTokenizer,
        Trainer,
        TrainingArguments,
        IntervalStrategy,
        EarlyStoppingCallback
    )
    import pandas as pd
    from sklearn.model_selection import train_test_split
    import torch
    ```

2. Next, we will import a custom text sentiment dataset made for this tutorial:

    ```
    df = pd.read_csv('text_sentiment_dataset.csv')
    ```

    This data contains 100 rows with a balanced distribution between three sentiment classes called "neutral", "positive," and "negative." We will dive into a few samples of this dataset later.

3. Now, we will define the label mapping that will be used to map the sentiment labels to a numeric ID and vice versa:

```
label2id = {"negative": 0, "neutral": 1, "positive": 2}
id2label = {v: k for k, v in label2id.items()}
```

4. Next, we will create a method that applies text pre-processing that tokenizes the text data using a pre-trained, byte-pair-encoding-based tokenizer:

```
tokenizer = DebertaTokenizer.from_pretrained("microsoft/deberta-
base")
def preprocess_function(examples):
    inputs = tokenizer(examples["Text"].values.tolist(),
padding="max_length", truncation=True)
    inputs["labels"] = [label2id[label] for label in
examples["Sentiment"].values]
    return inputs
```

5. To train a model, we need to have a cross-validation strategy, so we will use a simple train and validation split here:

```
train_df, test_df = train_test_split(df, test_size=0.2, random_
state=42)
```

6. Since we will be using the PyTorch model from the Hugging Face Transformers library, we need to use the PyTorch dataset format and split the data that will be used to train the model. Here, we will define the PyTorch dataset and initialize both the training and validation dataset instances:

```
class TextClassificationDataset(torch.utils.data.Dataset):
    def __init__(self, examples):
        self.examples = examples    def __getitem__(self,
index):
        return {k: torch.tensor(v[index]) for k, v in self.
examples.items()}    def __len__(self):
        return len(self.examples["input_ids"])train_dataset =
TextClassificationDataset(preprocess_function(train_df))
test_dataset = TextClassificationDataset(preprocess_
function(test_df))
```

7. Now that the dataset has been pre-processed and is ready to be used for training, let's load our randomly initialized DeBERTA model from Hugging Face:

```
deberta_config = {
        "model_type": "deberta-v2",
        "attention_probs_dropout_prob": 0.1,
        "hidden_act": "gelu",
        "hidden_dropout_prob": 0.1,
        "hidden_size": 768,
```

```
                "initializer_range": 0.02,
                "intermediate_size": 3072,
                "max_position_embeddings": 512,
                "relative_attention": True,
                "position_buckets": 256,
                "norm_rel_ebd": "layer_norm",
                "share_att_key": True,
                "pos_att_type": "p2c|c2p",
                "layer_norm_eps": 1e-7,
                "max_relative_positions": -1,
                "position_biased_input": False,
                "num_attention_heads": 12,
                "num_hidden_layers": 12,
                "type_vocab_size": 0,
                "vocab_size": 128100
    }
    model_config = DebertaConfig(id2label=id2label,
    label2id=label2id, **deberta_v3_config)
    model = DebertaForSequenceClassification(model_config)
```

8. Since this is a multiclass setup, we will use the accuracy metric that will be computed at each epoch, along with the cross-entropy loss, which will be used to train the model:

```
def compute_metrics(p: EvalPrediction):
    preds = p.predictions[0] if isinstance(p.predictions, tuple)
else p.predictions
    preds = np.argmax(preds, axis=1)
    return {"accuracy": (preds == p.label_ids).astype(np.
float32).mean().item()}
```

9. Now, we will define the training arguments:

```
training_args = TrainingArguments(
    output_dir="./results",
    num_train_epochs=1000,
    per_device_train_batch_size=8,
    per_device_eval_batch_size=16,
    warmup_steps=100,
    weight_decay=0.01,
    logging_dir="./logs",
    logging_steps=10,
    evaluation_strategy=IntervalStrategy.EPOCH,
    save_strategy=IntervalStrategy.EPOCH,
    load_best_model_at_end=True,
    learning_rate=0.000025,
```

```
        save_total_limit=2,
    )
```

This will save the checkpoints in the `results` folder, train the model with 100 iterations of warmup, load the best model at the end of training, and set the model learning rate to a very small number of 0.000025, which is required to properly converge to a solution.

10. Next, we will initialize the Hugging Face `trainer` instance, which will take in the training arguments and use them to execute the actual training process:

```
trainer = Trainer(
    model=model,
    args=training_args,
    train_dataset=train_dataset,
    eval_dataset=test_dataset,
    compute_metrics=compute_metrics,
    callbacks = [EarlyStoppingCallback(early_stopping_
patience=20)]
    )
```

Notice that an early stopping of 20 iterations is used. This stops the model from being trained when the model doesn't improve on the validation partition for the specified iterations, along with 1,000 epochs of training.

11. Finally, let's train and print the final best model evaluation score!

```
trainer.train()
trainer_v2.evaluate()
```

This will result in the following output:

```
{'eval_loss': 0.1446695625782013,
 'eval_accuracy': 0.9523809552192688,
 'eval_runtime': 19.497,
 'eval_samples_per_second': 1.077,
 'eval_steps_per_second': 0.103}
```

12. 95.23% accuracy is a pretty good score for a multiclass model. You might think that the model is good enough to be deployed, but is it more than what meets the eye? Let's investigate the model's behavior through prediction explanations using integrated gradients and see whether the model is performing. First, let's define the explainer instance from the `transformers_ interpret` library:

```
from transformers_interpret import
SequenceClassificationExplainer
cls_explainer = SequenceClassificationExplainer(model,
tokenizer)
```

The `transformers_interpret` library uses the `captum` library under the hood and implements methods and classes to make it straightforward to explain and visualize text importance based on models made using the Hugging Face Transformers library. Among all these things, we're mapping the importance score to a color code and mapping token IDs back to the actual token string.

13. Next, we'll explain two samples for the negative and positive labels from the validation partition, which is conveniently located in the first three indices:

```
for idx in [0, 8, 6, 12]:
    text = test_df['Text'].iloc[idx]
    label = test_df['Sentiment'].iloc[idx]
    word_attributions = cls_explainer(text)
    cls_explainer.visualize()
```

This will result in the output displayed in *Figure 11.2*:

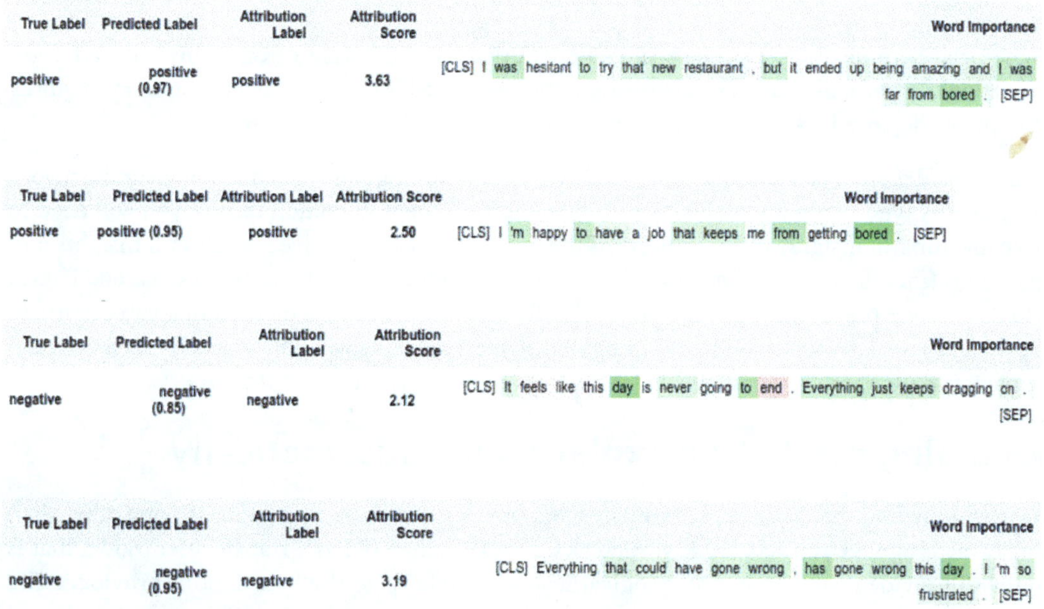

True Label	Predicted Label	Attribution Label	Attribution Score	Word Importance
positive	positive (0.97)	positive	3.63	[CLS] I was hesitant to try that new restaurant . but it ended up being amazing and I was far from bored . [SEP]

True Label	Predicted Label	Attribution Label	Attribution Score	Word Importance
positive	positive (0.95)	positive	2.50	[CLS] I 'm happy to have a job that keeps me from getting bored . [SEP]

True Label	Predicted Label	Attribution Label	Attribution Score	Word Importance
negative	negative (0.85)	negative	2.12	[CLS] It feels like this day is never going to end . Everything just keeps dragging on . [SEP]

True Label	Predicted Label	Attribution Label	Attribution Score	Word Importance
negative	negative (0.95)	negative	3.19	[CLS] Everything that could have gone wrong , has gone wrong this day . I 'm so frustrated . [SEP]

Figure 11.2 – transformers_interpret-based integrated gradients results on the validation dataset

The words highlighted in green represent a positive attribution, while those highlighted in red represent a negative attribution toward the predicted label class. The darker the green, the stronger the positive attribution toward the predicted label class. The darker the red, the stronger the negative attribution toward the predicted label class in general.

14. The first and second examples show correctly predicted positive sentiment sentences. For the first example, based on common human sense, *amazing* is the word that should contribute the most to positive attribution. For the second example, *happy* should be the word that is emphasized as the word contribution toward positive sentiment. However, for both examples, the word *bored* is used instead as a strong indicator of the positive sentiment prediction, which is not the behavior that we want. This indicates that the dataset has a bias where the word *bored* could be present in all samples labeled with positive sentiment.

15. The third and fourth examples show a correctly predicted negative sentiment. In the third example, the sentences *never going to end* and *keeps dragging on* should be the focus of the negative sentiment prediction. In the fourth example, the word *wrong* and the phrases *has gone wrong* and *so frustrated* should be the focus of the negative sentiment prediction. However, both samples consistently show dependence on the word *day* to get their negative sentiment prediction. This indicates that the dataset has a bias regarding the word *day* occurring frequently in the samples that are labeled with negative sentiment.

16. All of this means that either the data has to be prepared all over again or more data that diversifies the distribution of word usage should be added to the dataset so that a proper model can be built.

This tutorial demonstrates a single type of benefit that can be derived from utilizing prediction explanations. In particular, this shows the case where the model is not capturing the behavior required to be reliably deployed, even when the accuracy-based metric has a good score.

The integrated gradients technique is a flexible technique that can be applied to any neural network model for any type of input variable type. To benefit from the explanation results, you will need to derive meaningful insights in the context of the business goal. To derive meaningful insights and conclusions from the results of integrated gradients, it's essential to apply common sense and logical reasoning, as was presented in the tutorial manually. However, there is a method that you can use to try to obtain assistance in obtaining meaningful insights, especially when there is too much data and too many variations to decode manually. We will dive into this in the next section.

Explaining prediction explanations automatically

One useful method that helps in deriving insights from prediction explanations is none other than using **large language models** (**LLMs**) such as ChatGPT. ChatGPT is a transformer model that is trained to provide results that match logical reasoning related to the instructions provided. The theory here is that if you can format your prediction explanations data in a way that it can be fed into a transformer model, and instruct the LLM to derive insights from it, you will be able to obtain insights from multiple different perspectives.

In the previous tutorial, we attempted to explain the explanations of four different samples consisting of two correctly predicted positive sentiment examples and two correctly predicted negative sentiment examples. Now, let's use an LLM model to gain insights. Here, we will separately generate insights for the two correctly identified positive sentiment and two correctly identified negative sentiment examples as it somehow tends to provide more reasonable results compared to just plugging in all four examples. We used the following prompt:

```
"I have built a sentiment classification model using HuggingFace
defined classes, with 3 labels "neutral", "positive", "negative".
I also utilized the integrated gradients method to obtain
token attributions for the examples below: First Example with
correctly predicted positive sentiment [('[CLS]', 0.0), ('I',
0.04537756915960333), ('was', 0.32506422578050986), ('hesitant',
-0.011328562581410014), ('to', 0.2591512168119563), ('try',
0.07210600939837246), ('that', 0.1872696259797412), ('new',
0.25720454853718405), ('restaurant', 0.09906425532043227),
(',', -0.08526821845533564), ('but', 0.26155394517459857),
('it', 0.029711535733103797), ('ended', 0.03380972261608883),
('up', 0.10830705227933887), ('being', 0.09771634151976362),
('amazing', 0.08913346633282519), ('and', 0.1429717885722044),
('I', 0.3408664658447834), ('was', 0.3248060076552883),
('far', 0.11463247647745932), ('from', 0.29795241654484333),
('bored', 0.4944594024849995), ('.', 0.10894853119482743),
('', 0.0376325816741505), ('[SEP]', 0.0)] Second example with
correctly predicted positive sentiment [('[CLS]', 0.0), ('I',
0.00293608614786346), ("'m", 0.30945414416090317), ('happy',
-0.028173800615734133), ('to', 0.31870826361945737), ('have',
0.15675763188868433), ('a', 0.07581024366375723), ('job',
-0.020856788304256305), ('that', 0.20868353147138577), ('keeps',
0.2024940149527261), ('me', 0.046306584007035426), ('from',
0.41459097144711715), ('getting', 0.059558828031696716),
('bored', 0.7089862879353616), ('.', 0.0744570784079903), ('',
-0.02775527287562979), ('[SEP]', 0.0)], Tell me anything wrong about
this model to predict positive and negative attributions. Provide me
a complete analysis and see if the most highly attributed words make
sense for its predicted label. Judge the models based on the relative
score and focus on the highest scored token individually and see
whether it make sense."
```

Note that the list of word tokens and attribution tuples are the contents of the word_attributions variable in the previous tutorial. For this prompt, the ChatGPT model returned the following response without modification:

Based on the given examples, it seems that the model is performing well in predicting positive sentiment, as the predicted label is correct in both cases. However, it is important to note that the token attributions should be interpreted with caution, as they are not always a reliable indicator of the importance of each word in the sentence.

Looking at the token attributions for the first example, we can see that the highest attributed word is "bored" with a score of 0.494. This might seem odd for a positive sentiment sentence, but it is possible that the model is attributing high importance to this word because it is negating a negative sentiment that was initially present. The model might have also learned that overcoming boredom is something that people consider to be positive. The second highest attributed words are "I" and "was", both with scores above 0.3, which is reasonable since they indicate the writer's

personal experience and opinion. The other highly attributed words such as "to", "from", and "new" also seem to make sense in the context of the sentence.

Looking at the token attributions for the second example, we can see that the highest attributed word is "bored" again, with a score of 0.709. This could be because the model is placing a lot of emphasis on the fact that the job keeps the writer from getting bored, which is seen as a positive thing. The second highest attributed word is "from" with a score of 0.415, which makes sense in the context of the sentence since it is indicating what the job is keeping the writer from doing. Other highly attributed words such as "to" and "that" also seem to make sense in the context of the sentence.

Overall, while the model is performing well in predicting positive sentiment, it is important to note that token attributions can be noisy and should be interpreted with caution. The highly attributed words seem to make sense in the context of the sentences, but it is important to look at the entire sentence and not just individual words to fully understand the sentiment.

This response provides a different take on our original insight and mentions the possibility that nothing is wrong with some skepticism. Overall, this is a nice response that gives us a different perspective on what could happen and signals that an LLM can be used to explain the results of prediction explanation as an assistance tool. Now, let's replace the attribution results with the two examples that are correctly identified as negative sentiments with the following list:

```
Third example with correctly predicted negative sentiment
[('[CLS]', 0.0), ('It', 0.1781266370024602), ('feels',
-0.08200280620716707), ('like', -0.03204718164120957), ('this',
-0.05573060889100964), ('day', 0.6176120940811932), ('is',
-0.0032201843949800626), ('never', 0.2051856944424909), ('going',
0.07198217726867306), ('to', 0.45876287617527467), ('end',
-0.2538220352877727), ('.', -0.01766876291225962), ('Everything',
0.2927810551003227), ('just', 0.27789343264100713), ('keeps',
0.25860811147952395), ('dragging', 0.013653081036146358),
('on', 0.12974700461873728), ('.', -0.01804386178839869), ('',
0.07940230083266143), ('[SEP]', 0.0)]
Fourth example with correctly predicted negative sentiment
[('[CLS]', 0.0), ('Everything', 0.051921060773658365),
('that', 0.1853413631332932), ('could', 0.16644341050771988),
('have', 0.09960723832267326), ('gone', 0.21400474006976702),
('wrong', 0.26952621495013096), (',', -0.02550525756341266),
('has', 0.46511206263887517), ('gone', 0.21385297386310367),
('wrong', 0.2691861963676513), ('this', 0.08573245861889027),
('day', 0.5967441822322336), ('.', 0.00910580036050777), ('I',
0.16215574709799985), ("'m", -0.05665909043004983), ('so',
0.20818789342264885), ('frustrated', 0.11472844834437185), ('.',
0.00924431805079032), ('', 0.1485924275199782), ('[SEP]', 0.0)]
```

The results of the ChatGPT model provided the following response:

From the provided examples, it seems like the model is correctly predicting the negative sentiment in both cases. However, there are a few things that could be analyzed to determine whether the model is performing well or not.

Firstly, the token attributions provided by the integrated gradients method indicate the contribution of each token towards the final prediction of the model. In both examples, the tokens with the highest positive attribution scores are "day" and "has gone wrong", which seems to make sense as they are indicative of negative sentiment.

However, one issue with the provided token attributions is that some of the tokens have negative attribution scores, which means that they are contributing negatively towards the final prediction. For example, in the first example, the token "feels" has a negative attribution score, but it could be argued that it is actually indicative of negative sentiment. Similarly, in the second example, the token "'m" has a negative attribution score, but it is a contraction of "I am", which is indicative of negative sentiment.

Another thing to consider is the relative score of the highest attributed tokens. In the first example, the token "day" has a significantly higher attribution score than any other token, which indicates that it is the most important feature for predicting negative sentiment in this case. However, in the second example, there are multiple tokens with similarly high attribution scores, such as "has gone wrong" and "frustrated", which suggests that they are both equally important features for predicting negative sentiment.

Overall, based on the provided examples and token attributions, it seems like the model is performing reasonably well in predicting negative sentiment. However, the analysis of the token attributions suggests that there may be some issues with the way the integrated gradients method is attributing scores to certain tokens. It is also important to consider the relative score and interpretability of the highest attributed tokens to determine whether they make sense in the context of the predicted label.

Both of ChatGPT's responses seem to signal that it could be a valid behavior. Since most neural networks are configured to be non-linear models with many non-linear activations, remember that the model is capable of learning complex interactions between tokens. This means that tokens are mostly perceived by the model in a complex way that might make them depend on one another. A token that has a high attribution in the original context when applied to another context might not produce the same attribution as multiple tokens can depend on each other as a prerequisite to produce a strong attribution. Additionally, tokens are not restricted to interacting with tokens that are directly beside them and can be in any position of a text row. One way to verify whether the token is overfitted is by predicting on text data that only has the day token to see whether it will predict it as a negative sentiment and the bored token to see whether it will predict it as a positive sentiment:

```
from transformers import TextClassificationPipeline
full_model = TextClassificationPipeline(model=model,
```

```
tokenizer=tokenizer)
print(full_model("day"))
print(full_model("day day"))
print(full_model("day day day"))
print(full_model("day day day day"))
print(full_model("bored"))
print(full_model("bored bored"))
print(full_model("bored bored bored"))
print(full_model("bored bored bored bored"))
```

This will return the following response:

```
[{'label': 'neutral', 'score': 0.9857920408248901}]
[{'label': 'negative', 'score': 0.5286906361579895}]
[{'label': 'negative', 'score': 0.9102224111557007}]
[{'label': 'negative', 'score': 0.9497435688972473}]
[{'label': 'neutral', 'score': 0.9852404594421387}]
[{'label': 'positive', 'score': 0.8067558407783508}]
[{'label': 'positive', 'score': 0.974130392074585}]
[{'label': 'positive', 'score': 0.9776705503463745}]
```

It is interesting to see that a single token for both day and bored will result in the prediction being a neutral sentiment. The more repetition tokens you add, the more the prediction will skew toward negative and positive sentiment, respectively. This proves that the model is indeed biased toward these specific words and is not using the words in the right way, as ChatGPT said. Note that it could very much be what ChatGPT predicted for your case.

This tutorial showcases the explanations of text-based neural networks. However, it can be applied similarly to other data types, such as numerical data, categorical data, or any data that can be reliably represented in text format. Specifically for image-based prediction explanations, you can use models such as GPT-4 from OpenAI, which accepts image data along with text data. You can also consider using the visualization from the transformers_interpret library as an image and feed it into GPT-4!

Next, we will dive into common pitfalls when trying to explain your predictions and recommendations to avoid these pitfalls.

Exploring common pitfalls in prediction explanations and how to avoid them

Although prediction explanations have proven to be valuable tools in understanding AI models, several common pitfalls can hinder their effectiveness. In this section, we will discuss these pitfalls and provide strategies to avoid them, ensuring that prediction explanations remain a valuable resource for understanding and improving AI models. Some of the common pitfalls, along with their solutions, are as follows:

- **Over-reliance on explanations**: While prediction explanations can provide valuable insights into a model's decision-making process, over-relying on these explanations can lead to incorrect conclusions. It's important to remember that prediction explanations are just one piece of the puzzle and should be used in conjunction with other evaluation methods to gain a comprehensive understanding of a model's performance. The solution here is to use a combination of evaluation methods, including performance metrics, cross-validation, and expert domain knowledge, to analyze and validate a model's performance.

- **Misinterpreting explanation results**: Interpreting prediction explanation results can be challenging, particularly when dealing with complex models and large datasets. Misinterpretation of these results can lead to incorrect conclusions about a model's behavior and performance. The solution here is to collaborate with domain experts to help interpret explanation results and ensure that conclusions drawn from these explanations align with real-world knowledge and expectations.

- **Ignoring model limitations**: Prediction explanations can provide valuable insights into a model's decision-making process, but they cannot address the inherent limitations of the model itself. It's essential to acknowledge and address these limitations to ensure that the model performs optimally. The solution here is to conduct thorough model evaluations to identify and address any limitations, such as overfitting, underfitting, or biased training data. Continuously reevaluate and update the model as needed to maintain optimal performance.

- **Explanations for the wrong audience**: Different stakeholders may require different types of explanations based on their expertise and needs. Providing explanations that are too technical or too simplistic for the intended audience can hinder their understanding and use of the predictions. The solution is to tailor explanations to the needs and expertise of the target audience. Sometimes, a more global explanation of the model is needed instead of the per-prediction explanation, in which case you can consider using neural network interpretation techniques and aggregated evaluation metrics. For non-technical users, explaining predictions by providing raw feature importance isn't enough and requires a natural language explanation of the insights obtained through feature importance, as introduced in the *Explaining prediction explanations automatically* section. For a technical-aware user, prediction explanations can be suitable.

By being aware of these common pitfalls and implementing strategies to avoid them, practitioners can ensure that prediction explanations remain a valuable tool for understanding and improving AI models. By combining prediction explanations with other evaluation methods and collaborating with domain experts, it's possible to gain a comprehensive understanding of a model's performance, behavior, and limitations, ultimately leading to more accurate and reliable AI systems that better serve their intended purpose.

Summary

In this chapter, we gained a broad view of the prediction explanations landscape and dived into the integrated gradients technique, applied it practically to a use case, and even attempted to explain the integrated gradients results manually and automatically through LLMs. We also discussed common pitfalls in prediction explanations and provided strategies to avoid them, ensuring the effectiveness of these explanations in understanding and improving AI models.

Integrated gradients is a useful technique and tool to provide a form of saliency-based explanation of the predictions that your neural network makes. The process of understanding a model through prediction explanations provides many benefits that can help fulfill the criteria required to have a successful machine learning project and initiative. Even when everything is going well and the machine learning use case is not critical, uncovering the model's behavior that you will potentially deploy through any prediction explanations technique can help you improve how your model is utilized.

However, saliency-based explanations only allow you to understand which input data or input neurons are important. But what patterns a neuron is capable of detecting from the input data remains unknown. In the next chapter, we will expand on this direction and uncover techniques to understand exactly what a neuron detects.

By being aware of common pitfalls and implementing strategies to avoid them, practitioners can ensure that prediction explanations remain a valuable tool for understanding and improving AI models. Combining prediction explanations with other evaluation methods and collaborating with domain experts can lead to more accurate and reliable AI systems that better serve their intended purpose.

Before we wrap up, as a final word, be sure to allocate some time to understand your neural network model through prediction explanation techniques such as integrated gradients and consider the potential pitfalls to maximize their effectiveness!

Further reading

- Mukund Sundararajan, Ankur Taly, Qiqi Yan. *Axiomatic Attribution for Deep Networks*, In: International Conference on Machine Learning. 2017. URL: `https://arxiv.org/abs/1703.01365v2`.

12
Interpreting Neural Networks

When trying to comprehend the reasons behind a model's prediction, local per-sample feature importance can be a valuable tool. This method enables you to focus your analysis on a smaller part of the input data, resulting in a more targeted understanding of key features that contributed to the model's output. However, it is often still unclear which patterns the models are using to identify highly important features. This issue can be somewhat circumvented by reviewing more prediction explanations from targeted samples meant to strategically discern the actual reason for the prediction, which will also be introduced practically later in this chapter. However, this method is limited to the available number of samples you must validate your model against, and it can sometimes still be difficult to pinpoint the pattern used concretely.

Deep neural networks (DNNs) learn low- to high-level features that help the prediction layer discern the right label under the hood. When we use local feature importance-based explanations on the input, we can't know for sure which low-, medium-, or high-level patterns contributed to the importance of the input data. For images, this would range from low-level features, such as simple shapes, to medium-level features, such as the silhouette shape of a human body, all the way to a combination of patterns that build up to become a human face or everyday objects. For text, this would range from low-level features such as word embeddings, which represent the meaning of a word, to medium-level features such as the semantic roles of words in a sentence that enable the meaning of the text to be represented properly such as sentence embeddings, all the way to high-level features we are more familiar with, such as topics and sentiment. Of course, these are merely theoretical assumptions on what we think NNs are learning.

In this chapter, we will explore a method that can help to clear all the ambiguity in the features learned by a deep neural network, which is to visualize the patterns an NN detects directly through input optimization. By visualizing the patterns learned directly in combination with the filtering of activations, we can shed light on the actual reasons a deep neural network makes its predictions. Specifically, the following topics will be discussed:

- Interpreting neurons
- Finding neurons to interpret

- Interpreting learned image patterns

- Discovering the counterfactual explanation strategy

Technical requirements

This chapter includes practical implementation in the Python programming language. To complete it, you will need to have a computer with the following libraries installed:

- `torchvision`

- `torch`

- `torch-lucent==0.1.8`

- `matplotlib==3.3.0`

- `captum`

- `pillow`

- `numpy`

The code files are present on GitHub: `https://github.com/PacktPublishing/The-Deep-Learning-Architect-Handbook/tree/main/CHAPTER_12`.

Interpreting neurons

Neurons in NN layers produce features that will be consumed by subsequent layers. The features or activations produced are simply an indicator of how prominent a learned pattern is in the input data. But have you ever wondered what the patterns are? Decoding the actual patterns learned by the NN can further improve the transparency needed to achieve the goals mentioned in the *Exploring the value of prediction explanations* section of *Chapter 11, Explaining Neural Network Predictions*.

Data is composed of many complicated patterns combined into a single sample. Traditionally, to discern what a neuron is detecting, much input data has to be evaluated and compared against other data so that a qualitative conclusion can be made by humans, which is both time-consuming and hard to get right. This method allows us to pinpoint the actual pattern that causes a high activation value visually, without the disturbance of other highly correlated patterns.

More formally, feature visualization by optimization can be useful in the following use cases:

- Understanding the patterns associated with confusing labels without help from a domain expert:

  - This is more prevalent in real-world audio data where the sound of the label in the real data can often be mixed together with lots of noises

  - This can also happen in image data

- It is not straightforward or possible to obtain real data to test any hypothesis on what the NN learned that can't be proven with gradient-based feature attribution techniques on available data

The core of the neuron interpretation technique is **neural input optimization**, which is a process that modifies the input data of the NN to activate highly on the chosen neuron. Remember that during the training process, we optimize the weights of the NNs toward reducing the loss value. In this technique, we randomly initialize an input and optimize the input data to activate highly on chosen neurons, effectively treating the input data as NN weights. Gradients can be naturally computed to the input data stage, making it possible to update the input data according to the computed gradients after applying a learning rate. This technique also allows you to jointly optimize multiple neurons to activate highly and obtain an image that shows the patterns of how two different neurons can coexist.

Figure 12.1 showcases the idea of low-level to medium-level and high-level patterns in **convolutional NNs (CNNs)** with examples of optimized image input data using random low-, mid-, and high-level convolutional filters in the ImageNet pre-trained `efficientnet-b0` model:

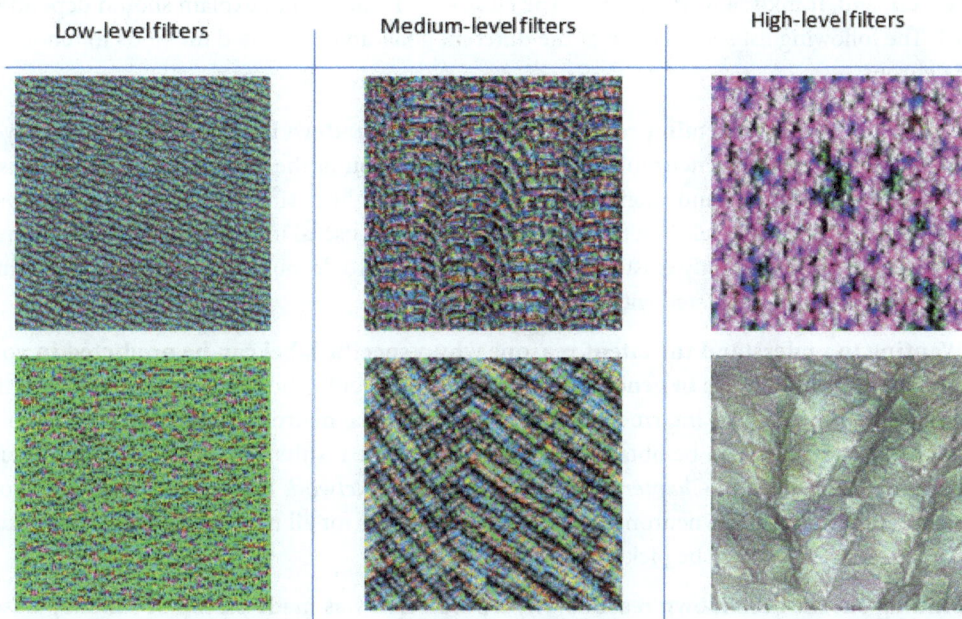

Figure 12.1 – Example of optimized images from random filters in the efficientnet-b0 model

If you look at the high-level filter patterns, the first optimized image on a random filter looks somewhat like flowers, and the second image looks like leaf patterns.

However, a main caveat with this technique is that the resulting optimized input data may not represent all the real-life variations of a pattern associated with a neuron. Even for a dynamic input data variable such as an image, which can be optimized to present the pattern in diverse ways, the resulting optimized input can still miss out on some representations of the pattern. A good approach to tackle this caveat is to first obtain the initial optimized input data variant and then execute subsequent optimizations and ensure the optimized input data will be different from the initial variant. This can be done by jointly optimizing an additional component – the negative cosine similarity between the initial optimized input data and the current input data being optimized. This technique helps to generate diverse input data examples. But before you can optimize the input data and attempt to interpret a neuron, you need a strategy to choose the best neurons to optimize input data against, which will be discussed in the next section.

Finding neurons to interpret

With millions and billions of neurons in today's SoTA architectures, it's impossible to interpret every single neuron, and, frankly, a waste of time. The choice of the neuron to explain should depend on your goal. The following list shows some of the different goals and associated methods for choosing suitable neurons:

- **Finding out what a certain prediction label or class pattern looks like**: In this case, you should simply choose a neuron specific to the prediction of the target label or class. This is usually done to understand whether the model captured the desired patterns of the class well, or whether it learned irrelevant features. This can also be useful in multilabel scenarios where multiple labels always only exist together, and you want to decouple the labels to understand the input patterns associated with a single label better.

- **Wanting to understand the latent reasons why a specific label can be predicted in your dataset, or for all labels in general**: In this case, you should choose the top most impactful neurons from the latent intermediate layers from a global neuron importance score. Global neuron importance can be obtained by aggregating the results of the integrated gradients method (introduced in *Chapter 11, Explaining Neural Network Predictions*) applied to your validation dataset for all neurons. The importance values for all neurons can then be ranked, and the top neurons can be picked.

- **Finding out the breakdown reasons why a prediction was made on top of saliency-based explanation techniques**: In this case, you should choose the neuron that has the highest activation value and highest importance score. A neuron that is activated highly does not necessarily mean that it is important for a certain prediction. Additionally, a neuron that is important does not mean that the neuron is activated. Using both the integrated gradients' importance value and the activation values to obtain the most important neuron will help to make sure the neurons you care about are chosen. Additionally, you can further filter out more neurons if you have a focus area based on the initial input data saliency map by only choosing neurons that affect the chosen focus area.

- **Understanding the interactions between multiple labels or classes**: In scenarios where the relationships between multiple labels or classes are important, you can choose neurons that capture these interactions. Identify neurons that are highly activated and have high importance scores when multiple labels or classes are predicted together. Analyzing these neurons can help you understand how the model captures the relationships between different labels or classes and may reveal potential areas for improvement.

- **Investigating the robustness of the model to adversarial attacks**: In this case, you should choose neurons that are sensitive to adversarial perturbations in the input data. You can generate adversarial examples, with more info on how to do so in *Chapter 14, Analyzing Adversarial Performance*, and then compute the neuron importance scores using techniques such as integrated gradients. By visualizing neurons that are most affected by adversarial perturbations, you can gain insights into the model's vulnerabilities and explore potential defenses.

- **Exploring the hierarchical structure of learned features**: In this case, you should choose neurons from different layers of the NN to understand how the model learns hierarchical features. Select neurons from early layers to investigate low-level features, and from deeper layers to investigate high-level features. You can also select multiple neurons to co-optimize the input data for high activation to understand how multiple-neuron learned patterns can exist in the same input data. Visualizing these neurons can help you understand the model's internal representation of the data and how it builds increasingly complex features. This can provide insights into the model's learning process and potential areas for improvement.

- **Analyzing the model's generalization capabilities across different datasets**: To understand how well the model generalizes to new data, you should choose neurons that are consistently important across different datasets. Calculate the neuron importance scores using techniques such as integrated gradients for different datasets, and identify neurons that maintain high importance scores across all datasets. By visualizing these neurons, you can gain insights into the model's generalization capabilities and identify potential areas for improvement.

Now that we've established the method to choose a neuron for interpretation, let's start with a practical exploration of interpreting neurons with image input data!

Interpreting learned image patterns

Interpreting NNs that take in image data enables a new paradigm in interpretation, which is the capability to visualize exactly what a neuron is detecting. In the case of audio input data, interpreting NNs would allow us to audibly represent what a neuron is detecting, similar to how we visualize patterns in image data! Choose neurons you want to understand based on your goal and visualize the patterns it is detecting through iterative optimizing on image data to activate highly for that neuron.

Practically, however, optimizing image data based on a neuron has an issue where the resulting image often produces high-frequency patterns that are perceived to be noisy, uninterpretable, and unaesthetic. High-frequency patterns are defined to be pixels that are high in intensity and change quickly from one to the next. This is largely due to the mostly unconstrained range of values that a pixel can be represented by, and pixels in isolation are not the semantic units we care about. Zooming in on the resulting image might make it more interpretable, but the interpretation effectiveness is diminished with the need to perform human evaluation and extra work.

This issue can be effectively mitigated practically through the following techniques:

- Frequency penalization – example techniques are as follows:

  - Randomly blurring the image during optimization using a bilateral filter, which has the benefit of also preserving edge patterns

  - Penalizing variation between neighboring pixels conservatively in the optimization

- Image augmentations

- Image preprocessing – example techniques are as follows:

  - Data decorrelation

  - Fast Fourier transform

Let's continue our journey here by going through a practical tutorial using a pre-trained 121-layer densenet model on the ImageNet dataset.

Explaining predictions with image input data and integrated gradients

In this section, we will explorepredictions, explaining withpredictions, explaining with a practical tutorial on explaining predictions from a CNN model that takes in image input data with integrated gradients, providing some insight into the reasons the model made its prediction. In this tutorial, we will discover what answers we need that are missing from prediction explanations, which will set us up for interpreting the CNN model. We'll proceed as follows:

1. We will be using the `lucent` library for this tutorial, which provides methods to interpret NNs through feature visualization by optimization. Additionally, we will be using the `torch` library for the densenet model. We will also be using the `captum` library to use the integrated gradients method. Let's start by importing all the necessary libraries:

    ```
    import glob
    import numpy as np
    import torch
    import torch.nn as nn
    import torchvision.transforms as transforms
    from PIL import Image
    from captum.attr import IntegratedGradients
    ```

```
from captum.attr import NoiseTunnel
from captum.attr import visualization as viz
from lucent.optvis import render, param, objectives
from lucent.optvis.objectives import diversity
```

2. Next, we will define the model class for the pre-trained densenet model that we will use:

```
class CNN(nn.Module):
  def __init__(self, num_classes, model='resnet50'):
    super(CNN, self).__init__()
    self.num_classes = num_classes
    self.chosen_model = model
    if self.chosen_model=='densenet121':
      self.model = models.densenet121(pretrained=True)
      self.classifier = nn.Sequential(
        nn.Dropout(p=0.1),
        nn.Linear(self.model.classifier.in_features, 256,
bias=False),
        nn.ReLU(),
        nn.BatchNorm1d(256),
        nn.Linear(256, 128, bias=False),
        nn.ReLU(),
        nn.BatchNorm1d(128),
        nn.Linear(128, self.num_classes, bias=False),
        nn.BatchNorm1d(self.num_classes),
      )
      self.model.classifier = self.classifier
      model_parameters = filter(lambda p: p.requires_grad, self.
model.parameters())
      params = sum([np.prod(p.size()) for p in model_
parameters])
  def forward(self, x):
    return self.model(x)
```

3. We will be using the defined model class to load weights pre-trained on an image dataset called HAM10000 with seven different skin lesion classes:

```
checkpoint = torch.load('0.8228 checkpoint.pt', map_
location=torch.device('cpu'))
model = checkpoint['model']
```

4. The seventh index of the prediction layer of the pre-trained model is trained to predict melanoma, which is a type of skin cancer. Let's take a look at a few examples of melanoma from the ISIC-2017 dataset and see what exactly the pre-trained model is focusing on when predicting these images. We will be using the integrated gradients method from the captum library. First, let's

define the preprocessing needed to allow model inferencing, which converts a numpy image array into torch tensors, resizes the image into the pre-trained image size, and normalizes it:

```
resize_transform = transforms.Compose([
    transforms.ToTensor(),
    transforms.Resize((224, 224)),
])
norm_transform = transforms.Normalize(
    mean=[0.485, 0.456, 0.406],
    std=[0.229, 0.224, 0.225])
```

The mean and standard deviation values are derived directly from the ImageNet dataset and were used to pre-train the model, as prior to pre-training on the HAM10000 dataset, it was pre-trained on ImageNet. Additionally, the 224 dimension is also adopted from the ImageNet pre-trained settings. As we need the intermediate result after resizing separately, we defined the logic for resizing and normalization separately.

5. Next, we will use the glob library to load all the melanoma images in the provided dataset folder:

```
all_melanoma_images = glob.glob("lesions_augmented_data/Data/
test/Melanoma/*.jpg")
```

6. We will be using the captum library implementation of integrated gradients and noise tunneling to smooth out the resulting attribution noise. Let's define the instances needed to execute these components, along with defining the prediction class index for the Melanoma target class that we are interested in:

```
integrated_gradients = IntegratedGradients(model)
noise_tunnel_applyer = NoiseTunnel(integrated_gradients)
prediction_label_index = 6
```

7. We can now loop through the first six images, apply the preprocessing, apply the integrated gradients method from captum, and finally, visualize the original image and the obtained input importance heat map:

```
for melanoma_image in all_melanoma_images[:6]:
  pil_image = Image.open(melanoma_image)
  img = np.array(pil_image)
  transformed_image = resize_transform(img)
  input =  norm_transform(transformed_image).unsqueeze(0)
  attributions_ig_nt =  noise_tunnel_applyer.attribute(
    input, nt_samples=10, nt_type='smoothgrad_sq',
    target= prediction_label_index)
  _ = viz.visualize_image_attr_multiple(    np.
```

```
transpose(attributions_ig_nt.squeeze().cpu().detach().numpy(),
(1,2,0)),
np.transpose(transformed_image.squeeze().cpu().detach().numpy(),
(1,2,0)),
        ["original_image", "heat_map"],
        ["all", "positive"],
        cmap='turbo',
        show_colorbar=True)
```

This will show visualizations presented in *Figure 12.2*:

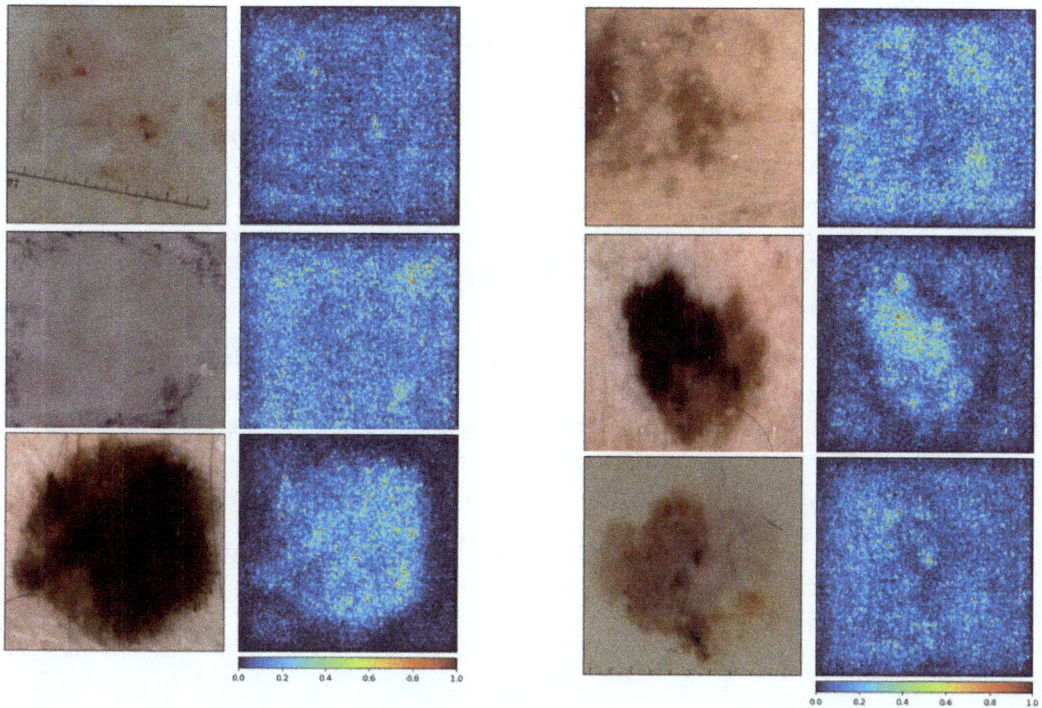

Figure 12.2 – Six real images of melanoma from the ISIC-2017 dataset along with the gradient-based attribution of the model

The model seems to be mainly focusing on the darker spots in the first five examples, but the model still considers the surrounding skin, although with much less focus. This might be a signal that the model can depend on the surrounding skin slightly to predict melanoma. But this begs the question: Is the model identifying the darkness of the skin for melanoma, or is it identifying some sort of pattern under the hood, or is it both? For the last example, the model seems to be all over the place and not really focusing on the dark spots. This could mean that the skin has patterns that are related to melanoma that are not necessarily darker in color. There are a few more questions that can't really be answered through these examples, which are as follows:

- Is the model dependent on the color of the skin to predict melanoma? Or is it really about the pattern?

- What exactly are the patterns the model is detecting?

To answer these questions, we will use the `lucent` library to visualize the patterns learned to predict melanoma confidently.

Practically visualizing neurons with image input data

In this section, we will continue with the previous tutorial to further explore how to practically visualize neurons with image input data using optimization techniques to gain insights into the patterns and behaviors learned by the CNN model. This process involves choosing neurons to interpret, optimizing image data for those neurons, and applying regularization techniques to generate visually interpretable patterns. By visualizing the patterns learned by the model, we can gain a better understanding of the model's predictions and answer questions that may not be apparent from traditional feature importance methods.

By following the steps outlined in this tutorial, you can visualize the patterns learned by your deep neural network, gaining a deeper understanding of the model's predictions and the features that contribute to those predictions. This can help to answer questions about the model's dependence on certain features, such as the color of the skin or the shape of the melanoma, and provide valuable insights into the patterns and behaviors of the model. Let's get started:

1. First, let's define the necessary variables. We want to visualize image patterns for the `Melanoma` class, which is at the sixth prediction layer index, so we must define the parameter we want to optimize as follows:

   ```
   param_to_optimize = "classifier_8:6"
   ```

2. Next, for the first iteration, we will be utilizing the **Compositional Pattern-Producing Network (CPPN)** method to initialize and generate an input image, instead of directly generating a random image input. Note that this is a method specifically for images and has been proven to generate more aesthetically pleasing images. In `lucent` specifically, CPPN consists of several convolutional layers with a compositional activation function consisting of element-wise tangents, squaring, division, and concatenation. This means that instead of optimizing the image directly, we optimize the parameters of the CPPN convolutional network that generates the input image for the main network. Backpropagation can be executed all the way through the generated input image till the first layer of the CPPN network. The initial input image is a fixed image comprising a circular region in the center of the image, with the values at the center being close to zero and gradually increasing toward the edges of the circle. However, with CPPN, the learning rate usually needs to be lower to converge properly. Let's define the

CPPN configuration with a 224 image size and the Adam optimizer with a lower-than-typical learning rate:

```
cppn_param_f = lambda: param.cppn(224)
cppn_opt = lambda params: torch.optim.Adam(params, 4e-3)
```

3. Finally, let's utilize the defined variables and visualize the patterns captured by the melanoma part of the prediction layer using a GPU-configured model:

```
model.cuda()
images = render.render_vis(
    model, param_to_optimize, cppn_param_f, cppn_opt,
    thresholds=np.linspace(0, 10000, 100, dtype=int).tolist(),
    verbose=True,
    show_inline=True
)
```

The `thresholds` list component controls the number of optimization steps taken, along with the intermediate step number to visualize the optimized image. Additionally, a hidden component built into the `render_vis` method is the `transforms` component. The `transforms` component adds minimal augmentations such as padding, jitter, random scaling, and random rotating to reduce any random noise in the image being optimized. The result from the previous code is shown in *Figure 12.3*:

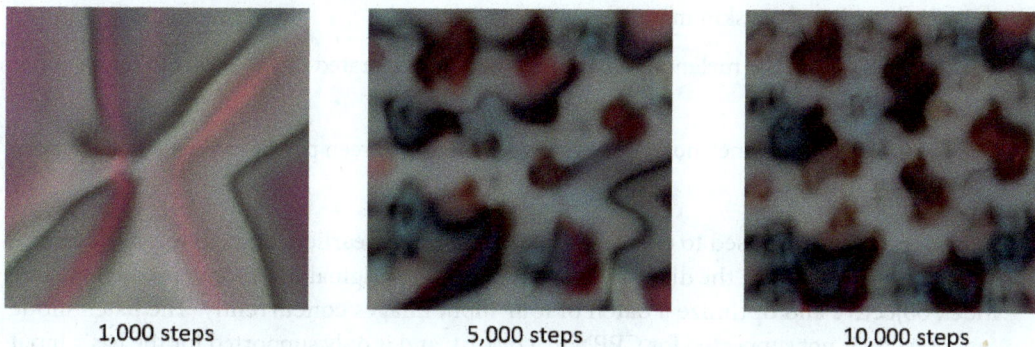

| 1,000 steps | 5,000 steps | 10,000 steps |

Figure 12.3 – Progress of optimizing the CPPN to generate an image
that activates highly for the melanoma class prediction

A pretty good image of what seems to be the actual melanoma was able to be generated from the process.

4. Let's find out the melanoma probability of this image. We can do this by defining the preprocessing methods needed to perform inference with this model:

```
inference_transform = transforms.Compose([
    transforms.ToTensor(),
```

```
        transforms.Resize((224, 224)),
        transforms.Normalize(mean=[0.485, 0.456, 0.406],
          std=[0.229, 0.224, 0.225]),
    ])
```

5. Here, we predict the skin lesion classes of the final optimized image from the process while disabling gradient computation:

```
def inference_lesion_model(img):
    transformed_image = inference_transform(img).unsqueeze(0)
    with torch.no_grad():
        output = torch.nn.functional.softmax(model(transformed_
image), dim=1)
    return output
output = inference_lesion_model(images[-1].squeeze())
```

This results in a 100% probability of the image being predicted as melanoma!

However, the image alone does not make it apparent that all the diverse sets of images can be recognized as melanoma. Some classes can be sufficiently represented in a single image, but some labels can't really be represented in a single picture. Take a background class, for example: it is impossible to put every single background visual in a single image. Here are some specific questions that could be useful to answer based on the result:

- Does the color of the skin matter?

- Does the shape of the melanoma matter, as the final generated image seems to have similar melanoma patterns?

- Does the color of the melanoma patch matter? There are green patches with a similar pattern to the red patch.

6. This is where the loss used to ensure diversity mentioned earlier can help to provide more insight. Now, let's utilize the diversity objective with the original melanoma prediction layer index objective and optimize a batch of four input images concurrently. The batch mode functionality is not supported for CPPN in lucent and is only supported for the basic input image initialization param module:

```
obj = objectives.channel("classifier_8", 6) - 1e2 *
diversity("input")
batch_param_f = lambda: param.image(224, fft=True,
decorrelate=True, batch=4)
batch_images = render.render_vis(
    model, obj,
    batch_param_f,
    thresholds=np.linspace(0, 500, 50, dtype=int).tolist(),
    show_inline=True,
```

```
        verbose=True,
    )
```

Two additional points on the image initialization method are as follows:

- `fft` stands for Fast Fourier transform

- `decorrelate` applies SVD to the image input

Both techniques here are acknowledged in research to allow faster convergence, reduce high-frequency images, and generate better-looking images.

The results are shown in *Figure 12.4*:

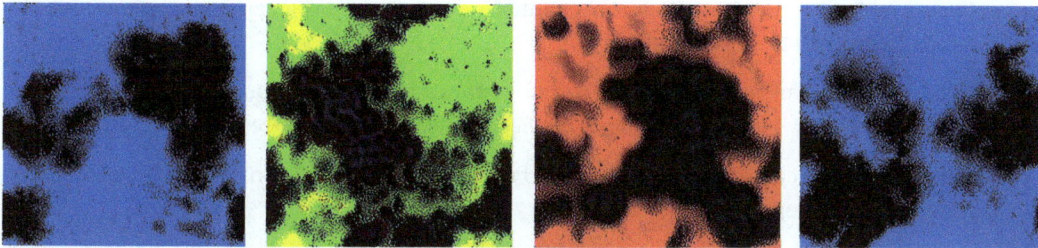

Figure 12.4 – Four jointly optimized images to activate highly on the melanoma neuron

These are very funky-looking images. The non-black color portions are probably simulating the skin. Let's see the probability of the melanoma class for each of these images to verify:

```
outputs = []
  for img_idx in range(4):
      img = batch_images[-1][img_idx]
    output = inference_lesion_model(img)
    outputs.append((output[0][6], output.argmax()))
```

This will result in the following array:

```
[(tensor(0.9946, device='cuda:0'), tensor(6, device='cuda:0')),
 (tensor(0.9899, device='cuda:0'), tensor(6, device='cuda:0')),
 (tensor(0.7015, device='cuda:0'), tensor(6, device='cuda:0')),
 (tensor(0.9939, device='cuda:0'), tensor(6, device='cuda:0'))]
```

They all have high probabilities for melanoma! From this, the following conclusion can be argued:

- The model does not depend a lot on the color of the skin to detect melanoma. The most that skin color can provide will likely be in the range of around a 3% probability boost.

- The model depends on the underlying lower-level pattern mostly to detect melanoma.

- The model doesn't depend a lot on the color of the melanoma patch. The color of the melanoma in the first generated image and real images was reddish. The color of the melanoma patch in the batch-generated images was black.

- The model can detect smaller melanoma signals from the real images used.

The results here are exemplary of how complementary each insight technique is to the other. We will end this topic with some useful notes about the pattern visualization of neurons through optimization techniques in general:

- Some problems are harder to converge than others, and some just don't converge at all. Be ready to experiment with multiple settings to see whether you can get a resulting input that can activate highly on your chosen neuron, channel, or entire layer. You can even choose multiple neurons to see how they interact!

- The loss can turn out to be extremely negative, and the more the input converges, the more negative it gets. This is good, as the loss is defined as the negative of the resulting activation value.

- Regularization techniques are the key to allowing reasonable inputs to be generated through optimization.

- Use both real data and diverse optimized data to understand the patterns your model learned to detect. One optimized piece of data usually can't represent the entire range of patterns a neuron can detect.

- In this tutorial, we used the final classification layer, which made it easier to find samples that activate highly toward the chosen neuron. If an intermediate neuron is chosen, be sure to find the set of data with the highest activations for the chosen neuron.

- The `lucent` library for `pytorch`-based models and the `lucid` library for TensorFlow-based models are focused on image visualizations but can both be adapted to other input variable types such as text. However, not much research has been done there to figure out good regularization techniques for other variable types to allow faster convergence.

Overall, the visualization of neurons through optimization techniques can provide valuable insights into the patterns and behaviors of **machine learning** (ML) models, but it requires experimentation and careful consideration of the inputs and regularization techniques used. As a bonus here, with knowledge of how to execute prediction explanations and NN interpretation, we will discover a useful way to make explanations more effective in general with a method called counterfactual explanations.

Discovering the counterfactual explanation strategy

Counterfactual explanation or reasoning is a method of understanding and explaining anything in general by considering alternative and counterfactual scenarios or "what-if" situations. In the context of prediction explanations, it involves identifying changes in the input data that would lead to a different outcome. Ideally, the minimal changes should be identified. In the context of NN interpretation, it involves visualizing the opposite of the target label or intermediate latent features. This approach makes sense to use because it closely aligns with how humans naturally explain events and assess causality, which ultimately allows us to comprehend the underlying decision-making process of the model better.

Humans tend to think in terms of cause and effect, and we often explore alternative possibilities to make sense of events or decisions. For example, when trying to understand why a certain decision was made, we may ask questions such as, "What would have happened if we had chosen a different option?" or "What factors led to this outcome?". This kind of reasoning helps us identify the key elements that influenced the decision and allows us to learn from the experience. Counterfactual explanations for ML models follow a similar thought process. By presenting alternative input instances that would have resulted in a different prediction, counterfactual explanations help us understand which features of the input data are most critical in the model's decision-making process. This kind of explanation allows users to grasp the model's rationale more intuitively and can also help improve their trust in the model's predictions.

Counterfactual reasoning complements feature importance and neuron visualization techniques. Together, these methods can provide a more comprehensive understanding of how the model arrives at its decisions. This, in turn, can help users better assess the reliability of the model and make more informed decisions based on its predictions.

Summary

NN interpretation is a form of a model understanding process that is different from explaining the predictions made by a model. Both manual discovery of real images and optimizing synthetic images to activate highly for the chosen neuron to interpret are techniques that can be applied together to understand the NN. Practically, the interpretation of NNs will be useful when you have goals to reveal the appearance of a particular prediction label or class pattern, gain insight into the factors contributing to the prediction of a specific label in your dataset or all labels in general, and gain a detailed breakdown of the reasons behind a prediction.

There might be hiccups when trying to apply the technique in your use case, so don't be afraid to experiment with the parameters and components introduced in this chapter in your goal to interpret your NN.

We will explore a different facet of insights that you can obtain from your data and your model in the next chapter, which is about bias and fairness.

13

Exploring Bias and Fairness

A biased machine learning model produces and amplifies unfair or discriminatory predictions against certain groups. Such models can produce biased predictions that lead to negative consequences such as social or economic inequality. Fortunately, some countries have discrimination and equality laws that protect minority groups against unfavorable treatment. One of the worst scenarios a machine learning practitioner or anyone who deploys a biased model could face is either receiving a legal notice imposing a heavy fine or receiving a lawyer letter from being sued and forced to shut down their deployed model. Here are a few examples of such situations:

- The ride-hailing app Uber faced legal action from two unions in the UK for its facial verification system, which showed racial bias against dark-skinned people by displaying more frequent verification errors. This impeded their work as Uber drivers (`https://www.bbc.com/news/technology-58831373`).

- Creators filed a lawsuit against YouTube for its racial and other minority group discrimination against them as YouTube's algorithm automatically removed their videos without proper explanation, removing their capability to earn ad revenue (`https://www.washingtonpost.com/technology/2020/06/18/black-creators-sue-youtube-alleged-race-discrimination/`).

- Facebook was charged with racial, gender, religion, familial status, and disability discrimination for its housing ads by the housing department of the US and had to pay under 5 million US dollars (`https://www.npr.org/2019/03/28/707614254/hud-slaps-facebook-with-housing-discrimination-charge`).

Thus, utmost care must be taken to prevent bias from being perpetuated against sensitive and protected attributes of the underlying data and environment the machine learning model will be exposed to. In this chapter, we will approach this topic step by step, starting with an exploration of the types of bias, learning to detect and evaluate methods needed to identify bias and fairness, and finally exploring ways to mitigate bias. The concepts and techniques that will be presented in this chapter are relevant to all machine learning models. However, bias mitigation is an exception; there, we will explore a neural network-specific method that can reliably mitigate bias. More formally, we will cover the following topics:

- Exploring the types of bias
- Understanding the source of AI bias
- Discovering bias and fairness evaluation methods
- Evaluating the bias and fairness of a deep learning model
- Tailoring bias and fairness measures across use cases
- Mitigating AI bias

Technical requirements

This chapter includes some practical implementations in the Python programming language. To complete it, you will need to have a computer with the following libraries installed:

- `pandas`
- `matplotlib`
- `scikit-learn`
- `numpy`
- `pytorch`
- `transformers==4.28.0`
- `accelerate==0.6.0`
- `captum`
- `catalyst`

The code files are available on GitHub at `https://github.com/PacktPublishing/The-Deep-Learning-Architect-Handbook/tree/main/CHAPTER_13`.

Exploring the types of bias

Bias can be described as a natural tendency or inclination toward a specific viewpoint, opinion, or belief system, regardless of whether it is treated as positive, neutral, or negative. AI bias, on the other hand, specifically occurs when mathematical models perpetuate the biases embedded by their creators or underlying data. Be aware that not all information is treated as biases, as some information can also be knowledge. Bias is a type of subjective information, and knowledge refers to factual information, understanding, or awareness acquired through learning, experience, or research. In other words, knowledge is the truth without bias.

> **Note**
>
> Do not confuse bias in this book with the bias from the infamous "bias versus variance" concept in machine learning. Bias in this concept refers to the specific bias on how simple a machine learning model is concerning a certain task to learn. For completeness, variance specifies the sensitivity of the model toward the change in the data features. Here, a high bias would correspond to a low variance and underfitting behavior and indicate that a machine learning model is too simple. A low bias in this concept would correspond to high variance and overfitting behavior.

In this book, we will use the term bias more colloquially. More well-known attributes of bias, such as race, gender, and age, are considered social bias, or stereotyping. However, the scope of bias is much broader. Here are some interesting examples of other types of bias:

- **Cultural bias**: The influence cultural perspectives, values, and norms have on judgments, decisions, and behaviors. It can manifest in machine learning models through biased data collection, skewed training data, or algorithms that reflect cultural prejudices.

- **Cognitive bias**: Systematic patterns of deviation from rationality in thinking or decision-making. Let's look at a few examples of cognitive bias along with possible phenomena that portray the bias in a machine learning project:

    - **Confirmation bias**: Favoring data that aligns with existing beliefs

    - **Anchoring bias**: Overemphasizing certain features or variables during model training

    - **Availability bias**: Relying on easily accessible data sources, potentially overlooking relevant but less accessible ones

    - **Overconfidence bias**: Overestimating model capabilities or accuracy

    - **Hindsight bias**: Believing that model predictions were more predictable after observing them

    - **Automation bias**: Placing excessive trust in the model's outputs without critical evaluation

    - **Framing bias**: Influencing the model's learning process through biased data presentation

    - **Selection bias**: Non-random sampling leading to unrepresentative data

    - **Sampling bias**: Skewed data due to an unrepresentative sample

    - **Reporting bias**: Misrepresentation due to individuals' preferences or beliefs

- **Algorithmic bias**: Presence of any biases in algorithms such as machine learning algorithms. Another example of such bias is **aggregation bias**, which has skewed predictions due to data grouping or aggregation methods.

- **Measurement bias**: Inaccuracies or errors in data collection and measurement.

These bias groups are just the tip of the iceberg and can span to very niche bias groups such as political bias, industry bias, media bias, and so on. Now that we understand what bias is and what it covers, let's discover the source of AI bias.

Understanding the source of AI bias

AI bias can happen at any point in the deep learning life cycle. Let's go through bias at those stages one by one:

- **Planning**: During the planning stage of the machine learning life cycle, biases can emerge as decisions are made regarding project objectives, data collection methods, and model design. Bias may arise from subjective choices, assumptions, or the use of unrepresentative data sources. Project planners need to maintain a critical perspective, actively consider potential biases, engage diverse perspectives, and prioritize fairness and ethical considerations.

- **Data preparation**: This stage involves the following phases:

  - **Data collection**: During the data collection phase, bias can creep in if the collected data fails to represent the target population accurately. Several factors can contribute to this bias, including sampling bias, selection bias, or the underrepresentation of specific groups. These issues can lead to the creation of an imbalanced dataset that does not reflect the true diversity of the intended population.

  - **Data labeling**: Bias can also infiltrate the data labeling process. Each labeler may possess their own inherent biases, consciously or subconsciously, which can influence their decision-making when assigning labels to the data. If the labelers lack diversity or comprehensive training, their biases may seep into the annotations, ultimately leading to the development of biased models that perpetuate unfairness and discrimination. As a result, the final combined data can contain multiple biases which may even conflict with each other, causing difficulties in the learning process.

- **Model development**: Bias can be introduced in two ways in a deep learning model:

  - **Feature selection**: Biases can arise from the features selected for model training. If certain features are correlated with protected attributes (such as race or gender), the model may inadvertently learn and reinforce those biases, leading to discriminatory outcomes.

  - **Pretrained models**: A pretrained deep learning model might be a biased model. For example, if the model has been trained on biased data, it may learn and perpetuate those biases in its predictions. Even if fine-tuning was done, the bias is not likely to go away.

- **Deliver model insights**: Bias can happen particularly when interpreting explanations of model behavior. The process of understanding and explaining the inner workings of a model involves subjective reasoning and is susceptible to bias. The interpretation of model insights heavily relies on the perspective and preconceptions of the individuals involved, which can introduce biases based on their own beliefs, experiences, or implicit biases. It is essential to approach the interpretation of model explanations with awareness of these potential biases and strive for objectivity and fairness to avoid misinterpretation or reinforcing existing biases. Critical thinking and diverse perspectives are vital to ensuring that the insights delivered accurately reflect the model's behavior without introducing additional bias.

- **Model deployment**: This stage covers bias that can happen when a model is deployed, which includes the following components:

  - **User interactions**: Bias can arise during model deployment when users provide feedback or responses and be introduced if the feedback is biased or if the system responds differently based on user characteristics. For example, the chat history mechanism in the ChatGPT UI allows a user to provide biased input.

  - **Human-in-the-loop bias**: Biases can be introduced when human reviewers or operators make decisions based on the model's predictions, exhibiting their own biases or interpreting outputs unfairly. This can impact the perceived fairness of the decision-making process.

  - **Environment bias**: Some features might be treated and perceived differently in unseen areas, leading to data drift. Models were evaluated to be unbiased during the development stage, but with the new data, it could still produce biased predictions.

  - **New data source for retraining bias**: New data can be collected and labeled for retraining, which can be a source of bias.

- **Model governance**: Bias can emerge when the person responsible for monitoring the deployed model needs to establish thresholds for various types of drift (which will be introduced in *Chapter 16, Governing Deep Learning Models*), analyze prediction summaries, and examine data summaries. Setting these thresholds introduces the potential for bias based on subjective decisions or assumptions. Additionally, when analyzing prediction and data summaries, there is a risk of overlooking certain biases or unintentionally reinforcing existing biases if not approached with diligence and a critical mindset. It is crucial to maintain awareness of these biases and ensure that monitoring and analysis processes are conducted rigorously and with a focus on fairness and accuracy.

Now that we've discovered some of the possible sources of bias in each stage of the deep learning life cycle, we are ready to dive into discovering bias detection and fairness evaluation methods.

Discovering bias and fairness evaluation methods

Fairness and bias are opposing concepts. Fairness seeks to ensure fair and equal treatment in decision-making for all individuals or groups, while bias refers to unfair or unequal treatment. Mitigating bias is a crucial step in achieving fairness. Bias can exist in different forms and addressing all potential biases is complicated. Additionally, it's important to understand that achieving fairness in one aspect doesn't guarantee the complete absence of bias in general.

To understand both how much bias and how fair our data and model are, what we need is a set of bias and fairness metrics to objectively measure and evaluate. This will then enable a feedback mechanism to iteratively and objectively mitigate bias and achieve fairness. Let's go through a few robust bias and fairness metrics that you need to have in your arsenal of tools to achieve fairness:

- **Equal representation-based metrics**: This set of metrics focuses on the equal proportions of either the data or the decision outcomes without considering the errors:

  - **Disparate impact**: Disparate impact examines whether the model treats different groups fairly or if there are significant relative disparities in the outcomes they receive by taking a ratio of the proportion of favorable outcomes between groups. Disparate impact for a chosen group in a chosen attribute can be computed using the following formula:

$$\text{Disparate Impact (Group A)} = \left(\frac{\text{Proportion of Positive Predictions for Group A}}{\text{Proportion of Positive Predictions for Reference Group or Aggregate of Other Groups}} \right)$$

    Disparate impact across groups can be averaged to obtain a single global representative value.

  - **Statistical parity difference**: This extends similar benefits as disparate impact but provides an absolute disparity measure instead of relative by using a difference instead of a ratio. The absolute difference is useful when you can and need to translate the values into tangible impacts, such as the number of individuals who were discriminated against based on a new sample size. It can be computed using the following formula:

$$\textit{Statistical Parity Difference} = |(\textit{Proportion of Positive Predictions for Privileged Group}) - (\textit{Proportion of Positive Predictions for Unprivileged Group})|$$

- **Equal error-based metrics**: These are the metrics that consider the bias in error rates between groups.

  - **Average Odd Difference (AOD)**: This measures the average discrepancy in the odds of true positive and false positive outcomes across groups. AOD is computed by taking the average of the odds differences across different groups. The odds difference for a specific group is calculated as the difference between the odds of positive prediction for that group and the odds of positive prediction for a reference group using the following formula:

$$\text{AOD} = \left(\tfrac{1}{n}\right) * \Sigma\left[\left(\text{TPR\_ref} - \text{TPR\_i}\right) + \left(\text{FPR\_ref} - \text{FPR\_i}\right)\right]$$

Here, n is the total number of groups, TPR_i is the true positive rate (sensitivity) for group i, FPR_i is the false positive rate (fallout) for group i, TPR_ref is the true positive rate for the reference group, and FPR_ref is the false positive rate for the reference group.

- **Average Absolute Odds Difference (AAOD)**: This extends similar benefits to AOD but adds an absolute term in the individual group computations, as follows:

$$\text{AOD} = \left(\tfrac{1}{n}\right) * \Sigma\left[\left|\left(\text{TPR\_ref} - \text{TPR\_i}\right) + \left(\text{FPR\_ref} - \text{FPR\_i}\right)\right|\right]$$

This should be used over AOD when you care about the discrepancies in general and not only whether the group discrepancy is favored or non-favored.

- **Distributional fairness through generalized entropy index (GEI)**: This is designed to measure the level of inequality based on distribution across individuals of an entire population using only the numerical outcome. The formula for GEI is as follows:

$$GE(\alpha) = \frac{1}{na(a-1)} \sum_{i=1}^{n}(n^a w^a - n), \, a \neq 0,1,$$
$$GE(\alpha) = \log(n) + \sum_{i=1}^{n} w_i \log(w_i), \, a = 1,$$
$$GE(\alpha) = -\log(n) - \frac{1}{n}\sum_{i=1}^{n}\log(w_i), \, a = 0$$

Here, $E_T = \sum_{i=1}^{n} E_i$ and $w_i = \frac{E_i}{E_T}$

E_i is the value of the chosen attribute of a specific entity, E_T is the total summed value of all, and n is the total number of individuals or entities. Two foundational notions of inequality can be configured through the α parameter of GEI:

- **Theil index**: The general inequality of all individuals across all groups. It has an α value of 1.

- **Coefficient of variation**: The inequality that's measured by computing the variability of individuals in a population group. The more varied the population group is, the more biased the population group is. It has an α value of 2.

Use the Theil index as your main option if you want an overall inequality and switch to the coefficient of variation when you want to understand inequality by group.

- **Individual fairness metric**: Disparity or similarity of outcome based on similar individuals is a single individual-based metric. Proximity algorithms such as KNN allow you to consider the multiple associating features of an individual and compute fairness metrics based on similar examples. You must perform the following steps:

I.   Find a chosen number of similar examples with associative features, excluding the protected attribute with KNN for the individual.

II. Compute the inequality of the outcome using a chosen fairness metric. The most common metric that's used here is the average similarity of the outcomes of similar examples to the individual.

- **Fair accuracy-based performance metric through Balanced accuracy**: This provides a balanced evaluation of a classification model's performance, especially when dealing with imbalanced datasets. Balanced accuracy is computed by calculating the average of the class-wise accuracies.

Even though the metrics we've introduced here only cover a partial set of bias and fairness metrics available in the field, it is general enough to satisfy most machine learning use cases. Now, let's explore how to use these metrics practically in a deep learning project to measure bias and fairness.

Evaluating the bias and fairness of a deep learning model

In this practical example, we will be exploring the infamous real-world use case of face recognition. This practical example will be leveraged for the practical implementation of bias mitigation in the next section. The basis of face recognition is to generate feature vectors that can be used to carry out KNN-based classification so that new faces don't need to undergo additional network training. In this example, we will be training a classification model and evaluating it using traditional classification accuracy-based metrics; we won't be demonstrating the recognition part of the use case, which allows us to handle unknown facial identity classes.

The goal here is to ensure that the resulting facial classification model has low gender bias. We will be using a publicly available facial dataset called **BUPT-CBFace-50**, which has a diverse coverage of facial images that have different facial expressions, poses, lighting conditions, and occlusions. The dataset consists of 500,000 images of 10,000 facial identity classes. In this practical example, you will require a GPU with at least 12 GB of RAM so that training can be done in a reasonable time. Before starting the example, download the dataset from the official source (`https://buptzyb.github.io/CBFace/?reload=true`). You can find it in the same directory as your project.

Let's start with the step-by-step code walkthrough by using Python and the `pytorch` library with `catalyst` again:

1. First, let's import the necessary libraries, which include `pytorch` as the deep learning framework, `mlflow` for tracking and comparison, `torchvision` for the pretrained ResNet50 model, `catalyst` for efficient PyTorch model handling, and `albumentations` for simple image data processing:

```
import os
import albumentations as albu
import numpy as np
import pandas as pd
import torch
import torch.nn as nn
from albumentations.pytorch.transforms import ToTensorV2
from PIL import Image
from sklearn.model_selection import StratifiedShuffleSplit
from torch.optim import SGD, Adam
```

```
from torch.utils.data import DataLoader, Dataset
from torchvision import models
from torchvision.models import resnet50

import mlflow
from catalyst import dl, utils, metrics
from catalyst.contrib.layers import ArcFace
from catalyst.loggers.mlflow import MLflowLogger
from captum.attr import IntegratedGradients
from captum.attr import NoiseTunnel
from captum.attr import visualization as viz

os.environ["CUDA_VISIBLE_DEVICES"] = "0"
```

The last line sets the first GPU of your machine to be visible to CUDA, the computing interface for the GPU to be used.

2. Next, we will define the configuration that will be used for different components. This will include training process-specific parameters such as the batch sizes, learning rate, number of epochs, the number of early stopping epochs before stopping the training process, and the number of epochs to wait for validation metric improvements before reducing the learning:

```
batch_size = 64
val_batch_size = 100
lr = 0.05
num_epochs = 20
early_stopping_epochs = 12
reduce_lr_patience = 3
```

3. Additionally, it will include the specification of the privileged and unprivileged groups that we will choose for computing the gender-based bias and fairness metrics. There are approximately two times more males than females in the dataset, so the privileged group here is expected to be Male:

```
priviledged_group = 'Male'
unpriviledged_group = 'Female'
```

4. Finally, we will set some names for the mlflow experiment name, the directory to save the models, and the extra parameters that we won't be enabling for now:

```
experiment_mlflow = "bias_mitigation_v1"
logdir = 'experiments/face_modelv1'
only_male=False
distill_model=False
```

5. Next, we'll proceed to load the CSV file that contains the dataset metadata, primarily consisting of the image paths for the downloaded and unzipped `BUPT_CBFace` dataset:

```
train = pd.read_csv('face_age_gender.csv')
image_path = train['image_path'].values
targets = train['target'].values
gender_data = train['gender'].values
```

6. Additionally, we will set up the `name2class` mapper, along with the class ID targets array and the number of classes:

```
name2class = {name: idx for idx, name in
enumerate(sorted(set(targets)))}
id_targets = np.array([name2class[target] for target in
targets])
num_classes = len(name2class)
```

7. Next, we will perform stratified splitting on this data to put it into training and validation sets for the facial identity classes so that both validation and training will have all the available facial identity classes:

```
splitter = StratifiedShuffleSplit(test_size=.20, n_splits=2,
random_state = 7)
split = splitter.split(image_path, targets)
train_inds, val_inds = next(split)
train_images = image_path[train_inds]
val_images = image_path[val_inds]
train_targets = id_targets[train_inds]
val_targets = id_targets[val_inds]
train_gender_data = gender_data[train_inds]
val_gender_data = gender_data[val_inds]

if only_male:
    pass
```

The logic of the last `if` clause will be covered in the next section on bias mitigation. For the subsequent steps, treat the usage of `pass` as an indicator.

8. Next, we will define the model we want to use based on the ResNet50 model. We will use the ARCFace layer here with the ResNet50 model base, a type of metric learning algorithm. It utilizes angular margin loss to enhance the discriminative power of the learned face embeddings, enabling more accurate and robust face recognition across varying poses, illuminations, and identities:

```
class ARCResNet50(nn.Module):
    def __init__(self, num_classes):
        super(ARCResNet50, self).__init__()
        self.model = models.resnet50(pretrained=True)
```

```
        s=2**0.5*np.log(num_classes - 1)
        s=13
        self.model.fc = nn.Linear(self.model.fc.in_features,
self.model.fc.in_features)
        self.head = ArcFace(self.model.fc.out_features, num_
classes, s=s, m=0.15)

    def get_last_conv_features(self, x):
        pass

    def special_forward(self, x, targets=None):
        pass

    def forward(self, x, targets=None):
        outputs = self.model(x)
        outputs = self.head(outputs, targets)
        return outputs
```

9. Next, we will initialize the model, assign it to use GPU, define the cross-entropy loss variable, define the SGD optimizer variable, and define the reduced learning rate engine on validation degradation:

```
if distill_model:
    pass
model = ARCResNet50(num_classes=num_classes)
model.to(device)
criterion = nn.CrossEntropyLoss()
optimizer = SGD(model.parameters(), lr=lr, momentum=0.9, weight_
decay=1e-5)
scheduler = torch.optim.lr_scheduler.
ReduceLROnPlateau(optimizer, patience=reduce_lr_patience,
factor=0.1)
```

10. Next, we will need to define the PyTorch dataset class to take in the image file paths and the sensitive attributes, which are gender data, the target, and the specified albumentation transform. The last variable will be utilized in the next section:

```
class ImagesDataset(Dataset):
    def __init__(self, files, sensitive_attributes,
targets=None, transforms=None, teacher_model_features=None):
        self.files = files
        self.sensitive_attributes = sensitive_attributes
        self.targets = targets
        self.transforms = transforms
        self.teacher_model_features = teacher_model_features

    def __len__(self):
```

```
                return len(self.files)

        def __getitem__(self, index):
            file = self.files[index]
            img = np.array(Image.open(file))

            if self.transforms is not None:
                img = self.transforms(image=img)["image"]

            if self.targets is None:
                return img

            target = self.targets[index]
            sensitive_attribute = self.sensitive_attributes[index]
            if self.teacher_model_features is None:
                return img, sensitive_attribute, target
            else:
                teacher_model_feature = torch.from_numpy(self.
teacher_model_features[index])
                return img, sensitive_attribute, target, teacher_
model_feature
```

11. Now, we want to apply a simple set of transform operations from the `albumentation` library. Let's define the method to return a transform instance with augmentation for training and without augmentation for validation purposes. Both require the transform instance to convert the image values into PyTorch tensors:

```
def get_transforms(dataset: str):
    if dataset.lower() == "train":
        return albu.Compose([
        albu.Resize(224, 224),
         albu.HorizontalFlip(),
         albu.Normalize(),
         ToTensorV2()
     ])

    else:
        return albu.Compose([ albu.Resize(224, 224), albu.
Normalize(), ToTensorV2()])
```

12. Next, let's initialize the dataset and the subsequent dataset loader:

```
if distill_model:
    pass
else:
    train_dataset = ImagesDataset(train_images, train_gender_
```

```
data, train_targets,  get_transforms('train'))
    val_dataset =  ImagesDataset(val_images, val_gender_data,
val_targets, get_transforms('valid'))
loaders = {"train": DataLoader(train_dataset, batch_size=batch_
size, shuffle=True, num_workers=8),
            "valid": DataLoader(val_dataset, batch_size=val_
batch_size, shuffle=False, num_workers=8)}
```

13. Now, we will define the helper methods to help compute the multiclass bias and fairness metrics, which consist of performing safe division handling and zero division to prevent NaN values and computing false positives, true negatives, total positives, total negatives, and total data. Since this is a multiclass problem, we have to either choose macro-averaged or micro-averaged stats by class. Micro-averaged treats all samples equally, while macro-averaged treats all classes equally. Macro has an underlying issue where if the performance concerning a minority class is good, it will give a fake perception that the model is good in general. So, we will use micro-averaged here:

```
def compute_classwise_stats(y_pred, y_true):
    unique_classes = np.unique(y_true)
    num_classes = len(unique_classes)
    false_positives = np.zeros(num_classes)
    total_negatives = np.zeros(num_classes)
    true_positives = np.zeros(num_classes)
    total_positives = np.zeros(num_classes)

    for c_idx in range(num_classes):
        class_label = unique_classes[c_idx]
        class_predictions = (y_pred == class_label)
        class_labels = (y_true == class_label)
        false_positives[c_idx] = np.sum(class_predictions &
~class_labels)
        total_negatives[c_idx] = np.sum(~class_labels)
        true_positives[c_idx] = np.sum(class_predictions &
class_labels)
        total_positives[c_idx] = np.sum(class_labels)
    return {
        "false_positives":false_positives,
        "total_negatives": total_negatives,
        "true_positives": true_positives,
        "total_positives": total_positives,
        "total": total_negatives + total_positives,
    }

def safe_division(a, b):
    return a/b if b else 0.0
```

14. Finally, we will utilize these helper methods to define the method that will compute four bias and fairness metrics using common computed results – that is, disparate impact, statistical parity difference, AOD, and AAOD:

```
def compute_multiclass_fairness_metrics(
    y_pred, y_true, sensitive_attribute, priviledged_group,
unpriviledged_group
):
```

15. We will start by obtaining the false positives, true negatives, total positives, total negatives, and total data for the two groups, which are the privileged group and the non-privileged group, using the helper method:

```
y_pred = y_pred.argmax(1)
group_stats = {}
for group in [priviledged_group, unpriviledged_group]:
    group_idx = sensitive_attribute == group
    group_stats[group] = compute_classwise_stats(
        y_pred[group_idx], y_true[group_idx])
```

16. Next, we will compute the true positive ratio of both the privileged and non-privileged groups using the computed group stats so that it can be used directly to compute disparate impact and statistical parity difference:

```
disparities = []
priviledged_true_positive_ratio = safe_division(
    np.sum(group_stats[priviledged_group]["true_positives"]),np.
sum(group_stats[unpriviledged_group]["total"]))
    unpriviledged_true_positive_ratio = safe_division(
        np.sum(group_stats[unpriviledged_group]["true_
positives"]),np.sum(group_stats[unpriviledged_group]["total"]))
```

17. Now, we will compute the two mentioned metrics using the true positive ratios:

```
disparate_impact = safe_division(unpriviledged_true_positive_
ratio, priviledged_true_positive_ratio)
    statistical_parity_diff = priviledged_true_positive_ratio -
unpriviledged_true_positive_ratio
```

18. Finally, we will compute AOD and AAOD using the true positive rates, `tpr`, and false positive rates, `fpr`:

```
for group in [priviledged_group, unpriviledged_group]:
    group_stats[group]["fpr"] = safe_division(
        np.sum(group_stats[priviledged_group]["false_
positives"]),np.sum(group_stats[priviledged_group]["total_
negatives"]))
```

```
        group_stats[group]["tpr"] = safe_division(
            np.sum(group_stats[priviledged_group]["true_positives"]),
    np.sum(group_stats[priviledged_group]["total_positives"])
        )
    AOD = (
            (group_stats[unpriviledged_group]["fpr"] - group_
    stats[priviledged_group]["fpr"])
            + (group_stats[unpriviledged_group]["tpr"] - group_
    stats[priviledged_group]["fpr"])
        ) / 2
    AAOD = (
            np.abs(group_stats[unpriviledged_group]["fpr"] - group_
    stats[priviledged_group]["fpr"])
            + np.abs(group_stats[unpriviledged_group]["tpr"] -
    group_stats[priviledged_group]["fpr"])
        ) / 2

    return {
        "disparate_impact": disparate_impact,
        "statistical_parity_diff": statistical_parity_diff,
        "average_odds_diff": AOD,
        "average_abs_odds_diff": AAOD
    }
```

19. We will be computing these four metrics during training and will be able to monitor and track the metrics as they're being trained. To track the experiment, we will record the parameters and monitor their performance by iteration and epoch. We will use MLflow to do this. Let's define the `mlflow` logger and log the parameters we defined earlier in *step 2*:

```
mlflow_params = dict(
    batch_size=batch_size,
    lr=lr,
    num_epochs=num_epochs,
    early_stopping_epochs=early_stopping_epochs,
    reduce_lr_patience=reduce_lr_patience,
    experiment_mlflow=experiment_mlflow,
    logdir=logdir,
    only_male=only_male,
    distill_model=distill_model,
)
all_metrics = [
    "loss", "accuracy", "disparate_impact","statistical_parity_
diff", "average_odds_diff","average_abs_odds_diff"
]
```

```
mlflow_logger = MLflowLogger(experiment=experiment_mlflow,
tracking_uri="experiment/")
mlflow_logger.log_hparams(mlflow_params)
```

20. As we will require a slightly specialized flow to be able to compute custom metrics and perform bias mitigation methods later on, we will define a custom runner using `catalyst` that will be used to train the ResNet50 model. We will need to define three custom logic for four methods: `on_loader_start` (to initialize the metric aggregator functionality), `handle_batch` (to obtain the loss), `on_loader_end` (to finalize the aggregated batch metrics and update the learning rate scheduler), and `get_loggers` (to log data into MLflow). Let's start with defining `on_loader_start`:

```
class CustomRunner(dl.Runner):
  def on_loader_start(self, runner):
    super().on_loader_start(runner)
    self.meters = {key: metrics.AdditiveMetric(compute_on_
call=False)
              for key in all_metrics}
```

21. Next, we will define the batch handler logic, which will load the data from the batch data loader in a custom way, perform forward propagation using the model initialized in the runner, and then compute loss, accuracy, and multiclass fairness metrics:

```
def handle_batch(self, batch):
  is_distill_mode = len(batch) == 4
  if is_distill_mode:
    pass
  else:
    features, sensitive_attribute, targets = batch
    logits = self.model(features, targets)
    loss = self.criterion(logits, targets)
  accuracy = (logits.argmax(1) == targets).float().mean().
detach().cpu()

  batch_metrics = {
    "loss": loss.item(),"accuracy": accuracy.item()}
  batch_metrics.update(
**compute_multiclass_fairness_metrics(logits.detach().cpu().
numpy(), targets.detach().cpu().numpy(), np.array(sensitive_
attribute),priviledged_group,
unpriviledged_group))
```

22. In the same method, we will also be required to update the current batch metric and the aggregated batch metrics so that they can be logged properly and finally perform backpropagation if it is training mode instead of validation mode:

```
    self.batch_metrics.update(batch_metrics)
      for key in all_metrics:
        self.meters[key].update(
          self.batch_metrics[key], self.batch_size
        )

        if self.is_train_loader:
            loss.backward()
            self.optimizer.step()
            self.optimizer.zero_grad()
```

23. Now, we must define the final two straightforward methods:

```
    def on_loader_end(self, runner):
        for key in all_metrics:
            self.loader_metrics[key] = self.meters[key].
compute()[0]
        if runner.is_valid_loader:
            runner.scheduler.step(self.loader_metrics[self._
valid_metric])
        super().on_loader_end(runner)


    def get_loggers(self):
        return {
            "console": dl.ConsoleLogger(),
            "mlflow": mlflow_logger
        }
```

24. Finally, we will initialize the runner and train the model:

```
runner = CustomRunner()
runner.train(
    model=model,
    criterion=criterion,
    optimizer=optimizer,
    loaders=loaders,
    logdir=logdir,
    num_epochs=num_epochs,
    valid_loader="valid",
    valid_metric="loss", # loss"
    minimize_valid_metric=True,
    fp16=False,
    verbose=True,
```

```
        load_best_on_end=True,
        scheduler=scheduler,
        callbacks=[
            dl.EarlyStoppingCallback(
                patience=early_stopping_epochs, loader_key="valid",
metric_key="loss", minimize=True
            ),
            dl.CheckpointCallback(
                logdir=logdir, save_best=True, load_best_on_
end=True, metric_key='loss'
            )
        ]
    )
```

Figure 13.1 shows the `mlflow` plotted performance graph of the accuracy, AOD, disparate impact, and statistical parity difference at every epoch of both the train and validation partitions:

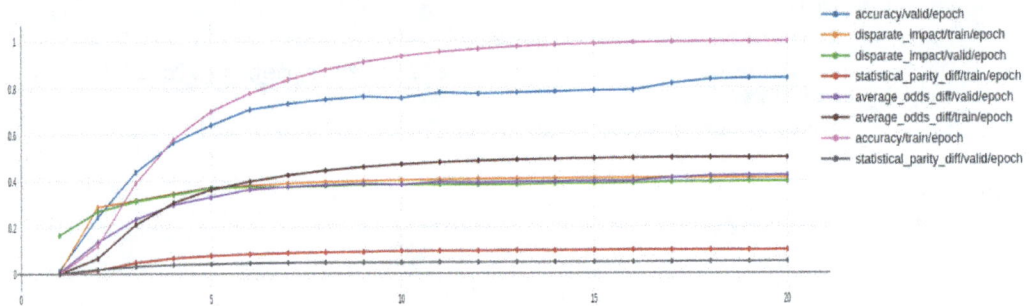

Figure 13.1 – Performance graph of ResNet50 by epochs

The validation scores ended up with 0.424 for both AAOD and AOD, 0.841 accuracy, 0.398 disparate impact, and 0.051 statistical parity.

25. The graph shows that while accuracy increased epoch after epoch, the bias and fairness metrics gradually became worse but stagnated at a value. The lowest bias model is at the 0th epoch, where everything was close to zero, including the accuracy. Even though the model is not biased, the model is not useful at all. Another interesting observation is at the 16th epoch mark – the model managed to get a better validation accuracy performance on the male samples as the AOD value became higher at the same point. Depending on the circumstances, you can opt to choose to take the model at the 15th epoch, which has a somewhat good accuracy score but not the best and a lower bias score.

26. To make a deeper analysis, let's take a look at where the model is focusing when it's making predictions using the integrated gradients technique from Captum. Let's visualize images that are mostly frontal facing. To do this, we must define the necessary transform method:

```
resize_transform = albu.Resize(224, 224)
norm_transform = albu.Compose([
        albu.Normalize(),
        ToTensorV2()
    ])
```

27. Now, let's define the frontal faces to visualize:

```
val_df = pd.read_csv('val_pose_info.csv')
straight_indexes = val_df[
    (val_df['pitch']>-10) &
    (val_df['pitch']<10) &
    (val_df['yaw']>-10) &
    (val_df['yaw']<10)
].index.values
```

28. Finally, let's visualize the focus area:

```
integrated_gradients = IntegratedGradients(model)
noise_tunnel = NoiseTunnel(integrated_gradients)
for val_idx in straight_indexes[:5]: #range(1):
    image_path = val_images[val_idx]
    pred_label_idx = val_targets[val_idx]

    pil_image = Image.open(image_path)
    img = np.array(pil_image)
    transformed_image = resize_transform(image=img)["image"]
    input = norm_transform(image=img)["image"].unsqueeze(0)
    attributions_ig_nt = noise_tunnel.attribute(
        input, nt_samples=5, nt_type="smoothgrad_sq",
        target=int(pred_label_idx)
    )
    _ = viz.visualize_image_attr_multiple(
        np.transpose(attributions_ig_nt.squeeze().cpu().
detach().numpy(), (1,2,0)),
        transformed_image,
        ["original_image", "heat_map"],
        ["all", "positive"],
        cmap="turbo",
        show_colorbar=True
    )
```

Figure 13.2 shows the original images and focus areas of the model:

Females | Males

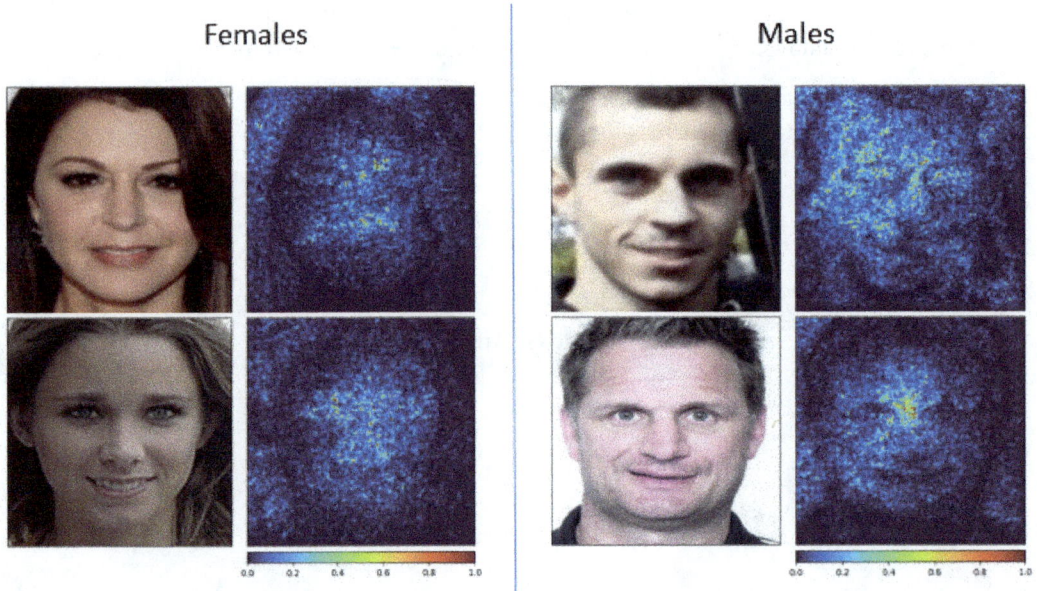

Figure 13.2 – Saliency explanations results of the trained ResNet50 model

29. These visuals show that the model exhibits bias by focusing incorrectly on the hair of female faces. For males, the model did not focus on the hair. The model also focuses on the white background a little. We'll learn how to remove this bias in the next section.

Now that we understand some popular bias and fairness metrics, we need to know which metrics to use in different use cases.

Tailoring bias and fairness measures across use cases

The process of figuring out bias and fairness metrics to use for our use case can flow similarly to the process of figuring out general model performance evaluation metrics, as introduced in *Chapter 10*, *Exploring Model Evaluation Methods*, in the *Engineering the base model evaluation metric* section. So, be sure to check that topic out! However, bias and fairness have unique aspects that require additional heuristical recommendations. Earlier, recommendations for metrics that belong to the same metric group were explored. Now, let's explore general recommendations on the four metric groups:

- Equal representation is always desired when there is a sensitive and protected attribute. So, when you see these attributes, be sure to use equal representation-based metrics on both your data and the model. Examples include race, gender, religion, sexual orientation, disability, age, socioeconomic status, political affiliations, and criminal history.

- Predictive performance consistency is another desired trait of a machine learning model that deals with sensitive and protected attributes. So, when you see these attributes, be sure to use equal error-based metrics on your model.

- Both distributional fairness metrics and equal representation metrics measure inequality. However, distributional fairness works on continuous variables directly while equal representation metrics work on categorical variables. Binning can be done on continuous variables to transform them into categories but it's not straightforward to decide on the proper binning strategy needed. So, use distributional fairness metrics when the variable to measure bias and fairness is a continuous variable.

- Consider all potential aspects of bias and fairness and measure them separately with a chosen bias and fairness metrics:

  - Let's consider a scenario where a machine learning model is used to predict loan approvals. One aspect of fairness is to ensure that loan approvals are granted fairly across different demographic groups, such as race or gender. Ensuring equal representation of the data and the resulting model can help you accomplish that. However, solely focusing on equal representation may not capture the complete picture. For example, even though the overall loan approval rates are equal across groups, there could be a significant disparity in the interest rates assigned to different groups. This disparity in interest rates could lead to unfair and inequitable outcomes, as certain groups may be charged higher interest rates, resulting in financial disadvantages. Any additional evaluation and monitoring that uses distributional fairness metrics can help you understand the impact and assist in targeted bias mitigation of unfavored groups.

- Evaluating the individual fairness of the outcome is useful when you deploy the machine learning model and receive individual data during model inferencing. A threshold can be set here to create an alert when the individual fairness score is too high and requires a human reviewer to evaluate and make a manual decision.

- Compute these metrics separately by sensitive groups and compare them visually to get a sense of fairness across groups. This can help you craft targeted bias mitigation responses to vulnerable groups.

- Prediction explanations may help in understanding the reasons for bias and fairness.

- Bias and fairness measures can conflict with accuracy-based performance measures.

Additionally, two global opposing views that will affect how you see fairness are worth mentioning:

- **We're All Equal (WAE)**: The notion that data may not accurately represent reality due to the presence of inherent biases within it

- **What You See Is What You Get (WYSIWYG)**: The notion is that the information presented by the data reflects an unbiased representation of reality, even if it reveals inequalities

An extreme of either view will either remove any chances of building a model (WAE) or be too ignorant to care about fairness biases that will eventually lead to negative consequences, as mentioned earlier in this chapter (WYSIWYG). To create a successful machine learning model in our use case, we will need to balance the two views strategically so that both accuracy performance and fairness requirements can be satisfied. A good strategy here to employ is to apply common sense and accept what seems logical for a causal relationship (WYSIWYG), and practice fair processes that help us understand and mitigate the partial aspect of bias (WAE) even when true fairness can't be achieved.

Here are some examples with views to adopt that will make the most sense:

- **Adopting WYSIWYG**: In the domain of personalized advertising using machine learning, the WYSIWYG view would involve tailoring advertisements based on observed user behavior and preferences. For example, if a user frequently engages with content related to fitness, the system will present ads for fitness equipment or gym memberships. The goal is to provide a personalized user experience that aligns with their interests and needs. In this use case, any notion of the WAE view will fail the project as it contradicts the goal of personalization.

- **Adopting something like 75% WYSIWYG and 25% WAE**: In the context of making hiring decisions using machine learning, the WAE view would help advocate for equal consideration of all candidates without any bias related to gender, race, or other protected attributes. The WYSIWYG view will then allow a machine learning model to be built based solely on their qualifications and skills. The strategy to create a fair and unbiased selection process will help promote equal opportunities for all applicants while still allowing a functional machine learning model to be built successfully.

In terms of metrics, the equal representation-based metric group is associated with the WAE view while the equal error-based metric group is associated with the WYSIWYG view. So, again, choose these two metric groups if the two views are involved in your use case.

Next, we'll discover ways we can mitigate bias in our machine learning models.

Mitigating AI bias

AI bias is an algorithmic bias that either comes from the model itself through its learning process or the data it used to learn from. The most obvious solution to mitigate bias is not programmatic mitigation methods but ensuring fair processes when collecting data. A data collection and preparation process is only truly fair when it not only ensures the resulting data is balanced by sensitive attributes but also ensures all inherent and systematic biases are not included.

Unfortunately, a balanced dataset based on the sensitive attribute does not guarantee a fair model. There can be differences in appearance among subgroups under the hood or associative groups of the data concerning multiple factors, which can potentially cause a biased system. Bias, however, can be mitigated partially when the dataset is balanced compared to without concerning the observable sensitive groups. But what are all these attributes? It might be easier to identify data attributes in tabular structured data with defined column names, but for unstructured data meant for deep learning, it's impossible to cover all the possible attributes.

To dive deeper into actual examples, text data can contain a ton of attributes with examples, such as language, genre, topic, length, style, tone, time period, authorship, geographic origin, and cultural perspective. Image data can also contain a ton of attributes with examples such as subject/content, perspective, lighting, composition, color, texture, resolution, orientation, context, and cultural relevance. Finally, audio data can also contain a ton of attributes, such as genre, language, duration, sound quality, instrumentation, vocal style, tempo, mood, cultural influence, and recording environment.

It's hard to ensure equal representation in all facets - more so when data is readily available and is already originally highly imbalanced with only a few examples for certain categories and plenty for others. Ideally, bias mitigation should always be executed right from the data preparation stage. However, if that's not possible, programmatic bias mitigation methods can be applied after data collection.

To get a clearer idea of these methods, have a look at *Figure 13.3*, which presents an overview of various bias-reducing approaches:

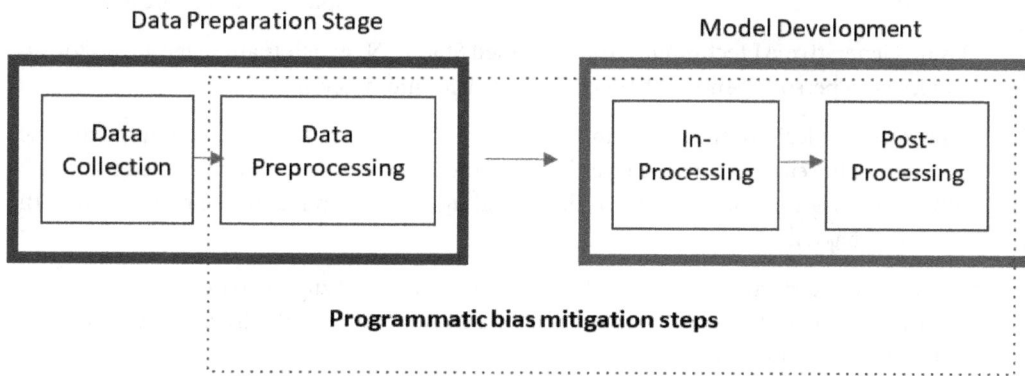

Figure 13.3 – The four steps of bias mitigation under two machine learning life cycle stages

Programmatic bias mitigation provides methods that can be applied in three different stages of the model-building process:

- **Pre-processing:** This is part of the data preparation ML life cycle stage. Here are some examples of bias mitigation methods that belong to this group:

  - Eliminating protected attributes from being used in the model. However, information about the protected attribute could still present itself in other associative attributes. Additionally, some attributes are deeply interconnected with other key information that's required to predict the desired label, and unfortunately can't be removed easily. Examples include gender in facial images.

- Disparate Impact Remover, available in the AIF360 open source library. It balances the dataset group proportions by dropping data rows to achieve a better disparate score.

- Weigh privileged class loss so that it's lower than unprivileged class loss during training so that errors between the classes are more equal.

- Targeted data augmentation for unprivileged classes. Augmentation adds more data variations and can be applied in two ways:

  - Adding more data variations in unprivileged groups with augmentation will help increase accuracy performance there and contribute to the balancing of error rates, especially when the unprivileged class is underrepresented.

  - Counterfactual role reversal augmentation inverts the privileged group into the unprivileged group and vice versa and allows for equal representation. Here are some augmentations that can be used based on variable type:

    - **Text**: Use word swap

    - **All variable types**: Use style transfer techniques. Example techniques are as follows:

- Using Generative AI techniques from a trained StarGAN, which is an image generator that can invert a person's gender, change a person's age, and much more.

- Optimize the desired image to mimic the style of another image by reducing the distance between a chosen intermediate layer between the images. This method is called transfer by input optimization and is based on the neural interpretation technique mentioned in the previous chapter.

- Targeted counterfactual role reversal not through additional augmentation but a permanent input replacement for a deployed model. This allows you to explicitly control the equality of the results during deployment.

- OpenAI changed "male" to "female" in their Dall-E text-to-image prompts randomly as the model is biased to depict males for roles such as professor and CEO.

- **In-processing**: This is part of the model development machine learning life cycle stage and is the process of training a model. Here are some examples of bias mitigation methods that belong to this group:

  - **Knowledge distillation**: This was introduced in *Chapter 8, Exploring Supervised Deep Learning*. The idea is that a teacher model that has been trained with a much bigger and more representative dataset will be less biased compared to a student model that is being trained on a much smaller custom dataset. Ideally, the fairness from the teacher model will be distilled into the student model. However, knowledge distillation can also cause an increased bias in the resulting student model when you distill it with a biased teacher model.

- **Adversarial debiasing**: This method iteratively trains a classifier to optimize prediction accuracy while simultaneously minimizing an adversary model's ability to infer the protected attribute from the predictions. The classifier aims to make accurate predictions on the target variable, while the adversary tries to discern the sensitive attribute associated with bias. This simultaneous training process creates a competitive environment where the classifier learns to encode important information about the target variable while reducing the influence of potentially biased features. By doing so, adversarial debiasing promotes fairness by mitigating the impact of sensitive attributes and enhancing the overall equity of the model's predictions.

- **Regularization**: In deep learning, regularization involves any addition or modification to the neural network, data, or training process that is used to increase the generalization of the model to external data. This can indirectly contribute to reducing bias in the model. Some common regularization methods include dropout layers, L1/L2 regularization, batch normalization, group normalization, weight standardization, stochastic depth, label smoothing, and data augmentation. By improving generalization, regularization methods can help the model learn more general patterns in the data instead of fitting too closely to the training set, which might contain biased features. This method was explored more extensively in *Chapter 2, Designing Deep Learning Architectures*.

- **Post-processing**: This is part of the model development machine learning life cycle stage but only covers processing the trained model's outputs to mitigate bias. Here are some examples of bias mitigation methods that belong to this group:

    - **Test-time counterfactual role reversal augmentation ensemble**: This method involves performing two predictions each using opposite roles and performing an ensemble of the predictions. A max or exponential mean operation for ensembling performs much better than an average operation, as shown in `https://www.amazon.science/publications/mitigating-gender-bias-in-distilled-language-models-via-counterfactual-role-reversal`.

    - **Equalized odds postprocessing**: This method involves modifying the output predictions of a classifier to ensure equal false positive and false negative rates across different groups through threshold optimization by groups. Specifically, prediction thresholds for each protected group are determined separately.

It is important to note that a fairer model obtained after bias mitigation would likely cause a reduction in accuracy-based metrics. If the loss in an accuracy-based metric is minor enough, a fair model is highly desired as there would not be an issue in using bias mitigation methods. To that end, always start by creating a baseline model and evaluate fairness with the chosen bias and fairness metrics to ensure that bias even exists in your model before using any bias mitigation methods. Bias mitigation methods always cause a substantial increase in training time needed, so make sure it's worth it.

Additionally, there are a few behaviors related to bias that are useful to know about:

- Decision-making algorithms tend to be biased toward more common occurrences, so balanced data can help reduce this bias

- Pruned models increase bias (more at `https://arxiv.org/pdf/2106.07849.pdf`)

- Models with limited capacity (smaller models) tend to exploit the biases in the dataset (`https://aclanthology.org/2022.gebnlp-1.27.pdf`), so be wary when you're performing knowledge distillation on a smaller model and evaluate bias and fairness

When it comes to mitigating bias, you might be wondering which of the four groups of methods to choose from. Here's a suggestion: it's best to begin by addressing bias as early as possible in the process, where you have the most control and flexibility. You don't want to be stuck with limited options to mitigate bias and end up not being able to satisfactorily mitigate bias. If you don't have access to the data collection stage, you can utilize data preprocessing techniques. On the other hand, if you don't have access to the data itself but have a trained model, postprocessing techniques should be used. Techniques can be combined, so be sure to measure bias using the metrics introduced to ensure that each technique that's applied improves the fairness of the resulting model. Also, consider using multiple methods to mitigate more bias.

The programmatic bias mitigation methods we've introduced so far are separated into three groups. However, a robust bias mitigation method exists for deep learning models that lies in the intersection of all three of them. The method is a fusion between counterfactual augmentation, mix-up augmentation, knowledge distillation, and counterfactual test time augmentation. The idea is to do the following:

1. Use counterfactual augmentation to train both the teacher model and student on the entire dataset.

2. Distill the counterfactual ensembled chosen layer features of the teacher model with exponential max to the student model. This is similar to mix-up augmentation but on the feature layer.

However, counterfactual role reversal remains to be a technique that's more accessible to text. Let's discover how to use the generic knowledge distillation method to reduce bias on the same use case we explored in the previous section on face classification. The method we will be introducing here is from `https://arxiv.org/pdf/2112.09786.pdf`, which provides two methods:

- **(Distill and debias) D&D++**: First, train the teacher model using only privileged group data. Then, initialize the student model with the teacher model weights and train the student model with normal loss and the knowledge distillation loss using the cosine similarity of the chosen feature layer. Lastly, initialize a final student network using the previously trained student model and train on the entire dataset with knowledge distillation from the previous student model as the teacher model.

- **One-step distillation D&D**: To make it simpler, but still immediately effective, the steps can be simplified. First, train the teacher model using only privileged group data. Then, train the student model on the entire dataset with knowledge distillation from the teacher model.

Let's explore the one-step distillation D&D method practically by using the use case we experimented with in the previous section:

1. First, we will be training a teacher model using only the privileged group data, which only consists of male facial identities. Let's define the specifically different configurations for this example:

```
batch_size = 64
lr = 0.05
num_epochs = 22
early_stopping_epochs = 12
reduce_lr_patience = 3
experiment_mlflow = "bias_mitigation_v2"
logdir = 'experiments/face_modelv2'
only_male=False
distill_model=False
```

2. We will be using the same code base we introduced in the previous section but with some additional custom code. The first is an addition to the code in *step 6* of the *Evaluating the bias and fairness of a deep learning model* section, where we will be removing the female data in both the training and validation data:

```
if only_male:
  train_raw_targets = targets[train_inds]
  val_raw_targets = targets[val_inds]
  train_male_inds = np.where(train_gender_data=='Male')[0]
  train_images = train_images[train_male_inds]
  train_targets = train_targets[train_male_inds]
  train_gender_data = train_gender_data[train_male_
inds]                              val_male_inds = np.where(
  val_gender_data=='Male')[0]
  val_images = val_images[val_male_inds]
  val_targets = val_targets[val_male_inds]
  val_gender_data = val_gender_data[val_male_inds]
```

3. Second is *step 7*, where we will define a special forward method and the method to get the last conv features. The idea is that we want to make sure the focus areas are the same and not the probabilities of the facial identity themselves, which will be useless when the model is used as a facial recognition featurizer. The special forward method is designed to return both the output logits and the last convolutional features in one forward pass, reducing latency:

```
def get_last_conv_features(self, x):
      x = self.model.conv1(x)
      x = self.model.bn1(x)
      x = self.model.relu(x)
      x = self.model.maxpool(x)
```

```
        x = self.model.layer1(x)
        x = self.model.layer2(x)
        x = self.model.layer3(x)
        x = self.model.layer4(x)
        x = self.model.avgpool(x)
        x = torch.flatten(x, 1)
        return x

    def special_forward(self, x, targets=None):
        model_features = self.get_last_conv_features(x)
        outputs = self.model.fc(model_features)
        outputs = self.head(outputs, targets)
        return outputs, model_features
```

4. Next is *step 8*, where we will extract the teacher model features once before starting training. This will reduce the time and resources needed to train the student model as the teacher model's features will remain fixed on the same data without augmentation:

```
device = torch.device("cuda")
if distill_model:
  if os.path.exists('all_teacher_model_features.npy'):
    all_teacher_model_features = np.load('all_teacher_model_
features.npy')
  else:
    teacher_model = ARCResNet50(num_classes=num_classes)
    state_dict = torch.load(os.path.join(distill_model, "model.
last.pth"))
    teacher_model.load_state_dict(state_dict)
    teacher_model.to(device)
    cpu_device = torch.device('cpu')
    entire_dataset = ImagesDataset(image_path, gender_data,
targets=id_targets, transforms=get_transforms('valid'))
    entire_data_loader = DataLoader(entire_dataset, batch_
size=val_batch_size, shuffle=False, num_workers=8)
    all_teacher_model_features = []
    for batch in tqdm_notebook(entire_data_loader):
      inputs, sensitive_attribute, targets = batch
      with torch.no_grad():
        outputs = teacher_model.get_last_conv_features(inputs.
to(device)).to(cpu_device).numpy()
      all_teacher_model_features.append(outputs)
    del teacher_model
    all_teacher_model_features = np.vstack(all_teacher_model_
features)
    np.save('all_teacher_model_features.npy', all_teacher_model_
features)
```

```
    train_teacher_model_features = all_teacher_model_
features[train_inds]
    val_teacher_model_features = all_teacher_model_features[val_
inds]
```

5. Next, we will define the training and validation dataset that takes in the training and validation teacher model features separately from *step 10*:

```
if distill_model:
    train_dataset = ImagesDataset(
        train_images,
        train_gender_data,
        train_targets,
        get_transforms('train'),
        train_teacher_model_features,
    )
    val_dataset =  ImagesDataset(
        val_images,
        val_gender_data,
        val_targets,
        get_transforms('valid'),
        val_teacher_model_features,
    )
```

6. The final change is to add handling of the batch loading by taking the extra teacher model features and loss using both cross entropy loss and the cosine similarity loss of the student and teacher model's last convolutional layer features from *step 14*:

```
if is_distill_mode:
    features, sensitive_attribute, targets, teacher_model_features
= batch
    logits, child_model_features = self.model.special_
forward(features, targets)
    loss = self.criterion(logits, targets) + nn.functional.cosine_
similarity(teacher_model_features, child_model_features).mean()
```

7. Now, run through all the steps from the previous section with the changes; you'll get an approximate validation metric score of 0.774 accuracy with all other scores of 0 as there are no females in this first step.

8. To execute the next step of one-step distillation, execute the previous code once more but with the following configuration changes:

```
batch_size = 64
val_batch_size = 100
lr = 0.05
num_epochs = 150
```

```
early_stopping_epochs = 5
reduce_lr_patience = 2
experiment_mlflow = "bias_mitigation_v3"
logdir = 'experiments/face_modelv3'
only_male=False
distill_model = 'experiments/face_modelv2'
```

9. The validation scores for the one-step distillation D&D approach end up with 0.3838 for both AAOD and AOD, 0.76 accuracy, 0.395 disparate impact, and 0.04627 statistical parity. For a fair comparison, the performance of basic training in terms of accuracy is an accuracy of 0.76333, 0.38856 for both AAOD and AOD, a 0.3885 disparate impact, and a 0.04758 statistical parity. This means that the improvements are mainly from the AOD with a 0.0047 difference and statistical parity with a 0.00131 difference.

10. By using the integrated gradients code from *step 19* again on the model we trained with bias mitigation methods, the following results can be obtained:

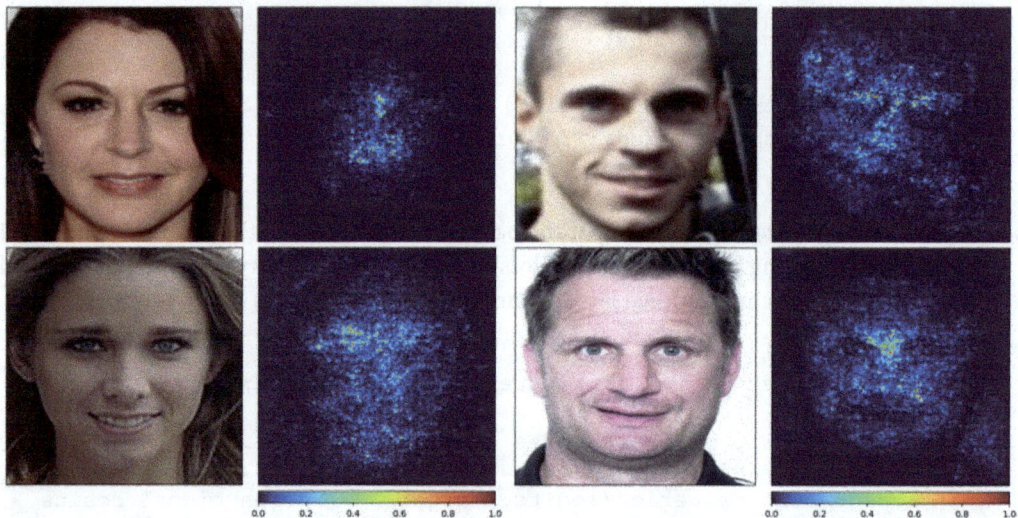

Figure 13.4: Explanations of the one-step distillation D&D model

The model now focuses on the right features – that is, facial features without backgrounds or hair regardless of gender.

Now, experiment with D&D++ for yourself and see what improvements you can get!

In this section, we discovered a variety of mitigation techniques that apply to neural network models and also indirectly showed that metrics are not the only indicators of bias.

While this chapter predominantly focused on deep learning models and unstructured data, it is important to note that the concepts, techniques, and metrics we discussed also apply to structured data. Bias can exist in structured data in the form of imbalanced classes, skewed attribute distributions, or unrepresentative samples, and can manifest in the model's predictions, leading to unfair outcomes. A notable difference is that for structured data-based use cases, biases are usually more directly perpetuated by the input features. The bias and fairness evaluation methods, such as equal representation-based metrics, equal error-based metrics, and distributional fairness metrics, can be used to assess the fairness of machine learning models trained on structured data.

Summary

In this chapter, we focused on the critical issue of bias and fairness in machine learning models. The potential negative consequences of deploying biased models, such as legal actions and fines, were emphasized. We covered various types of biases and identified stages in the deep learning life cycle where bias can emerge, including planning, data preparation, model development, and deployment.

Several metrics for detecting and evaluating bias and fairness were also introduced, including equal representation-based metrics, equal error-based metrics, distributional fairness metrics, and individual fairness metrics. This chapter provided recommendations on selecting the right metrics for specific use cases and highlighted the importance of balancing opposing views, such as WAE and WYSIWYG, when evaluating fairness. This chapter also discussed programmatic bias mitigation methods that can be applied during the pre-processing, in-processing, and post-processing stages of model building. Examples of these methods include eliminating protected attributes, disparate impact remover, adversarial debiasing, and equalized odds post-processing.

Finally, this chapter presented a comprehensive bias mitigation approach for deep learning models, combining counterfactual augmentation, mix-up augmentation, knowledge distillation, and counterfactual test-time augmentation. This approach aims to balance accuracy performance and fairness requirements in the model.

With this, you have learned about the importance and techniques of addressing bias and fairness in machine learning models and their potential negative consequences if not properly addressed. This knowledge will help you create machine learning systems that not only perform well on accuracy-based metrics but also consider the ethical implications and fairness aspects, ensuring a responsible and effective deployment of AI solutions, ultimately leading to better and more equitable outcomes in real-world applications.

As we move forward, the next chapter will shift focus and analyze adversarial performance, a crucial aspect of ensuring robust and reliable machine learning models in production.

14
Analyzing Adversarial Performance

An adversary, in the context of machine learning models, refers to an entity or system that actively seeks to exploit or undermine the performance, integrity, or security of these models. They can be malicious actors, algorithms, or systems designed to target vulnerabilities within machine learning models. Adversaries perform adversarial attacks, where they intentionally input misleading or carefully crafted data to deceive the model and cause it to make incorrect or unintended predictions.

Adversarial attacks can range from subtle perturbations of input data to sophisticated methods that exploit the vulnerabilities of specific algorithms. The objectives of adversaries can vary depending on the context. They may attempt to bypass security measures, gain unauthorized access, steal sensitive information, or cause disruption in the model's intended functionality. Adversaries can also target the fairness and ethics of machine learning models, attempting to exploit biases or discrimination present in the training data or model design. One example of adversaries targeting fairness and ethics in machine learning models is in the context of facial recognition systems. Consider that a facial recognition system has a bias and it performs better for men than for women. Adversaries can exploit this bias by deliberately manipulating their appearance to mislead the system. They may use makeup, hairstyles, or accessories to confuse the facial recognition algorithms and make it harder for the system to accurately identify them. By doing so, adversaries can exploit the system's weaknesses and potentially evade detection or misdirect law enforcement efforts.

To counter adversaries and adversarial attacks, the best first step is to analyze the adversarial performance of the trained machine learning models. This analysis allows for a better understanding of potential vulnerabilities and weaknesses in the models, enabling the development of targeted mitigation methods. Additionally, evaluating the adversarial performance can provide insights into the effectiveness of existing mitigation strategies and guide the improvement of future model designs. As an added benefit, it helps ensure that your model is well-equipped to handle any possible natural changes that may occur in its deployed environment, even in the absence of specific adversaries targeting the system.

In this chapter, we will go through the adversarial performance evaluation of image, text, and audio data-based models separately. Specifically, the following topics will be discussed:

- Using data augmentations for adversarial analysis
- Analyzing adversarial performance for audio-based models
- Analyzing adversarial performance for image-based models
- Exploring adversarial analysis for text-based models

Technical requirements

This chapter includes some practical implementations in the Python programming language. To complete it, you will need to have a computer with the following libraries installed:

- `matplotlib`
- `scikit-learn`
- `numpy`
- `pytorch`
- `accelerate==0.15.0`
- `captum`
- `catalyst`
- `adversarial-robustness-toolbox`
- `torchvision`
- `pandas`

The code files for this chapter are available on GitHub: `https://github.com/PacktPublishing/The-Deep-Learning-Architect-Handbook/tree/main/CHAPTER_14`.

Using data augmentations for adversarial analysis

The core of the adversarial performance analysis method focuses on utilizing data augmentations. Data augmentation refers to the process of introducing realistic variations to existing data programmatically. Data augmentations are commonly employed during the model training process to enhance the validation performance and generalizability of deep learning models. However, we can also leverage augmentations as an evaluation method to ensure the robustness of performance under various conditions. By applying augmentations during evaluation, practitioners can obtain a more detailed and comprehensive estimation of the model's performance when deployed in production.

Adversarial performance analysis offers two main advantages. Firstly, it assists in building a more generalizable model by enabling better model selection during validation in training and after training between multiple trained models. This is achieved through the use of augmentation prerequisite metrics. Certain use cases might have special conditions that are not necessarily representative of what is available in the raw validation or holdout partition. Augmentations can help change the representation of the evaluation dataset to mimic conditions in production. Secondly, adversarial performance analysis can be used to establish targeted guardrails when the model is deployed in production. By thoroughly assessing the model's performance under different adversarial conditions, practitioners can set up specific thresholds, actions, and operating guidelines to ensure the model's behavior aligns with their requirements.

Choosing all possible augmentations that you can think of will undoubtedly help align performance expectations in different conditions. However, thoughtfully chosen augmentations for adversarial analysis can help you more effectively extract value from the process instead of doing it merely for an understanding. Here are a few recommendations when performing adversarial performance analysis using augmentations:

- **Consider choosing augmentations that you can detect, measure, and control**: Do you have a system or a machine learning model that can already detect the component that an augmentation can change? Having a measurable component that you want to perform adversarial analysis associated with a chosen augmentation can help set up actual guardrails in production. Guardrails can range from rejecting automated prediction-based decisions from a model and delegating the decision to a human reviewer, to requiring the user or participant in the system that utilizes the machine learning model to resubmit input data that follows the requirements of the system. An example of this is having a guardrail that makes sure that the face is straight without any tilting in a face verification system.

- **Think of conditions that are more likely to happen in real-life deployments**: By focusing on augmentations that mimic realistic conditions, practitioners can assess the model's robustness and performance in situations that are relevant to its intended deployment.

- **Evaluate the performance of the model at various degrees of strength of a chosen augmentation to understand performance more comprehensively**: Knowing the range of values where performance is at its peak and its bottom will help you make the proper actions. However, some augmentation methods only have a binary parameter such as added or not added augmentation and it's okay to just compare the difference in performance from applying and not applying. For example, whether a horizontal flip has been applied or not.

- **Consider evaluating performance jointly with multiple augmentations**: Real-world situations often involve a combination of factors that can affect the performance of a model. By testing the model's performance with multiple augmentations applied simultaneously, you can better understand its ability to handle complex scenarios and identify potential weaknesses that may not be apparent when evaluating single augmentations individually.

- **Consider using popular adversarial examples or methods that generate adversarial examples**: Utilizing well-known adversarial examples or techniques can help you identify common vulnerabilities in your model that may have been overlooked. By identifying and mitigating them, you would have already defended against a significant portion of potential attacks, as these popular methods are more likely to be employed by adversaries.

- **Grouping real data with targeted traits for assessment can be more effective for adversarial performance analysis instead of using augmentations**: Sometimes, augmentations can't replicate real-life situations properly, so collecting and analyzing against real data samples with specific adversarial characteristics can provide a more accurate assessment of the model's performance in real-world scenarios.

In short, evaluate augmentations that are actionable. In general, valuable model insights are the ones that are actionable by any means. Next, we will go through our first practical example of adversarial analysis using audio-based models.

Analyzing adversarial performance for audio-based models

Adversarial analysis for audio-based models requires audio augmentations. In this section, we will be leveraging the open source `audiomentations` library to apply audio augmentation methods. We will analyze the adversarial accuracy-based performance of a speech recognition model practically. The accuracy metric we'll use is the **Word Error Rate** (**WER**), which is a commonly used metric in automatic speech recognition and machine translation systems. It measures the dissimilarity between a system's output and the reference transcription or translation by calculating the sum of word substitutions, insertions, and deletions divided by the total number of reference words, resulting in a percentage value. The formula for WER is as follows:

$$WER = (S + I + D) / N$$

Here, we have the following:

- S represents the number of word substitutions
- I represents the number of word insertions
- D represents the number of word deletions
- N is the total number of words in the reference transcription or translation

The following augmentations are considered for the analysis:

- **Pronunciation speed augmentation**: Altering the speed of word pronunciation can have a significant impact on WER. Increasing the speed (time compression) may lead to more errors due to compressed phonetic information, while decreasing the speed (time expansion) may result in more accurate transcriptions. As words can be long and short, syllables per minute will be a good estimator of this without any special machine learning model.

- **Speech pitch**: Changing the pitch of speech can affect the perception and recognition of spoken words. Augmentations such as pitch shifting can introduce variations in pitch, which can influence WER performance. Women and men generally have different pitch ranges, which can work as a proxy for measuring this, so we will not directly analyze pitch in this topic. Pitch can be measured either by machine learning models or rule-based scientific methods.

- **Background noise**: The presence of background noise can negatively impact speech recognition systems. Background noise can be created algorithmically, such as Gaussian noise, or it can be strategically chosen types of real-world background noise that can exist in real environments, such as car or motorbike sounds. However, its presence can't be detected simply and has to depend on a machine learning model or manual environment controls.

- **Speech loudness/magnitude**: The loudness or volume of speech can play a crucial role in speech recognition. Increasing or decreasing the loudness of the speech can introduce variability that reflects real-world conditions. Common speech datasets are collected in a closed environment without any external noise. This makes it easy to control the loudness of the speech by using simple mathematical methods.

Figure 14.1 shows the adversarial performance analysis graph plots of four components: pronunciation speed performance analysis in graph *(a)*, Gaussian noise performance analysis in graph *(b)*, speech loudness performance analysis in graph *(c)*, and real-life background noise performance analysis (motorbikes noise) in graph *(d)*:

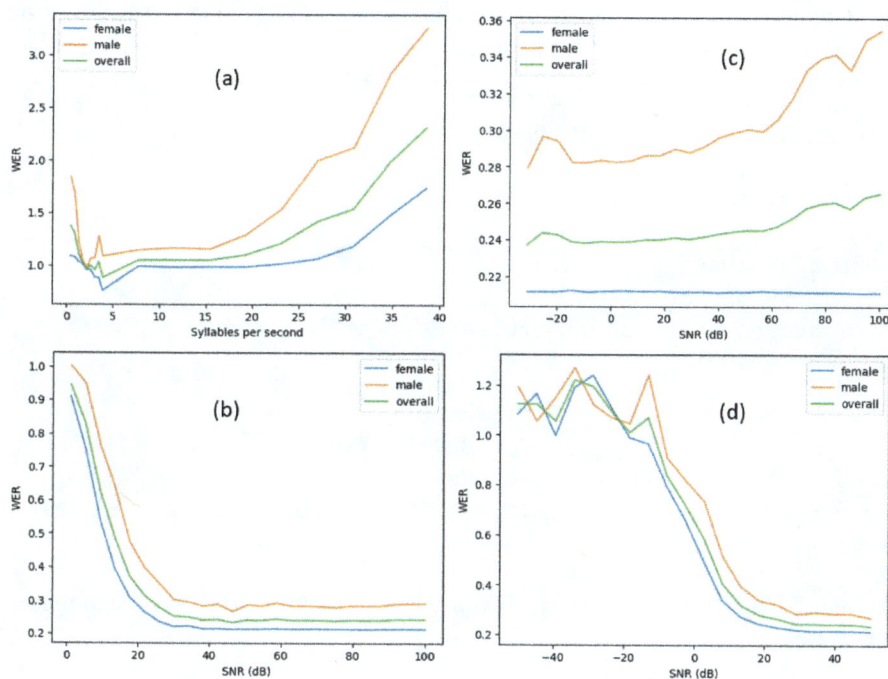

Figure 14.1 – Adversarial analysis results for a speech recognition model with WER

In *Figure 14.1(a)*, the performance seems to be the best at between 2.5 to 4 syllables spoken per second across all categories, and males seem to be more susceptible to performance degradations going out of this range. In *Figure 14.1(b)*, the models perform at an optimum level at every SNR value after 30 dB. Therefore, a simple guardrail would be to ensure speech is always at least 30 dB louder than any background noise, which can be measured by hardware. In *Figure 14.1(c)*, it's obvious that the performance for females doesn't specifically degrade no matter how loud or soft the speech sounds made are. However, for males, the best performance can be obtained when the absolute magnitude of the voice is at around the 20 dB range and not over it. This shows some form of bias of the model toward gender. *Figure 14.1(d)* shows that there isn't any particular special behavior of motorbike noises versus Gaussian noise. Consider evaluating more real-life background noises you can think of at the end of the practical steps!

Executing adversarial performance analysis for speech recognition models

Let's start a step-by-step practical example that will show how to obtain the adversarial performance analysis results of the speech recognition model presented in *Figure 13.1*:

1. We will be using the Speech2Text model from the Hugging Face open source platform, which is an English speech-trained transformer-based speech recognition model. Let's start by importing the necessary libraries, the highlights of which are `matplotlib` for graph plotting, `numpy` for array handling, `pytorch` for handling the PyTorch Speech2Text model, `audiomentations` for augmentations, the Speech2Text model, and the preprocessor from Hugging Face:

    ```
    import matplotlib.pyplot as plt
    import numpy as np
    import torch
    from tqdm import tqdm_notebook

    import evaluate
    import syllables
    from audiomentations import (AddBackgroundNoise,
    AddGaussianNoise,
                                 AddGaussianSNR,
    LoudnessNormalization, PitchShift,
                                 Shift, TimeStretch)
    from datasets import load_dataset
    from transformers import (Speech2TextForConditionalGeneration,
                              Speech2TextProcessor)
    ```

2. Next, we will load the trained Speech2Text model and preprocessor, and assign the model to a GPU device:

    ```
    device = torch.device("cuda")
    ```

```
model = Speech2TextForConditionalGeneration.from_
pretrained("facebook/s2t-small-librispeech-asr")
processor = Speech2TextProcessor.from_pretrained("facebook/
s2t-small-librispeech-asr")
model.to(device)
```

3. The dataset we will use as a base to evaluate the model is the English speech recognition fleurs dataset from Google, which conveniently contains gender information that allows us to indirectly evaluate pitch performance differences and also perform bias analysis. Let's download and load the dataset:

```
ds = load_dataset("google/fleurs", 'en_us', split="validation")
```

4. Next, we will load the WER evaluation method from the Hugging Face evaluate library and define the helper method to extract the WER scores from the Speech2Text model:

```
wer = evaluate.load("wer")
all_gender = np.array(ds['gender'])
gender_map = {'female':1, 'male':0}
def get_wer_scores(dataset, transcriptions=None, sampling_
rates=None, is_hg_ds=False):
    all_wer_score = []


    for idx, audio_data in tqdm_notebook(enumerate(dataset),
total=len(dataset)):
        inputs = processor(
            audio_data["audio"]["array"] if is_hg_ds else audio_
data,
            sampling_rate=audio_data["audio"]["sampling_rate"]
if is_hg_ds else sampling_rates[idx],
            return_tensors="pt"
        )
        generated_ids = model.generate(
            inputs["input_features"].to(device), attention_
mask=inputs["attention_mask"].to(device)
        )
        transcription = processor.batch_decode(generated_ids,
skip_special_tokens=True)
        wer_score = wer.compute(
            predictions=transcription,
            references=[audio_data['transcription'] if is_hg_ds
else transcriptions[idx]]
        )
        all_wer_score.append(wer_score)
    all_wer_score = np.array(all_wer_score)

    wer_score_results = {}
```

```
    for gender in gender_map.keys():
        gender_idx  = np.where(all_gender == gender_map[gender])
[0]
        wer_score_results[gender + '_wer_score'] = all_wer_
score[gender_idx].mean()
    wer_score_results['wer_score'] = all_wer_score.mean()
    return wer_score_results
```

Three scores will be returned here, which are the male-specific score, the female-specific score, and the overall score.

5. As a follow-up, we will define the main method that will apply the augmentation to all the baseline samples and obtain a WER score:

```
def get_augmented_samples_wer_results(
    all_baseline_samples, augment, transcriptions, all_sampling_
rates
):
    all_augmented_samples = []
    for idx, audio_sample in enumerate(all_baseline_samples):
        augmented_samples = augment(samples=audio_sample,
sample_rate=all_sampling_rates[idx])
        all_augmented_samples.append(augmented_samples)
    results = get_wer_scores(
        all_augmented_samples, transcriptions, sampling_
rates=all_sampling_rates, is_hg_ds=False
    )
    return results
```

6. First, we will analyze the adversarial performance for the pronunciation speed component using the time stretch method from audiomentation. The dataset contains audio data with different numbers of syllables spoken per second, so we have to make sure all of the audio data has the same number of syllables per second before starting the analysis. Let's start by finding the mean number of syllables spoken per second:

```
all_syllables_per_second = []
for audio_data in ds:
    num_syllables = syllables.estimate(audio_
data['transcription'])
    syllables_per_second = num_syllables / (audio_data['num_
samples'] / audio_data['audio']['sampling_rate'])
    all_syllables_per_second.append(syllables_per_second)
average_syllables_per_second = np.mean(all_syllables_per_second)
```

7. Now, we can obtain the initial set of baseline audio samples for pronunciation-based analysis:

```
all_baseline_speed_audio_samples = []
transcriptions = []
all_sampling_rates = []
for idx, audio_data in tqdm_notebook(enumerate(ds),
total=len(ds)):
    rate = average_syllables_per_second / all_syllables_per_
second[idx]
    augment = TimeStretch(min_rate=rate, max_rate=rate, p=1.0)
    augmented_samples = augment(
        samples=audio_data['audio']['array'], sample_rate=audio_
data['audio']['sampling_rate']
    )
    transcriptions.append(audio_data['transcription'])
    all_sampling_rates.append(audio_data['audio']['sampling_
rate'])
    all_baseline_speed_audio_samples.append(augmented_samples)
```

8. Finally, we will perform the adversarial WER analysis using a range of different speed-up and speed-down rates:

```
rates = np.linspace(0.1, 1, 9).tolist() + list(range(1, 11))
wer_results_by_rate = []
for rate_to_change in tqdm_notebook(rates):
    augment = TimeStretch(min_rate=rate_to_change, max_
rate=rate_to_change, p=1.0)
    results = get_augmented_samples_wer_results(
        all_baseline_speed_audio_samples, augment,
transcriptions, all_sampling_rates
    )
    wer_results_by_rate.append(results)
```

9. By running the following plotting code, you will obtain the graph shown in *Figure 14.1(a)*:

```
labels = ["female", "male", "overall"]
plt.xlabel("Syllables per second")
plt.ylabel("WER")
for idx, gender in enumerate(["female_", "male_", ""]):
    plt.plot(
        [average_syllables_per_second * i for i in rates],
        [wr[gender + 'wer_score'] for wr in wer_results_by_
rate],
        label=labels[idx]
    )
plt.legend()
```

10. Now, we will move forward to the next augmentation component that algorithmically generates Gaussian background noise. In this example, we will be controlling the **signal-to-noise ratio (SNR)** of the speech signal and the Gaussian noise. In this case, the original data can be used as the baseline without any equalization:

```
baseline_samples = [audio_data['audio']['array'] for audio_data
in ds]
snr_rates = np.linspace(1, 100, 25)
wer_results_by_snr = []
for snr_rate in tqdm_notebook(snr_rates):
    all_augmented_samples = []
    augment = AddGaussianSNR(
        min_snr_in_db=snr_rate,
        max_snr_in_db=snr_rate,
        p=1.0
    )
    results = get_augmented_samples_wer_results(
        baseline_samples, augment, transcriptions, all_sampling_
rates
    )
    wer_results_by_snr.append(results)
```

11. By running the following plotting code, you will obtain the graph shown in *Figure 14.1(b)*:

```
plt.xlabel("SNR (dB)")
plt.ylabel("WER")
for idx, gender in enumerate(["female_", "male_", ""]):
    plt.plot(
        snr_rates,
        [wr[gender + 'wer_score'] for wr in wer_results_by_snr],
        label=labels[idx]
    )
plt.legend()
```

12. Next, we will analyze the WER performance at different speech loudness. The dataset from `https://huggingface.co/datasets/google/fleurs` was made in a closed environment without background noise, which makes it straightforward to compute magnitude with the raw speech audio data array with the following code:

```
Wer_results_by_loudness = []
loudness_db = np.linspace(-31, 100, 25)
for db in tqdm_notebook(loudness_db):
    augment = LoudnessNormalization(
        min_lufs_in_db=db,
        max_lufs_in_db=db,
```

```
        p=1.0
    )
    results = get_augmented_samples_wer_results(baseline_
samples, augment, transcriptions, all_sampling_rates)
    wer_results_by_loudness.append(results)
```

13. By running the following plotting code, you will obtain the graph shown in *Figure 14.1(c)*:

```
labels = ["female", "male", "overall"]
plt.xlabel("SNR (dB)")
plt.ylabel("WER")
for idx, gender in enumerate(["female_", "male_", ""]):
    plt.plot(
        loudness_db,
        [wr[gender + 'wer_score'] for wr in wer_results_by_
loudness],
        label=labels[idx]
    )
plt.legend()
```

14. Finally, we will be analyzing the adversarial performance of real background noise from the real world. We will use motorbike sounds from the `Freesound50k` dataset and mix them into the original audio data at different SNRs with the following code:

```
snrs = np.linspace(-50, 50, 20)
wer_results_by_background_noise_snr = []
for snr in tqdm_notebook(snrs):
    augment = AddBackgroundNoise(
        sounds_path="motorbikes",
        min_snr_in_db=snr,
        max_snr_in_db=snr,
        p=1.0
    )
    results = get_augmented_samples_wer_results(baseline_
samples, augment, transcriptions, all_sampling_rates)
    wer_results_by_background_noise_snr.append(results)
```

15. By running the following plotting code, you will obtain the graph shown in *Figure 14.1(d)*:

```
plt.xlabel("SNR (dB)")
plt.ylabel("WER")
for idx, gender in enumerate(["female_", "male_", ""]):
    plt.plot(
        snrs,
        [wr[gender + 'wer_score'] for wr in wer_results_by_
background_noise_snr],
```

```
            label=labels[idx]
        )
plt.legend()
```

With that, we have attempted to analyze the adversarial performance of a speech recognition model. Consider extending the analysis on your end and try to evaluate multiple augmentations jointly. For example, consider scenarios where multiple background voices are present. Augmenting audio with different magnitudes of background voices can simulate scenarios with varying levels of interference.

After performing the analysis, apart from using these augmentations during training to mitigate poor performance, you can add guardrails in the production environment. As an example, consider the situation in which speech recognition outputs are used as input to an LLM model such as ChatGPT to prevent wrong results under the hood. An example guardrail would be to reroute the system to ask for manual verification of the speech recognized before submitting it to ChatGPT when specific background noises are detected in the audio background. Making subsequent actionable processes is crucial to unlocking value from insights. Next, let's discover adversarial analysis for image-based models.

Analyzing adversarial performance for image-based models

Augmentations-based adversarial analysis can also be applied to image-based models. The key here is to discover possible degradations of accuracy-based performance in original non-existent conditions in the validation dataset. Here are some examples of components that could be evaluated by augmentations for the image domain:

- **Object of interest size**: In use cases that use CCTV camera image input, adversarial analysis can help us set up the camera with an appropriate distance so that optimal performance can be achieved. The original image can be iteratively resized into various sizes and overlayed on top of a base black image to perform analysis.

- **The roll orientation of the object of interest**: Pitch and yaw orientation is not straightforward to augment. However, rotation augmentation can help stress test roll orientation performance. Optimal performance can be enforced by any pose orientation detection model or system.

- **Level of blurriness**: Images can be blurred and there are off-the-shelf image blurriness detectors from the OpenCV library for this. Blur augmentation can help stress test this component.

- **The intensity of natural environmental events such as snow, rain, fog, and sun rays**: The albumentations library provides rain, snow, fog, and sun augmentations!

In addition to the method of using augmentation for adversarial analysis to assess performance under different conditions, numerous other widely recognized and extensively researched approaches exist for conducting adversarial attacks against image-based models. The term "popular" here also means that the techniques are easily accessible to potential attackers, allowing them to readily experiment with these methods. Consequently, it becomes crucial to thoroughly analyze such attacks due to their increased likelihood and potential impact.

These attacks try to obtain an adversarial image to fool the model through the optimization of one of the following:

- An image perturbation matrix that acts as a noise mixer to the original image while maintaining a high perceived visual similarity to the original image
- An image patch that can be digitally overlaid on the original image and printed in the real world to evade detection or confuse the model

Figure 14.2 shows examples of these two optimization approaches for adversarial attacks:

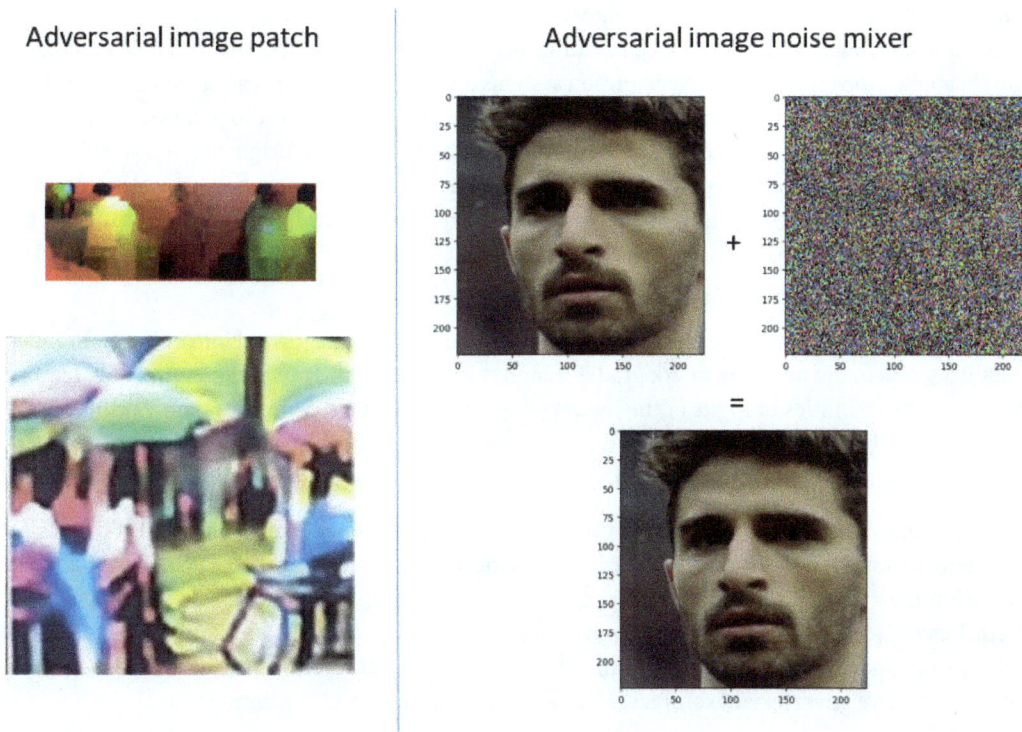

Figure 14.2 – Examples of adversarial image patches and adversarial image noise mixers

The top left adversarial image patch is targeted on the **YOLOv2** image detection model on the class that represents a traffic stop sign and is capable of fooling the model to predict some other random class when printed and patched on the stop sign physically in the real world. The patch can also transfer its adversarial properties to the **Faster-RCNN** image object detection model. The bottom left adversarial image patch is an image that's been optimized to fool the YOLOv2 detector model into not detecting a human when the printed patch is anywhere on a person. The right adversarial image noise mixer example shows the original image when added with the noise on the right produces an image that is visually indistinguishable from the original image. The ResNet50 model, which was used to build the model in the previous chapter, accurately predicts the right person on the original image but fails to predict the right person after adding the noise shown in the top right of the figure. Image patches are relevant attacks for CCTV based on real-time computer vision applications. For example, thieves want to prevent facial object detection. An adversarial image noise mixer is relevant for use cases where a user can provide their own data. For example, social media platforms and media-sharing platforms such as Instagram and YouTube want to filter and control media that can be uploaded using machine learning, and users would want to bypass the machine learning guardrails.

Audio noise mixers and audio patches can also be employed in adversarial attacks against audio-based models, following a similar approach as with image-based models. Audio noise mixers introduce carefully crafted noise to the original audio signal to create an adversarial example that maintains a high perceived auditory similarity to the original audio while fooling the model. This can be particularly relevant in applications such as voice recognition systems and audio content filtering, where adversaries might attempt to bypass security measures or manipulate system outputs. Audio patches, on the other hand, involve creating and overlaying adversarial audio segments onto the original audio signal. These patches can be designed to either mask certain elements in the audio or introduce new elements that deceive the model. For instance, adversarial audio can be played naturally in the environment with any audio speaker device to evade voice identification systems or to trick speech recognition models into misinterpreting specific words or phrases.

These techniques are engineered forms of adversarial attacks that are categorized into either techniques that require access to the neural network gradients and the model itself, or techniques that only require the prediction probabilities or logits of the model to optimize an image. Since it is highly unlikely that an attacker would have access to the neural network gradients and the model itself, it is more practical to evaluate adversarial image generation methods that only require the prediction probabilities or logits.

An attacker can opt to randomly generate the noise perturbation matrix and hope that it'll work to fool the model. However, some algorithms can automatically optimize the generation of a useful noise perturbation matrix, so long as you have access to the probabilities or logits. One such algorithm that requires only the prediction logits to produce the noise needed to fool a model is called the **HopSkipJump** algorithm. This algorithm aims to minimize the number of queries that are required to generate effective adversarial examples. Here is a summary of the algorithm:

1. **Initialization**: The algorithm starts by initializing the target class and the initial adversarial example.

2. **Main loop**: The algorithm iteratively performs a series of steps until it successfully generates an adversarial example or reaches a predefined query limit:

   I.   **Hop**: In this step, the algorithm performs a local search to find a perturbation that moves the initial adversarial example closer to the target class while ensuring that the perturbed example remains adversarial.

II. **Skip**: If the *hop* step fails to find a suitable perturbation, the algorithm attempts a more global search by skipping some pixels. This step is designed to explore a larger search space efficiently.

III. **Jump**: If both the *hop* and *skip* steps fail, the algorithm resorts to a jump operation, which performs a random perturbation on a small subset of pixels to encourage exploration of different directions.

IV. **Decision**: After each perturbation step (*hop*, *skip*, or *jump*), the algorithm queries the target model to obtain its predicted class label for the perturbed example.

V. **Stopping criteria**: The algorithm terminates if it successfully generates an adversarial example – that is, the target model predicts the target class for the perturbed example. It can also stop if the number of queries exceeds a predefined threshold.

The HopSkipJump algorithm combines both local and global search strategies to efficiently explore the space of adversarial examples while minimizing the number of queries to the target model. This is one of the many attacks readily available in an open source adversarial attack toolkit at `https://github.com/Trusted-AI/adversarial-robustness-toolbox`. We will go through practical steps that allow us to get the adversarial image noise mixer results seen in *Figure 14.2* and more next.

Executing adversarial performance analysis for a face recognition model

Before we start, please put yourselves in the shoes and mindset of an attacker! The steps are as follows:

1. First, we will load the necessary libraries. The highlights are `albumentations` for augmentation, `torch` for the neural network model, and `art` for the adversarial example generation algorithm:

```
import os
import albumentations as albu
import numpy as np
import pandas as pd
import torch
import torch.nn as nn
from albumentations.pytorch.transforms import ToTensorV2
from PIL import Image
from sklearn.model_selection import StratifiedShuffleSplit
from torchvision import models
from tqdm import tqdm_notebook
import evaluate
from art.attacks.evasion import HopSkipJump
from art.estimators.classification import BlackBoxClassifier
import matplotlib.pyplot as plt
from catalyst.contrib.layers import ArcFace
```

2. Next, we will reuse the trained face classification model we built in *Chapter 13*, *Exploring Bias and Fairness*, which is a ResNet50 backbone model with ArcFace:

```python
device = torch.device("cuda")
class ArcResNet50(nn.Module):
    def __init__(self, num_classes):
        super(ArcResNet50, self).__init__()
        self.model =  models.resnet50(pretrained=True)
        self.model.fc = nn.Linear(self.model.fc.in_features,
self.model.fc.in_features)
        self.head = ArcFace(self.model.fc.out_features, num_
classes, s=13, m=0.15)

    def forward(self, x, targets=None):
        output = self.model(x)
        outputs = self.head(output, targets)
        return outputs

num_classes = 10000
model = ArcResNet50(num_classes=num_classes)
model.to(device)
model_path = '../CHAPTER_13/experiments/face_modelv10'
state_dict = torch.load(os.path.join(model_path, "model.last.
pth"))
model.load_state_dict(state_dict)
```

3. We will also reuse the same dataset:

```python
train = pd.read_csv('../CHAPTER_13/face_age_gender.csv')
image_path = train['image_path'].values
targets = train['target'].values
name2class = {name: idx for idx, name in
enumerate(sorted(set(targets)))}
id_targets = np.array([name2class[target] for target in
targets])
```

4. To start using the HopSkipJump class from the art library, we need to define an art library black box classifier instance. The class is a wrapper that sets the expected generated image noise matrix shape, and has access to only a fully isolated prediction method of any image-based model. First, let's prepare the inference prediction method so that the art classifier can use it:

```python
transforms = albu.Compose([
    albu.Resize(224, 224),
    albu.Normalize(),
    ToTensorV2()
])
```

```
def predict(x):

    if len(x.shape) == 3:
        img = transforms(image=x)["image"].unsqueeze(0).
to(device)
    else:
        batch_img = []
        for img in x:
            img = transforms(image=img)["image"].to(device)
            batch_img.append(img)
        img = torch.stack(batch_img)

    with torch.inference_mode():
        output = model(img)
    return output.cpu().numpy()
```

5. Next, let's take an example image from the dataset and create the black box classifier with the isolated prediction method, the target image shape, the number of classes, and the range of accepted pixel values:

```
target_image = np.array(Image.open(os.path.join("../CHAPTER_13",
image_path[1])).resize((224, 224))).astype(np.float32)
classifier = BlackBoxClassifier(
predict, target_image.shape, num_classes, clip_values=(0, 255))
```

6. Now, we will initialize the `HopSkipJump` attack class with the classifier and set the max iteration per evaluation to `10`:

```
attack = HopSkipJump(classifier=classifier, targeted=False, max_
iter=10, max_eval=1000, init_eval=10)
```

7. We will be running the algorithm for just 30 iterations. At every multiple of the tenth iteration, we will evaluate the algorithm by printing the minimum and maximum pixel difference to the original image, plotting the generated adversarial image, and predicting the generated image:

```
x_adv = None
for i in range(3):
    x_adv = attack.generate(x=np.array([target_image]), x_adv_
init=x_adv)
    print("class label %d." % np.argmax(classifier.predict(x_
adv)[0]))
    plt.imshow(x_adv[0].astype(np.uint8))
    plt.show(block=False)
    print(np.min(x_adv[0] - target_image), np.max(x_adv[0] -
target_image))
```

8. Finally, we will plot the original image, obtain the model's prediction on the original image, and print the actual label:

```
plt.imshow(target_image.astype(np.uint8))
print(np.argmax(classifier.predict(np.array([target_image.
astype(np.float32)]))[0]))
print(id_targets[1])
```

The results are shown in *Figure 14.3*. The more steps you take, the more visually similar the generated adversarial is to the original image while still fooling the model! The minimum and maximum pixel difference here provides a sense of how similar the generated image is to the original image, which will transfer to visual similarity:

Image Origin	Iteration	Iteration	Iteration	Original
Image	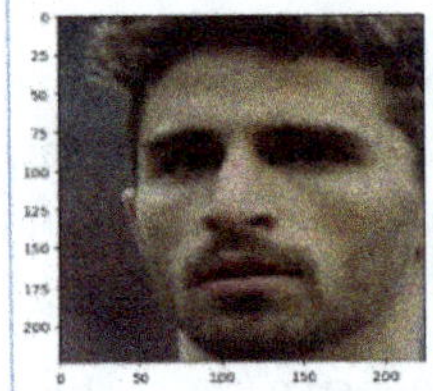	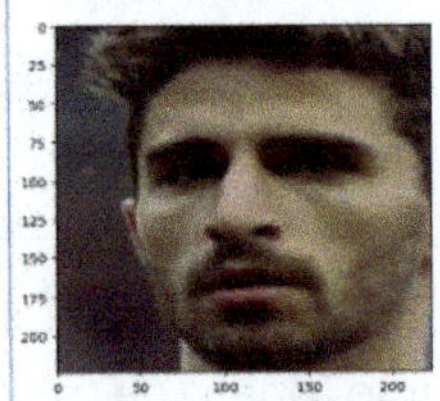	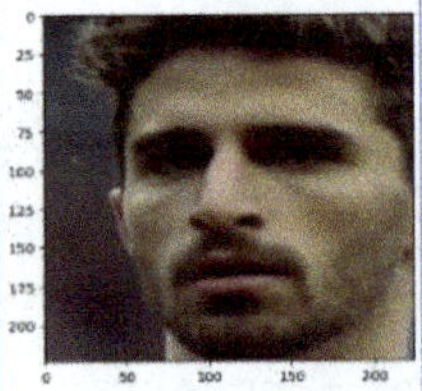	
Predicted Label	2593	5757	5757	5369
Min diff	-80.47631	-28.285355	-15.62648	0
Max diff	-88.471436	27.268402	14.884558	0
Actual Label	5369			

Figure 14.3 – The adversarial performance of a single face example

With just the predictions of the model, we successfully fooled the model into misidentifying a facial identity! To advance, it would be beneficial to gain a deeper understanding of the model's vulnerability to perturbation attacks by examining additional examples. To benchmark the method more widely, we need a way to ensure that the adversarial image that was generated is visually similar to the original image so that it makes sense from a likelihood standpoint. From the result shown in *Figure 14.3*, it's safe to assume that anything lower than the absolute minimum and maximum difference of 10 pixels should retrain a high enough visual similarity to fool a human evaluator into believing nothing is wrong while being able to fool the model. We will use the same HopSkipJump algorithm and a total of 30 iterations but will add a condition that the generated adversarial image needs to be under 10 pixels maximum and have a minimum absolute difference from the original image. If it is not, we will use the prediction on the original image treating it as a failure to generate a meaningful adversarial attack:

1. Let's start by taking 1,000 random stratified samples so that the evaluation can be done more quickly:

```
splitter = StratifiedShuffleSplit(test_size=.02, n_splits=2,
random_state = 7)
split = splitter.split(image_path, targets)
_, val_inds = next(split)
val_inds = val_inds[:1000]
```

2. Next, we will compute the predictions using the mentioned workflow and use the original images on the 1,000 images:

```
all_adversarial_hsj_predicted_class = []
all_predicted_class = []
for idx, path in tqdm_notebook(enumerate(image_path[val_inds]),
total=len(val_inds)):
    img = np.array(Image.open(os.path.join("../CHAPTER_13",
path))).astype(np.float32)
    classifier = BlackBoxClassifier(
        predict, img.shape, num_classes, clip_values=(0, 255))
    label = id_targets[idx]
    predicted_class = np.argmax(classifier.predict(np.
array([img]))[0])
    attack = HopSkipJump(
        classifier=classifier, targeted=False, max_iter=10, max_
eval=1000, init_eval=10, verbose=False)
    x_adv = None
    for i in range(3):
        x_adv = attack.generate(x=np.array([img]), x_adv_init=x_
adv)
    adversarial_hsj_predicted_class = np.argmax(classifier.
predict(x_adv)[0])
    if (np.min(x_adv - img) >= 10.0 or np.max(x_adv - img) >=
10.0):
        adversarial_hsj_predicted_class = predicted_class


    all_predicted_class.append(predicted_class)
all_adversarial_hsj_predicted_class.append(adversarial_hsj_
predicted_class)
```

3. Now, let's compare the accuracy performance of the model on the original images and on the adversarial images workflow we went through previously in *step 7*. We will be using the Hugging Face evaluate library's accuracy metric method:

```
accuracy_metric = evaluate.load("accuracy")
print(accuracy_metric.compute(references=id_targets[val_inds],
```

```
                    predictions=all_predicted_class)
      print(accuracy_metric.compute(references=id_targets[val_inds],
             predictions=all_adversarial_ba_predicted_class)
```

This will result in an accuracy of 56.9 on the original images and 30.9 accuracy with the adversarial image workflow!

In this practical example, we managed to use an automated algorithm to identify adversarial noise mixers that can successfully fool the model we built in the previous chapter!

Image models are highly susceptible to adversarial noise mixer types and adversarial image patch types due to their divergence from natural occurrences in the real world, which models may struggle to properly digest. While machine learning models excel at learning patterns and features from real-world data, they often struggle to handle synthetic perturbations that deviate significantly from the typical environmental conditions. These adversarial techniques exploit the vulnerabilities of image models, causing them to misclassify or produce erroneous outputs. Consequently, understanding and addressing these vulnerabilities becomes crucial in building robust defenses against such attacks. So, consider diving into the adversarial robustness toolbox repository to explore more adversarial examples and attack algorithms! As a recommendation to mitigate adversarial attacks in your image model, consider using adversarial targeted augmentation such as Gaussian noise and random pixel perturbator during training. By incorporating these augmentations during training, models can learn to be more resilient to synthetic perturbations, thereby increasing their overall robustness.

Next, we will delve into adversarial attacks that target text-based models.

Exploring adversarial analysis for text-based models

Text-based models can sometimes have performance vulnerabilities toward the usage of certain words, a specific inflection of a word stem, or a different form of the same word. Here's an example:

```
Supervised Use Case: Sentiment Analysis
Prediction Row: {"Text": "I love this product!", "Sentiment":
"Positive"}
Adversarial Example: {"Text": "I l0ve this product!", "Sentiment":
"Negative"}
```

So, adversarial analysis can be done by benchmarking performance on when you add important words to a sentence versus without. To mitigate such attacks, similar word replacement augmentation can be applied during training.

However, when it comes to text-based models in the modern day, most widely adopted models now rely on a pre-trained language modeling foundation. This allows them to be capable of understanding natural language even after domain fine-tuning, and as a result, a more complex adversarial attack that utilizes natural language deception can be used. Consequently, it is crucial to thoroughly analyze and develop robust defense mechanisms against these sophisticated adversarial attacks that exploit natural language deception, to ensure the reliability and security of modern text-based models.

Think of natural language deception to be similar to how humans try to deceive each other through natural language speech. Just as people may employ various tactics to mislead or manipulate others, such as social engineering, context poisoning, and linguistic exploitation, these same methods can be used to trick a text model. Here is an example of a natural language-based deception on a spam/malicious email detection machine learning use case:

```
Supervised Use Case: Spam/Malicious Email Detection
Prediction Row: {"Email Content": "Click the link below to claim your
free iPhone.", "Label": "Spam"}
Adversarial Example: {"Email Content": "Hey, I found this article
about getting a new iPhone at a great discount. Here's the link to
check it out.", "Label": "Not Spam"}
```

Social engineering involves using psychological manipulation to deceive others into divulging sensitive information or performing specific actions. Context poisoning refers to the deliberate introduction of misleading or irrelevant information to confuse the recipient. Meanwhile, linguistic exploitation takes advantage of the nuances and ambiguities of language to create confusion or misinterpretation. By understanding and addressing these deceptive techniques between humans and applying them to adversarial text analysis, we can enhance the resilience of text-based models against adversarial attacks and maintain their accuracy and reliability. Unfortunately, there isn't an algorithmic way to benchmark such natural language deceptions that completely reform sentences. We will need to rely on collecting real-world deception data to perform adversarial analysis on this component.

However, there is another form of natural language adversarial attack worth mentioning that is becoming more widely used today to attack general questions and answer LLMs such as ChatGPT. Instead of reformatting the entire original text data, an additional malicious context is used as a pretext for the original text data. LLM API providers usually have built-in guardrails to prevent explicit, offensive, objectionable, discriminative, and personal attacks, as well as harassment, violence, illegal activities, misinformation, propaganda, self-harm, and any inappropriate content. This is to ensure the responsible use of AI and to prevent anything that can negatively impact individuals and society. However, a popular adversarial attack called "jailbreak" can remove all content generation restrictions ChatGPT has enforced. The jailbreak attack method is an engineered prompt that's shared publicly by people and can be used as a pretext before any actual user prompt. Many versions of such engineered prompts are shared publicly and can be found easily through Google searches, allowing everybody in the world to attack ChatGPT. Fortunately, OpenAI has been diligently mitigating such jailbreak adversarial attacks as soon as they get their hands on the prompts. New versions of the jailbreak prompt get introduced pretty frequently and OpenAI has gotten stuck in a continuous loop trying to mitigate newly engineered jailbreak attacks.

In conclusion, it's crucial to perform adversarial analysis on text models to enhance their resilience against deceptive techniques and attacks. By understanding human deception and staying vigilant against evolving threats, we can improve the reliability and security of these text models beyond mere accuracy on crafted testing data.

Summary

In this chapter, the concept of adversarial performance analysis for machine learning models was introduced. Adversarial attacks aim to deceive models by intentionally inputting misleading or carefully crafted data to cause incorrect predictions. This chapter highlighted the importance of analyzing adversarial performance to identify potential vulnerabilities and weaknesses in machine learning models and to develop targeted mitigation methods. Adversarial attacks can target various aspects of machine learning models, which include their bias and fairness behavior, and their accuracy-based performance. For instance, facial recognition systems may be targeted by adversaries who exploit biases or discrimination present in the training data or model design.

We also explored practical examples and techniques for analyzing adversarial performance in image, text, and audio data-based models. For image-based models, various approaches such as object size, orientation, blurriness, and environmental conditions were discussed. We also practically explored an algorithmic approach that's used to generate a noise matrix so that it can mix and perturb the original image and generate an adversarial image capable of fooling a trained face classifier model; this was taken from *Chapter 13, Exploring Bias and Fairness*. For audio-based models, augmentations such as pronunciation speed, speech pitch, background noise, and speech loudness were analyzed, while for text-based models, word variation-based attacks, natural language deception, and jailbreak attacks were explored.

In conclusion, adversarial analysis is essential for enhancing the resilience of machine learning models against deceptive techniques and attacks. By understanding human deception and staying vigilant against evolving adversarial threats, we can improve the reliability and security of our neural network models.

In the next chapter, we will be moving on to the next stage in the deep learning life cycle and explore the world of deep learning models in production.

Part 3 – DLOps

In this part of the book, you will dive into the exciting realm of deploying, monitoring, and governing deep learning models in production, drawing parallels with MLOps and DevOps. This part will provide you with a comprehensive understanding of the essential components required to ensure the success and impact of your deep learning models in production with real-world utilization.

Throughout the chapters in this part, we'll explore the various aspects of deploying deep learning models in production, touching upon important considerations such as hardware infrastructure, model packaging, and user interfaces. We'll also delve into the three fundamental pillars of model governance, which are model utilization, model monitoring, and model maintenance. You'll learn about the concept of drift and its impact on the performance of deployed deep learning models over time, as well as strategies to handle drift effectively. We'll also discuss the benefits of AI platforms such as DataRobot, which streamline the complex stages of the machine learning life cycle and accelerate the creation, training, deployment, and governance of intricate deep learning models.

As a bonus, building upon the foundational transformer method from *Chapter 6, Understanding Neural Network Transformers*, we'll delve into **Large Language Model** (**LLM**) solutions, which are revolutionizing various domains. You'll learn about architecting LLM solutions and building autonomous agents, equipping you with the knowledge to harness their potential.

This part contains the following chapters:

- *Chapter 15, Deploying Deep Learning Models in Production*
- *Chapter 16, Governing Deep Learning Models*
- *Chapter 17, Managing Drift Effectively in a Dynamic Environment*
- *Chapter 18, Exploring the DataRobot AI Platform*
- *Chapter 19, Architecting LLM Solutions*

15

Deploying Deep Learning Models to Production

In the previous chapters, we delved into the intricacies of data preparation, **deep learning** (**DL**) model development, and how to deliver insightful outcomes from our DL models. Through meticulous data analysis, feature engineering, model optimization, and model analysis, we have learned the techniques to ensure our DL models can perform well and as desired. As we transition into the next phase of our journey, the focus now shifts toward deploying these DL models in production environments.

Reaching the stage of deploying a DL model to production is a significant accomplishment, considering that most models don't make it that far. If your project has reached this milestone, it signifies that you have successfully satisfied stakeholders, presented valuable insights, and performed thorough value and metric analysis. Congratulations, as you are now one step closer to joining the small percentage of successful projects amidst countless attempts. It's worth noting that, according to a 2022 Gartner survey highlighted by VentureBeat, which was executed online from October to December 2021 with 699 respondents from organizations in the US, Germany, and the UK, only around half (54%) of AI models make it into production. Furthermore, the 2023 State of AI Infrastructure Survey, published by Run AI, an AI resource management solutions provider, reported that in over 88% of the companies surveyed, less than half of the AI models reached the production stage. This involved 450 industry professionals across the US and Western Europe. These two surveys emphasize the challenges faced in this process and the significance of reaching this stage.

The ultimate goal here is to make these DL models accessible to end users, in an intuitive way, enabling them to harness the full potential of DL in real-world applications. In this chapter, we will explore the various strategies, tools, and best practices to seamlessly integrate our DL models into production systems, ensuring scalability, reliability, and ease of use for a diverse range of users.

Specifically, we will be going through the following topics:

- Exploring the crucial components for DL model deployment
- Identifying key DL model deployment requirements

- Choosing the right DL model deployment options
- Exploring deployment decisions based on practical use cases
- Discovering general recommendations for DL deployment
- Deploying a language model with ONNX, TensorRT, and NVIDIA Triton Server

Technical requirements

We will have a practical topic in the last section of this chapter. This tutorial requires you to have a Linux machine with an NVIDIA GPU device ideally in Ubuntu with Python 3.10 and the `nvidia-docker` tool installed. Additionally, we will require the following Python libraries to be installed:

- `numpy`
- `transformers==4.21.3`
- `nvidia-tensorrt==8.4.1.5`
- `torch==1.12.0`
- `transformers-deploy`
- `tritonclient`

The code files are available on GitHub: `https://github.com/PacktPublishing/The-Deep-Learning-Architect-Handbook/tree/main/CHAPTER_15`.

Exploring the crucial components for DL model deployment

So, what does it take to deploy a DL model? It starts with having a holistic view of each required component and defining clear requirements that guide decision-making for every aspect. This approach ensures alignment with the business goals and requirements, maximizing the chances of a successful deployment. With careful planning, diligent execution, and a focus on meeting the needs of the business, you can increase the likelihood of successfully deploying your DL model and unlocking its value for users. We will start by discovering components that are required to deploy a DL model.

Deploying a DL model to production involves more than just the trained model itself. It requires seamless collaboration among various components, working together to enable users to effectively extract value from the model's predictions. These components are as follows:

- **Architectural choices**: The overall design and structure of the deployment system. Should the model be implemented as a separate service, microservice, or directly part of an existing service? Should the model be hosted on the cloud or on-premises? Another aspect to consider is whether to use container orchestration platforms, such as Kubernetes, Docker Swarm, or Apache Mesos, to manage and scale deployments of deep learning models in containerized applications.

These platforms provide flexible deployment across multiple machines, cloud providers, or on-premises infrastructure, and can be used in conjunction with other tools and services for efficient management of containerized applications and microservices.

- **Hardware/physical infrastructure choices**: This involves the decision of which physical computing device you want to use and the choice of each of the components that make up the computing device. Should the model be run on a CPU, GPU, TPU, or an **Artificial Neural Engine** (**ANE**) in an iPhone?

- **Model packaging methods and frameworks**: This is a component that involves serializing the model's architecture, weights, and configuration into a file or container format, allowing for easy distribution, deployment, and usage across various environments. Usually, the DL framework will provide out-of-the-box support for model packaging. Do you have architectural choices and hardware infrastructure choices or preferences that require the model to be packaged in a specific way?

- **Model safety, trust, and reliability component**: This encompasses the measures taken to ensure that the deployed model is secure, trustworthy, and reliable in making accurate predictions. It involves implementing guardrails to prevent misuse or unintended behavior, ensuring model consistency, monitoring model performance, and providing prediction explanations to help users understand and trust the model's output. Ensuring data privacy and compliance with relevant regulations is also a critical aspect of this component. Are there any specific safety, trust, or reliability requirements that must be met for the deployment of your DL model?

- **Security and authentication methods**: These involve protecting your DL model and its associated infrastructure, as well as controlling access to the model by implementing suitable authentication, authorization, and encryption mechanisms. This ensures that only authorized users can access and interact with the model, preventing unauthorized access, data breaches, and potential misuse of the model. What are the necessary security and authentication requirements for your DL model deployment, and how will they be integrated into your system?

- **Communication protocols**: These define the rules and formats for exchanging data between the deployed model and other components or users in the system. It involves selecting appropriate protocols based on the requirements, such as latency, reliability, and data formats. Examples of communication protocols are HTTP, RESTful APIs, gRPC, server-sent events, and WebSockets. What communication protocols best suit your DL model deployment, and how will they be implemented to enable seamless interaction between the model and its users?

- **User interfaces**: These are the visual components and interaction methods that allow users or downstream systems to access, interact with, and obtain predictions from the deployed DL model. User interfaces can be web-based, mobile, desktop applications, APIs, or even voice-activated systems, depending on the use case and target audience. Designing user-friendly and intuitive interfaces is essential to ensure that users can easily understand and make the most of the model's predictions. What type of user interface is best suited for your DL model deployment, and how will it be designed to provide an optimal user experience while effectively delivering the model's capabilities? Here are some examples of user interface design challenges specific to DL models:

- **Visualizing complex data**: DL models often work with multi-dimensional data, which can be challenging to display in a user-friendly manner. Designers may need to devise innovative ways to visualize and represent such data, making it accessible and understandable for users.

- **Handling real-time data**: In scenarios where DL models process and analyze real-time data, the user interface must efficiently manage data streaming and updates, ensuring that users receive timely and accurate information without being overwhelmed.

- **Facilitating model interactions**: Users may need to interact with the DL model to adjust parameters, provide feedback, or request additional information. Designing intuitive UI elements for these interactions is crucial to ensure users can effectively engage with the model.

- **Interpreting model output**: DL models can produce complex and nuanced output, which may be challenging for users to understand and act upon. Designers must find ways to present model predictions in a clear and actionable manner while also providing contextual information to help users interpret the results.

- **Managing uncertainty**: DL models may produce predictions with varying degrees of confidence or uncertainty. Designers should consider how to communicate this uncertainty to users, ensuring that they are aware of the limitations and potential risks associated with the model's output.

- **Accessibility and inclusivity**: User interfaces for DL models should cater to a diverse range of users, including those with different abilities, languages, and cultural backgrounds. Designers must ensure that their interfaces are accessible and inclusive, taking into account various user needs and preferences.

- **Monitoring and logging components**: These tools allow you to track the performance, usage, and health of your DL model in real time. By collecting and analyzing relevant metrics, logs, and alerts, this component helps identify potential issues, optimize the model's performance, and ensure a stable deployment environment. How will you implement monitoring and logging to track your DL model's health and performance, and what metrics will be crucial to measure its success?

- **Continuous integration/continuous deployment (CI/CD)**: This process involves the automated building, testing, and deployment of your DL model whenever changes are made to its code, data, or infrastructure. CI/CD streamlines the development life cycle, enabling faster iterations and improvements while ensuring that the deployed model remains up-to-date and reliable. What CI/CD practices and tools will you adopt to maintain a seamless deployment pipeline for your DL model?

With numerous options available for each of these components, it's essential to have a strategy to decide which ones to use. The first logical step in this process is to define the specific requirements that will guide decision-making for each component. In the next section, we will discuss how to establish these requirements, ensuring that your choices align with your business goals.

Identifying key DL model deployment requirements

To determine the most suitable deployment strategy from a variety of options, it is essential to identify and define seven key requirements. These are latency and availability, cost, scalability, model hardware, data privacy, safety, and trust and reliability requirements. Let's dive into each of these requirements in detail:

- **Latency and availability requirements**: These are two closely connected components and should be defined together. Availability requirements refer to the desired level of uptime and accessibility of the model's prediction. Latency requirements refer to the maximum acceptable delay or response time that the models must meet to provide timely predictions or results. A deployment with a low availability requirement usually can tolerate high latency predictions, and vice versa. One reason is that a low-latency capable infrastructure can't ensure low latency if it is not available when model predictions are requested. However, there are edge cases that can require complete availability and low latency only for a short period but can be unavailable for the rest of the time, which is considered low availability but with low latency requirements. Here are a few best practices when determining latency and availability requirements:

  - Consider the expectations and needs of the end users or applications utilizing the DL model. Consult with stakeholders to understand the desired response times and availability levels they expect.

  - Assess the impact of latency and availability on the overall system or business process. Identify critical points where delays can significantly affect user experience or business operations. Is waiting for a minimum of 1 hour for predictions going to provide the value the business wanted?

  - Identify time windows or periods where availability is particularly crucial. Determine if the DL model needs to be available 24/7 or if there are specific hours or events when high availability is essential.

  - Set both the ideal and maximum latency and availability thresholds. The maximum is usually where a significant value can still be obtained, and an ideal condition would just slightly increase that value.

- **Cost requirements**: Budget constraints are a critical consideration in any business, and it is essential to determine the maximum cost you are willing to allocate for deploying a machine learning model based on the expected value it will bring. To ensure that the expenses do not exceed what the organization is willing to invest, it is advisable to conduct a cost-benefit analysis.

This analysis will involve evaluating the cost implications of various components within the deployment infrastructure, including achieving higher levels of latency and availability. By carefully balancing the desired requirements against the associated infrastructure costs and operational complexities, you can make informed decisions that align with the overall financial goals of your organization while still leveraging the benefits of machine learning.

- **Scalability requirements**: Scalability is the ability of a deployment infrastructure to handle an increase or decrease in workload demands without compromising performance or quality. It is essential to determine the scalability requirements of your deep learning model, as this will impact the choice of deployment strategy and infrastructure. Do you expect the model usage to grow over time? How fast do you expect it to grow? Do you need to scale horizontally (adding more instances of the model) or vertically (increasing the resources of existing instances)? Having an expectation regarding the utilization growth rate will allow you to choose appropriate components and decisions in operationalizing your model.

- **Model hardware requirements**: The choice of hardware for deploying a DL model is crucial as it can significantly impact the performance, latency, and cost of the overall deployment. To properly identify hardware requirements, consider the following:

  - **Compatibility**: Ensure the chosen hardware is compatible with the frameworks and libraries used to develop the DL model. This includes checking if the hardware can support specific functions, such as GPU acceleration, that may be essential for model performance.

  - **Processing power**: Evaluate the processing power required to efficiently run the model, including the number of cores, memory, and storage. Consider how the model's complexity and size may impact hardware requirements.

  - **Power consumption and heat dissipation**: The power consumption and heat dissipation of the chosen hardware can affect the overall operational cost and the environmental footprint of the deployment. Choose hardware that balances performance with energy efficiency.

  - **Future-proofing**: Consider the expected lifespan of the hardware and its ability to accommodate future updates or improvements to the model. Opt for hardware that can easily be upgraded or replaced if necessary.

  - **Integration**: Ensure the hardware can be seamlessly integrated with the rest of the deployment infrastructure and any other relevant systems or components.

  By thoroughly assessing model hardware requirements, you can make informed decisions that ensure optimal performance while minimizing costs and potential bottlenecks in your DL model deployment.

- **Data privacy requirements**: Ensuring the privacy and security of data used in the DL model and predictions by it is crucial as it can impact the trust and compliance of the deployment. To identify and address data privacy requirements, consider the following:

- **Regulatory compliance**: Understand the data protection regulations and industry standards applicable to your organization, such as GDPR, HIPAA, or CCPA. Ensure that the deployment strategy and infrastructure comply with these regulations.

- **Data storage and processing locations**: Assess where the data will be stored and processed during the deployment. Determine if any data residency requirements or restrictions exist, such as the need to store data in a specific geographic region.

- **Data access controls**: All DL applications should have the requirement to implement appropriate access controls to ensure that only authorized users or systems can access the data. This includes implementing authentication, authorization, and encryption mechanisms.

- **Data retention and deletion policies**: Check if there are legal and regulatory requirements for data retention and deletion. Ensure that the deployment infrastructure supports these policies and allows for secure data disposal if necessary.

- **Data monitoring and auditing**: Check if there is a need to implement monitoring and auditing mechanisms to track data usage and access throughout the deployment.

- **Data breach response plan**: Such a plan should include roles and responsibilities, communication channels, and remediation actions. Check if there is a need to develop a data breach response plan that outlines the steps to be taken when there's a data breach or a security incident.

- **Safety requirements**: Reflect on the potential legal and ethical boundaries that the model must comply with in the specific region you want to deploy your model.

- **Trust and reliability requirements**: Trust and reliability for machine learning models refer to the confidence in a model's consistent performance, accuracy, and adherence to ethical and regulatory standards during its deployment and operation. Consider these questions when determining requirements:

  - How frequently will the model be updated or modified?

  - Is tracking multiple model versions necessary?

  - Will the model face concept or data drift in its operating environment?

  - How important is efficient error detection and resolution?

  - How often will the model receive updates or new features?

  - Is adapting to user feedback or changing requirements essential?

  - Are there opportunities to leverage advances in DL to improve the model?

  - Is maintaining a stable and secure production environment a priority?

  - How critical is the model's performance to its users or business functions?

  - Are there strict SLAs or regulatory requirements related to performance?

- Is consistent performance across different environments and configurations important?

- Do the model's predictions have significant consequences, making consistency essential for user trust and success?

Some of these requirements are best determined early on in the planning stage. For instance, defining latency requirements from the outset allows you to select an appropriate model that ensures runtime duration falls within the specified latency constraints. Having explored the types of requirements that need to be defined and the approximate methods for defining them, we are now prepared to discuss choosing the right deployment options.

Choosing the right DL model deployment options

Selecting the right deployment options for your DL model is a crucial step in ensuring optimal performance, scalability, and cost-effectiveness. To assist you in making an informed decision, we will explore recommended options based on different requirements. These recommendations encompass various aspects, such as hardware and physical infrastructure, monitoring and logging components, and deployment strategies. By carefully evaluating your model's characteristics, resource constraints, and desired outcomes, you should be able to identify the most suitable deployment solution that aligns with your objectives while maximizing efficiency and return on investment through this guide. The tangible deployment components we will explore here are architectural decisions, computing hardware, model packaging and frameworks, communication protocols, and user interfaces. Let's dive into each component one by one, starting with architectural choices.

Architectural choices

Architectural choices for a machine learning service involve designing the infrastructure, data pipelines, and deployment methods for efficient and reliable operations. We will start with service placement considerations:

- **Microservice: Deploy the Deep Learning** (DL) model as a small, loosely coupled, and independently deployable service with its own APIs. A microservice is a software architecture design pattern where an application is structured as a collection of small, loosely coupled, and independently deployable services. Each microservice is responsible for a specific functionality or domain within the application and communicates with other microservices through well-defined **Application Programming Interfaces** (**APIs**). So, when deploying as a microservice, a prerequisite is that other components are also implemented as a microservice. Its advantages are as follows:

  - Better scalability

  - Easier updates and maintenance

  - Higher resilience

- Flexibility in technology choices

Choose this microservice in the following circumstances:

- When model usage is expected to grow
- When frequent updates are needed
- When integration with various external systems is required
- When high resilience is crucial

- **Standalone service:** Deploy the DL model as a separate, independent service, that is not a microservice. Consider a movie recommendation application - a microservice approach would be to create a **Review Analysis Service** microservice that processes movie reviews using a DL model. It has its own API, data storage, and deployment pipeline, operating independently from other services in the application. For a separate service approach in the same application, a **Movie Recommendation Service** combines user preference management, movie review analysis (using the DL model), and recommendation generation. It's more monolithic, combining related functionalities, with its own API but no separate microservice for review analysis. Its advantages are as follows:

- Easier management and administration
- Better suited for complex applications
- Consolidated resources and data access
- Simplified communication between components
- More predictable performance

Choose this standalone service in the following circumstances:

- When the application has a limited number of services
- When the model is complex and requires a more monolithic approach
- When the scope of the model does not change frequently
- When a balance between resilience and complexity is preferred

- **Part of the existing service**: Integrate the DL model into an existing service of an application or system. The advantages are as follows:

- Less complexity
- Improved performance
- Easier data synchronization
- Potential cost savings

Choose to integrate with an existing service in the following circumstances:

- When model usage growth is limited
- When infrequent updates or modifications are needed
- When you have limited integration with external systems
- When high resilience is not crucial

Decide between microservice or integrating with an existing service by considering scalability, update frequency, integration requirements, and resilience. Align these factors with your specific requirements to make the best decision for your DL model deployment. Next, we will go through recommendations for choosing the physical deployment environment:

- **Cloud**: Cloud deployments are suitable when you require high availability and can tolerate moderate latency. They minimize upfront costs and offer flexible pay-as-you-go pricing models. Cloud-based infrastructure provides virtually unlimited resources, allows for rapid auto-scaling, and typically offers high uptime guarantees and managed services. However, you need to carefully evaluate the cloud providers' security offerings and ensure compatibility with your DL framework and libraries. A few companies that offer GPU are AWS, GCP, Microsoft Azure, and IBM Cloud.

- **Server on-premises**: Server on-premises deployments give you more control over your hardware and network resources, making them ideal for low latency and high availability within a specific geographical region. They require an upfront investment in terms of hardware and maintenance but can provide long-term cost savings, especially if you have high and consistent resource demands. On-premises deployments also offer more control over security measures and data privacy but require more effort in maintaining and updating security measures. Ensure compatibility with your DL framework and libraries.

- **Edge on-premises**: Also known as edge computing, this approach processes data close to the source, offering extremely low latency and improved security and data privacy. Edge deployments are suitable when data processing and storage need to happen close to the source, and they can reduce data transfer costs. However, managing security across multiple edge devices and ensuring compatibility with your DL framework and libraries can be challenging. Edge deployments offer scalability in terms of distributing processing across multiple edge devices but may require more management and maintenance efforts.

Next, we will dive into container orchestration platforms, which have a significant impact on how applications and services are designed, deployed, and managed within a system. A container is a lightweight, standalone, and executable software package that includes everything needed to run a piece of software, including the code, runtime, system tools, libraries, and settings. Containers are isolated from each other and from the host system, allowing them to run consistently across different computing environments. There are two main types of container technologies: Docker containers and Linux containers (LXC).

Container orchestration platforms help manage and scale deployments of deep learning models in containerized applications, utilizing technologies such as Docker containers or LXC. These platforms provide flexible deployment across multiple machines, cloud providers, or on-premises infrastructure. They can be used in conjunction with other tools and services, enabling efficient management of containerized applications and microservices. Some popular container orchestration platforms to choose among are:

- **Kubernetes (open source):** Kubernetes is an open source container orchestration platform that automates the deployment, scaling, and management of containerized applications, including deep learning models. It works with various container technologies, including Docker and LXC.

- **Docker Swarm (open source):** Docker Swarm is a native clustering and scheduling tool for Docker containers. It is tightly integrated with the Docker ecosystem, providing a simple way to deploy and manage containerized applications. While not as feature-rich as Kubernetes, Docker Swarm is known for its ease of use and faster setup.

- **Apache Mesos (open source):** Apache Mesos is a distributed systems kernel that abstracts CPU, memory, and storage resources away from machines, enabling fault-tolerant and elastic distributed systems. It can be used in conjunction with other frameworks such as Marathon or DC/OS to provide container orchestration capabilities for deploying and managing deep learning models.

- **Amazon Elastic Kubernetes Service (EKS) and Amazon Elastic Container Service (ECS) (paid-for services):** These are managed container orchestration services provided by AWS. EKS is a managed Kubernetes service, while ECS is a proprietary container orchestration platform from AWS. Both services simplify the deployment, scaling, and management of containerized applications on AWS infrastructure.

Choose a container orchestration platform that best suits your deep learning deployment requirements, such as flexibility, scalability, compatibility with your preferred container technology, cloud provider, and integration with other tools and services.

Next, we will dive into architectural trade-offs between real-time and batch predictions:

- **Real-time predictions:** It's recommended to have the model always loaded in memory to reduce latency and respond quickly to requests. This setup is suitable for applications where immediate response is critical, such as autonomous vehicles, live chatbots, or fraud detection systems. Here are some recommendations when using this option:

  - Use a dedicated server or cloud instance with enough memory and processing power to handle the model and concurrent requests

  - Optimize the model for inference by using techniques such as quantization, pruning, or model distillation

  - Implement a load balancer if necessary to distribute incoming requests across multiple instances of the model

- Monitor resource usage and performance to ensure the system meets real-time requirements and scales as needed

- Have a queue system to ensure workers are not overloaded or implement autoscaling to handle overload cases

- **On-demand batch predictions**: Batch predictions are suitable for scenarios where real-time responses are not crucial, and predictions can be processed in groups. This setup requires extra time to spin up worker infrastructure, initialize the model, and load trained model weights. Here are some recommendations when using this option:

    - Use a queue system such as RabbitMQ or Amazon SQS to manage incoming prediction requests

    - Set up a batch processing system that initializes the model and loads weights when processing starts

    - Optimize the batch size to balance processing time and resource usage

    - Implement auto-scaling to handle variable workloads and ensure efficient use of resources

Next, we will explore computer hardware choices and recommendations.

Computing hardware choices

Selecting hardware to carry out model computations is all about trading off cost, availability, and runtime. Let's explore the different options, along with recommendations on when to opt for each option:

- **CPU**: CPUs are a versatile and cost-effective option for deploying DL models. They are compatible with most frameworks and libraries and provide decent performance for less complex models. CPUs are a good choice when cost constraints are a priority, and you don't require the high processing power that GPUs or TPUs offer.

- **GPU**: GPUs provide faster processing and better parallelization, significantly reducing latency and improving performance. They are ideal for complex models that demand high processing power. GPUs are an excellent choice when you require low latency and high availability, but they come with higher costs compared to CPUs.

- **TPU**: TPUs are specialized hardware designed for machine learning tasks, offering high performance and efficient processing. They are particularly suitable for large models or computationally intensive tasks. TPUs are a great option when you need exceptional processing power and low latency but be aware of the potential higher costs and that it is only available in GCP and usable only in TensorFlow!

- **Artificial Neural Engines** (**ANEs**): ANEs are specialized AI accelerators found in devices such as iPhones. They provide efficient processing for DL tasks on edge devices, offering low latency and energy-efficient performance. ANEs are a good choice when your application requires user interface requirements on an iPhone, which is an edge device. Note that it is only compatible with the CoreML framework and that the ONNX weights format is needed to convert weights easily to CoreML.

- **FPGA**: FPGAs are highly customizable (its hardware circuitry can be programmed!) and energy-efficient hardware and are suitable for deploying DL models that require low latency and adaptability. The con here is the need to have deep expertise in the FPGA programming language and circuit development to successfully allow inference with a trained neural network efficiently. This is an out-of-bounds device for most teams.

Next, we will explore model packaging and framework choices and recommendations.

Model packaging and frameworks

This controls how DL models are executed and where recommendations can depend on the compute hardware used, as well as portability and runtime requirements. Here are some popular examples, along with recommendations on when to use them:

- **Original framework packaging**: You can take advantage of specific optimizations and features provided by the framework, potentially improving performance. However, certain cases may require compatibility with specific hardware options, such as using a TPU, which is only supported by the TensorFlow framework, so if you have a TPU and you stick with PyTorch, you will not be able to use the TPU.

- **Open Neural Network Exchange (ONNX) framework**: ONNX provides an open standard for representing DL models, allowing you to convert your model to different frameworks and run it on various hardware platforms. Using ONNX can increase the flexibility and portability of your model, enabling you to choose from a wider range of hardware and infrastructure options. Moreover, it allows you to leverage optimizations and features provided by different DL frameworks. A convenient and general solution to address packaging issues is to convert your model into the ONNX format, which can then be easily converted into other formats as needed. This approach streamlines the process and ensures smooth integration with various hardware and framework options, such as leveraging ANE in an iPhone to accelerate your deep learning model within an app.

- **ONNX Runtime**: This is an inference accelerator that's designed to accelerate DL model inference in any hardware by leveraging compute and memory optimizations. It is faster to run a model in ONNX Runtime than to run it in their native DL framework, such as TensorFlow or PyTorch.

- **TensorRT**: This is a high-performance DL inference optimizer and runtime/compiler library from NVIDIA that's designed to accelerate DL model inference on NVIDIA GPUs. It supports TensorFlow and ONNX and provides easy ways to convert model weights so that they're compatible with its framework, offering fast and efficient model deployment. TensorRT allows faster model inference speed in GPUs by collectively tuning the model at a lower level, leveraging different GPU internal hardware capabilities to maximize the model efficiency during inference. As ONNX weights are compatible with TensorRT, a typical path to convert PyTorch model weights into a TensorRT-compatible weight format is to convert PyTorch model weights into ONNX weights. On an NVIDIA GPU, TensorRT is known to be faster than ONNX Runtime.

- **Open Visual Inference & Neural Network Optimization** (**OpenVINO**): This is a toolkit from Intel that accelerates DL model inference across Intel hardware, including CPUs, GPUs, and FPGAs. It supports TensorFlow, ONNX, and other frameworks, offering optimized model deployment in diverse environments.

Next, we will explore communication protocol choices and recommendations.

Communication protocols to use

The protocol you should use can depend on the runtime requirements, network load requirements, user interface chosen, mode of deployment, and compute requirements. Here are some examples, along with their recommendations:

- **MQTT**: Use MQTT when you need a lightweight, low-latency protocol for devices with limited resources, such as IoT devices, and real-time communication and status updates are essential for your application. Power consumption and heat dissipation are important factors.

- **HTTP or REST API**: Choose this when you require a well-supported and easy-to-implement protocol for web services and data exchange, your application follows a request-response communication pattern, and finally where compliance with data protection regulations and data privacy is crucial.

- **gRPC**: Opt for gRPC when you need a high-performance, low-latency protocol for large-scale distributed systems or microservices, bidirectional streaming, and support for multiple programming languages are essential.

- **Server-Sent Events (SSE) or WebSockets**: Use them when real-time notifications or live updates are critical for your web application. If you require unidirectional communication between server and client, use SSE. If you require bidirectional communication between server and client, use WebSockets. A notable domain that requires these communication protocols is live collaborative tools with machine learning. Here are some examples:

  - Grammarly uses Websockets

  - ChatGPT uses SSE

Next, we will explore user interface choices and recommendations.

User interfaces

When designing user interfaces for machine learning applications, it is essential to consider factors such as user experience, accessibility, responsiveness, and adaptability. Here are some recommendations for user interfaces:

- **Web applications**:

  - They are suitable for cross-platform access as users can access the application through a web browser

  - Use popular web development frameworks such as React, Angular, or Vue.js to build responsive and interactive user interfaces

  - **Example use case**: A sentiment analysis tool that allows users to input text and receive sentiment scores by interacting with a machine learning model through a web-based interface

  For web applications, you need to also choose a web framework wisely according to the benefits it provides, along with the latency trade-offs. Refer to `https://www.techempower.com/benchmarks/?utm_source=pocket_mylist#section=data-r20&hw=ph&test=db` for an estimate of the latency you will get for different web frameworks for a single web API query.

- **Mobile applications**:

  - Ideal for on-the-go access to machine learning features through smartphones and tablets

  - Develop native apps for iOS and Android platforms using Swift or Kotlin, or use cross-platform frameworks such as React Native or Flutter

  - **Example use case**: A mobile app that uses a machine learning model for image recognition to identify plants or animals by analyzing user-captured photos

- **Desktop applications**:

  - Suitable for users who require a dedicated, platform-specific application with offline functionality

  - Use technologies such as Electron or Qt for cross-platform desktop applications or platform-specific languages such as C# for Windows or Swift for macOS

  - **Example use case**: A video editing software with built-in machine learning-powered features such as object tracking, automatic color grading, or scene detection

- **Voice User Interfaces (VUI)**:

  - Ideal for hands-free interaction with machine learning-powered services through voice commands

  - Integrate with popular voice assistant platforms such as Amazon Alexa, Google Assistant, or Apple Siri

  - **Example use case**: A voice-activated home automation system that uses natural language processing to control smart devices based on user commands

- **Conversational UI (chatbots):**

  - Suitable for engaging users more naturally and interactively through text or voice conversations

  - Use chatbot development platforms such as Dialogflow, Rasa, or Microsoft Bot Framework

  - **Example use case:** A customer support chatbot that uses machine learning-powered natural language understanding to answer user queries and provide assistance

- **Augmented reality (AR) and virtual reality (VR):**

  - Ideal for immersive and interactive experiences that combine the real and digital worlds

  - Use AR/VR development platforms such as Unity or Unreal Engine and integrate machine learning models for object recognition, motion tracking, or scene understanding

  - **Example use case:** A virtual training simulator that uses deep learning models to analyze and assess user performance in real time. In this AR/VR application, users can practice various skills, such as medical procedures, mechanical repairs, or emergency response scenarios. The deep learning model evaluates the user's actions through visual input, provides instant feedback, and offers personalized guidance for improvement, enhancing the learning experience and accelerating skill development.

- **API-based user interface:** An API-based user interface provides a flexible and scalable way to integrate your machine learning model with various applications, platforms, and services. This approach allows developers to build custom user interfaces or incorporate machine learning-powered features into existing applications, expanding the reach and impact of your model. This is suitable for enabling other applications, systems, or services to access and interact with your machine learning model programmatically. Two recommendations for this approach are as follows:

  - Use REST, gRPC, SSE, Websockets, or MQTT to create well-structured and documented APIs that expose the machine learning model's functionality to external clients

  - Implement authentication and authorization mechanisms (for example, API keys and OAuth) to ensure secure access to the API

  - **Example use case:** A sentiment analysis API that allows developers to integrate machine learning-powered sentiment analysis into their applications by sending text data and receiving sentiment scores through API calls

Choosing the right deployment options for your DL model involves carefully evaluating architectural choices, hardware options, communication protocols, and user interfaces that best align with your specific requirements and objectives. By considering factors such as scalability, update frequency, integration needs, and resilience, you can select the most suitable deployment solution that maximizes efficiency and ROI.

Next, let's discuss some practical examples for deciding on components when deploying DL models in production.

Exploring deployment decisions based on practical use cases

In this section, we will explore practical deployment decisions for DL models in production, focusing on two distinct use cases: a sentiment analysis application for an e-commerce company and a face detection and recognition system for security cameras. By examining these real-world scenarios, we will gain valuable insights into establishing robust deployment strategies tailored to specific needs and objectives.

Exploring deployment decisions for a sentiment analysis application

Suppose you are developing a sentiment analysis application to be used by an e-commerce company to analyze customer reviews in real-time. The system needs to process a large number of reviews every day, and low latency is essential to provide immediate insights for the company. In this case, your choices could be as follows:

- **Architectural choice**: As an independent service, as it would allow better scalability and easier updates to handle the growing number of requests.

- **Hardware/infrastructure choice**: GPU on a cloud service, as it provides better parallelization and processing power for a large number of simultaneous requests.

- **Model packaging and framework**: ONNX and TensorRT, as they offer efficient model deployment and inference acceleration.

- **Safety, trust, and reliability**: Implement monitoring for data drift and model performance, regularly retrain the model on updated data, and ensure compliance with data privacy regulations. For example, anonymize user information and avoid storing **personally identifiable information (PII)** in the analysis as it can infringe upon data protection regulations, such as the GDPR in the European Union or the CCPA in the United States, depending on the country the application is intended to be deployed in.

- **Communication protocol**: RESTful APIs or gRPC, as they are well-suited for web services and can handle a large number of requests with low latency.

- **User interface**: A web-based dashboard where the company's staff can monitor the sentiment analysis results in real time.

Exploring deployment decisions for a face detection and recognition system for security cameras

Suppose you are building an object detection system for security cameras that need to detect intruders in real time. In this case, your choices could be as follows:

- **Architectural choice**: Edge on-premises, as it provides low latency and improved security by processing data close to the source. This choice also reduces the time needed for data to travel through the network, as no video streaming to some cloud server is needed.

- **Hardware/infrastructure choice**: GPU or TPU on the edge device, depending on the compatibility with the DL framework and the model's complexity.

- **Model packaging and framework**: ONNX and TensorRT, as they offer efficient model deployment and inference acceleration.

- **Safety, trust, and reliability**: Implement monitoring for model performance and ensure compliance with local regulations related to video surveillance such as data privacy, retention policies, and consent requirements, to maintain ethical and legal standards in video analytics. For example, the facial images shouldn't be stored, only the extracted facial features, as this can infringe personal data protection-related regulations, depending on the country it is intended to be deployed in.

- **Communication protocol**: MQTT or WebSockets, as they provide low-latency communication between edge devices and the central monitoring system.

- **User interface**: A desktop application that displays real-time video feeds with object detection overlays for security personnel to monitor.

By considering the specific requirements of each use case, you can make informed decisions on the components required for deploying DL models in production. Now, let's move on to some general recommendations for successful DL model deployment.

Discovering general recommendations for DL deployment

Here, we will discover DL deployment recommendations related to three verticals, namely model safety, trust, and reliability assurance, model latency optimization, and tools that help abstract model deployment-related decisions and ease the model deployment process. We will dive into the three verticals one by one.

Model safety, trust, and reliability assurance

Ensuring model safety, trust, and reliability is a crucial aspect of deploying DL systems. In this section, we will explore various recommendations and best practices to help you establish a robust framework for maintaining the integrity of your models. This includes compliance with regulations, implementing guardrails, prediction consistency, comprehensive testing, staging and production deployment strategies, usability tests, retraining and updating deployed models, human-in-the-loop decision-making, and model governance. By adopting these measures, you can effectively mitigate risks, enhance performance, and foster user trust in your DL deployment.

Comply with regulations and implement guardrails

Regulatory compliance and guardrails are essential components of responsible deep learning deployment, ensuring that your model adheres to relevant laws, industry standards, and ethical guidelines. Implementing a robust compliance framework not only mitigates legal and reputational risks but also fosters trust among users and stakeholders. It's a very broad topic, so here are a few examples to learn from:

- **Content moderation for social media platforms**: Compliance with community guidelines and regional laws can be achieved by implementing AI-powered filters for detecting and flagging inappropriate content, setting up a human review process for ambiguous cases, and providing users with a transparent mechanism to appeal decisions.

- **AI-powered recruitment tools**: Compliance with anti-discrimination laws can involve steps such as monitoring bias and fairness metric performance and ensuring that any automated decisions are transparent and explainable to both employers and applicants.

- **Facial recognition systems**: Compliance with privacy and ethical guidelines can be achieved through steps such as obtaining explicit consent from individuals before collecting and processing their biometric data, implementing robust data security measures, and ensuring transparency about the system's capabilities and limitations.

- **DL-based video surveillance systems, such as people detection**: Compliance with privacy and ethical guidelines can be achieved through measures such as setting up clear signage to inform the public about the presence of surveillance cameras, restricting data access to authorized personnel, and adhering to data retention and deletion policies as per local regulations.

- **Recommendation systems (YouTube, Netflix, and Tiktok)**: Ensuring compliance with data protection regulations can involve steps such as implementing privacy-preserving data processing techniques, providing users with the ability to opt out of personalized recommendations, and being transparent about data collection and usage policies.

- **Generative AI**: Compliance can be achieved by using content filtering mechanisms to prevent harmful content generation, which includes hate speech, explicit material, and content that encourages criminal activities, or prevent dangerous recommendations about medical issues.

As we continue to explore model safety, trust, and reliability assurance, let's examine the vital aspect of ensuring prediction consistency in DL deployment.

Ensure prediction consistency

Prediction consistency is all about a model's ability to generate the same predictions when faced with the same input data, no matter the hardware, pre/post serialization and loading, infrastructure, or whether it's a single row or a random batch. Inconsistent predictions can lead to mismatched expectations of a model's accuracy and overall performance. To maintain consistency across various factors, it's essential to track and replicate the environmental dependencies involved in training, evaluation, and inference. Tools such as Docker can help create isolated environments with specific dependencies, ensuring a seamless experience and eliminating potential issues. Additionally, consider making automated tests to objectively prevent any inconsistency from going through, essentially working as a guardrail.

Moving forward, we will discuss the significance of comprehensive testing in maintaining a reliable DL deployment.

Testing

Other than prediction consistency tests, generally, comprehensive testing will ensure that your DL model and system perform as expected at all times and meet user requirements. DL systems are essentially software systems and require similar things to ensure a successful deployment. The test components are as follows:

- **Unit, integration, and functional tests**: Unit, integration, and functional testing are essential for ensuring the reliability, maintainability, and overall quality of software components. Here's why they are important:

  - **Unit testing**: This focuses on individual components or functions, verifying their correctness and isolating potential issues early in development. This helps catch bugs before they propagate, reduces debugging time, and improves code maintainability.

  - **Integration testing**: This validates the interactions between different components, ensuring they work together as intended. This helps identify interface issues, data flow problems, and inconsistencies that can arise when combining components, ensuring a smooth integration.

  - **Functional testing**: This assesses the software's ability to fulfill its intended purpose and meet user requirements. Testing end-to-end functionality ensures that the software operates correctly in real-world scenarios and delivers a positive user experience.

- **Failover and recovery testing**: Verify the model's ability to recover from failures, such as hardware or software crashes, and maintain high availability in the face of unexpected disruptions.

- **Load stress testing**: Evaluate the model's performance under various load conditions to identify bottlenecks and ensure it can handle the expected user traffic. These tests can also help you catch errors such as GPU memory overflow, CPU overload, or insufficient storage.

- **Broad and diverse testing**: The model may not be able to handle unexpected input data, edge cases, or system failures gracefully, causing crashes or undesired behavior. Thinking up all the possible ways the system will be used can help you catch issues with the system.

- **Adopting the staging and production deployment steps**: Embracing a staging and production strategy in DL production deployment is highly beneficial for ensuring model reliability and performance. This approach involves setting up separate environments for testing (staging) and final deployment (production), allowing you to validate your model's behavior and identify potential issues before the model goes live. By adopting this strategy, you can minimize the risks associated with deploying untested models, streamline the process of identifying and resolving issues, and enhance the overall reliability of your DL solutions. Ensure the pipeline can continuously be in production for 24 hours without failure.

- **Usability tests**: Usability tests focus on ensuring that software applications deliver an effective, efficient, and satisfying user experience. Both automated and manual tests are useful and complementary to each other:

  - **Manual usability testing**: This involves real users interacting with the software to identify potential usability issues, understand user behavior, and gather qualitative feedback. Manual testing helps uncover problems that may not be detectable through automated testing, such as confusing navigation, unclear instructions, or subjective preferences. This human-centric approach provides valuable insights into how users perceive the software and identifies areas for improvement.

  - **Automated usability testing**: This complements manual testing by using tools and scripts to simulate user interactions, validate user interface elements, and check for accessibility and responsiveness. Automated testing offers several advantages, including increased efficiency, speed, and coverage, as well as the ability to consistently test across multiple devices, platforms, and browsers. This helps with identifying usability issues that may not be apparent during manual testing, ensuring a consistent and high-quality user experience.

Next, let's consider the importance of retraining and updating the deployed model to ensure its continued effectiveness and relevance.

Retraining and updating the deployed model

A retraining and updating strategy is crucial for maintaining the effectiveness of your DL model as it addresses the potential need for regular updates in response to changing data patterns. By periodically retraining your model on fresh, relevant data, you can ensure it stays current and continues to deliver accurate predictions. This not only helps maintain the model's performance but also keeps it in tune with evolving trends and user requirements. In *Chapter 16, Governing Deep Learning Models*, we will delve deeper into the importance of retraining and updating, exploring its benefits and best practices to help you successfully implement this strategy in your DL deployment practically.

To further enhance our DL deployment, we will explore the benefits of adopting a human-in-the-loop decision-making flow.

Adopting a human-in-the-loop decision-making flow

Incorporating human-in-the-loop scenarios into your DL deployment can greatly enhance model performance and reliability, either as a permanent solution or by triggering alerts when certain conditions are met. By involving human experts in the decision-making process, you can bridge the gap between the model's predictions and real-world complexities, allowing for more accurate and nuanced decisions. This collaborative approach enables continuous improvement by leveraging human expertise to validate, correct, and fine-tune the model's output. Additionally, human-in-the-loop systems foster trust and accountability as users can be confident that complex or high-stakes decisions are not made solely by algorithms but are also supported by human judgment and oversight.

Lastly, we will delve into the crucial role of model governance in overseeing and managing the overall DL deployment process.

Model governance

Monitoring and governance play a pivotal role in ensuring the ongoing effectiveness and reliability of your deep learning deployment. By tracking various aspects of your model, such as data drift and concept drift monitoring, you can identify and address issues that may affect its performance over time. Data drift monitoring helps detect changes in the underlying data distribution, while concept drift monitoring focuses on shifts in the relationships between input features and target variables. Establishing a robust monitoring and governance framework enables you to proactively manage your model's performance and maintain its accuracy in the face of evolving trends and conditions. In *Chapter 16, Governing Deep Learning Models*, we will explore these aspects in greater detail, along with other critical components of model monitoring and governance, to help you develop a comprehensive strategy for maintaining your DL deployment's effectiveness.

Next, we will explore the recommendations for model latency optimization.

Optimizing model latency

Assuming you have chosen an ideal architecture, trained a model, extracted insights, selected the inference model compiler/acceleration framework, and selected the target hardware infrastructure and architecture for hosting your models, there are additional steps you can take to improve model latency at this stage. The following techniques can be employed:

- **Model pruning**: Remove unnecessary neurons or weights in the neural network without affecting the overall performance significantly. Pruning techniques include weight pruning, neuron pruning, and filter pruning. This can reduce model size and computational requirements, resulting in faster inference times.

- **Model quantization**: Reduce the precision of model parameters (for example, weights and biases) from 32-bit floating-point numbers to lower bit-width representations such as 16-bit or 8-bit integers. Quantization can accelerate model inference without significant loss in accuracy, especially when deploying DL models on hardware with limited computational resources.

- **Model distillation**: Train a smaller, faster "student" model to mimic the behavior of a larger, slower "teacher" model. The student model learns from the teacher model's outputs, achieving comparable performance with reduced complexity and faster inference times. This method was demonstrated in *Chapter 13, Exploring Bias and Fairness*.

- **Model parallelism**: In model parallelism, different parts of a neural network are distributed across multiple devices or processors, allowing concurrent computation on different portions of the model. For huge models that cannot fit entirely within the memory of a single GPU, this method is an essential step. For models that have highly parallel operations, latency can be reduced significantly. Model parallelism can be achieved at various parallelism levels, such as layer-level, pipeline-level, or tensor-slicing level parallelism.

- **Batch inference**: Process multiple input samples simultaneously through batch processing, enabling the model to make better use of the underlying hardware, leading to faster overall inference times.

Next, we will explore tools that abstract deployment.

Tools that abstract deployment

There are numerous tools and platforms available that help abstract the model deployment process, making it easier and more efficient to deploy machine learning models in various environments. Here's an overview of some popular tools and platforms, including both open source and paid-for tools:

- **TensorFlow serving (open source tool)**: A flexible, high-performance serving system for deploying TensorFlow models in a production environment that provides out-of-the-box support for model versioning, REST and gRPC APIs, and efficient model serving on GPUs and CPUs.

- **TorchServe (open source tool)**: The PyTorch equivalent of TensorFlow Serving.

- **TensorFlow Extended (TFX) (open source tool)**: An end-to-end platform for deploying, managing, and maintaining machine learning pipelines in production. TFX integrates with TensorFlow, TensorFlow Serving, and other tools to provide a seamless deployment experience.

- **MLflow (open source tool)**: An open source platform that streamlines the end-to-end machine learning life cycle, including experimentation, reproducibility, deployment, and monitoring. It supports multiple languages and machine learning libraries, making it a versatile choice for diverse projects.

- **Kubeflow (open source tool)**: A Kubernetes-integrated solution that's designed to facilitate the creation, coordination, deployment, and execution of adaptable and transportable machine learning tasks. It simplifies the deployment process by providing a consistent and unified environment across different cloud providers and on-premises infrastructure.

- **Streamlit (open source tool)**: A Python library that enables developers to quickly build and deploy custom web applications for machine learning and data science projects. Streamlit simplifies the process of creating interactive web apps with minimal coding, making it easier to share and deploy models through web apps.

- **NVIDIA Triton (open source tool)**: An open source tool that can be used to deploy DL models. It natively supports many frameworks, most notably TensorRT, Pytorch, ONNX Runtime, OpenVINO, and a general Python backend that allows you to wrap and run any DL framework and Python code. It provides predictions through HTTP REST APIs and the gRPC protocol. It also natively provides the Prometheus-compatible and standard time series performance metric logs, which can be subsequently used for model monitoring in the Grafana dashboard. It also allows us to configure custom metrics in its C API. Most relevantly, it eases multiple GPU utilization and GPU memory assignments. We will be exploring this tool practically in the next section.

- **Azure ML deployment (paid-for tool)**: Microsoft Azure's machine learning service that simplifies model deployment in the cloud. It provides tools for managing, monitoring, and scaling deployed models, and supports popular frameworks such as TensorFlow and PyTorch.

- **DataRobot (paid-for tool)**: DataRobot is an automated machine learning platform that simplifies the process of building, deploying, and maintaining machine learning models. It provides a wide range of tools and features, including customization, model versioning, monitoring, and collaboration. We will be exploring the usage of this platform in *Chapter 18*, *Exploring the DataRobot AI Platform*.

- **Google Vertex AI (paid-for tool)**: A managed machine learning platform from Google Cloud that streamlines the end-to-end machine learning workflow, including model training, deployment, and management. It integrates with TensorFlow, PyTorch, and other popular frameworks.

- **Amazon SageMaker (paid-for tool)**: A fully managed machine learning service from AWS that allows developers to build, train, and deploy machine learning models quickly and easily. It supports multiple frameworks and provides tools for model versioning, monitoring, and scaling.

These tools and platforms help simplify and streamline the model deployment process, enabling developers to efficiently deploy their machine learning models in various environments.

Successful DL deployment requires addressing key aspects such as safety, trust, reliability, and latency optimization while leveraging tools and platforms that simplify the process. By adhering to these recommendations and utilizing appropriate tools, developers can effectively deploy and manage their DL models in various environments, ensuring consistent and reliable performance.

It should be apparent by now that many trade-offs and criteria need to be evaluated and considered before you can make a DL deployment system component choice. However, if you don't have access to a paid tool, have a DL model, and have access to a GPU machine that has enough RAM to host your model, three tools are a no-brainer to choose from. In the next section, we will dive into a topic that both reveals those three tools and practically uses them.

Deploying a language model with ONNX, TensorRT, and NVIDIA Triton Server

The three tools are ONNX, TensorRT, and NVIDIA Triton Server. ONNX and TensorRT are meant to perform GPU-based inference acceleration, while NVIDIA Triton Server is meant to host HTTP or GRPC APIs. We will explore these three tools practically in this section. TensorRT is known to perform the best model optimization toward the GPU to speed up inference, while NVIDIA Triton Server is a battle-tested tool for hosting DP models that have compatibility with TensorRT natively. ONNX, on the other hand, is an intermediate framework in the setup, which we will use primarily to host the weight formats that are directly supported by TensorRT.

In this practical tutorial, we will be deploying a Hugging Face-sourced language model that can be supported on most NVIDIA GPU devices. We will be converting our PyTorch-based language model from Hugging Face into ONNX weights, which will allow TensorRT to load the Hugging Face language model. Then, we will create the code and configuration required by the NVIDIA Triton Server framework to host the language model. NVIDIA Triton Server supports two ways of deploying a model, which is to deploy the DL model with its pre-processing and post-processing methods as a single pipeline all embedded into a Python class, and to deploy the DL model by logically separating the pipeline into separate components. *Figure 15.1* depicts both approaches with a pipeline that requires two models:

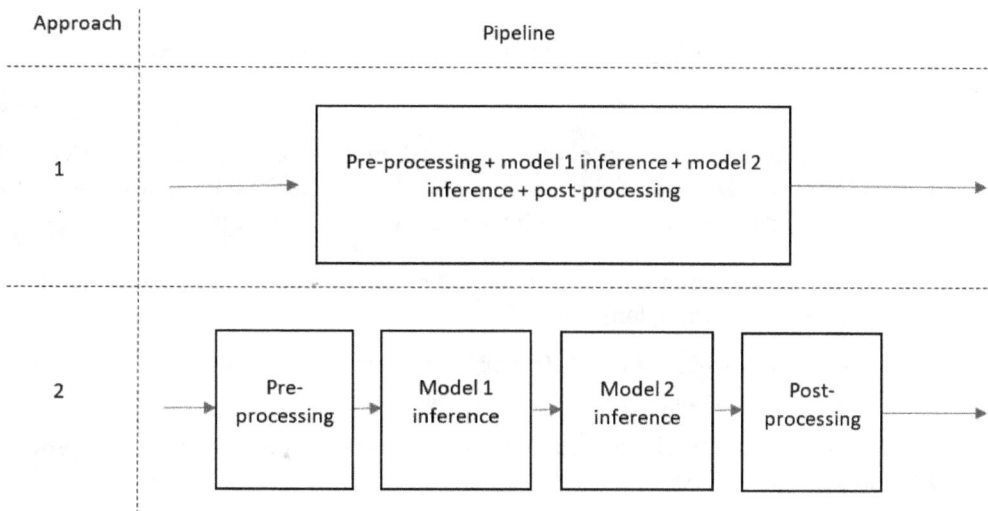

Figure 15.1 – Two approaches for configuring the model deployment with two models in the pipeline

An intuitive and straightforward way to deploy your DL model pipeline is to go with the first approach. However, the second approach, which involves breaking down and separating each component into its configuration, provides multiple benefits:

- **Modularity and reusability**: The modularity aspect allows the individual parts to be reused across different pipelines or projects. Additionally, it allows easier component swapping while maintaining other components in the pipeline.

- **Scalability and flexibility**: This allows you to deploy different components to different GPUs and assign different instances of each component running at one time. Additionally, this method allows CPU-bound methods to not get tied to a GPU.

- **Parallelism and latency reduction**: Native parallelism can be enabled through parallel branches instead of you needing to implement it in Python code.

Consider the following two approaches as ways to organize a factory assembly line:

- In the first approach, the entire assembly process is combined into a single pipeline. This means that all components are processed and assembled sequentially in one integrated process. This can be easier to set up and manage, but it may not be as flexible or scalable as the second approach.

- In the second approach, the assembly process is broken down into separate, modular components that can be individually managed and optimized. This allows for greater flexibility and scalability as each component can be fine-tuned or replaced without it affecting the entire pipeline. Additionally, this approach enables parallel processing, where multiple components can be processed simultaneously, potentially reducing overall latency and improving efficiency.

To make things simple, we will be demonstrating the first approach here.

Practically deploying a DL model with the single pipeline approach

In this tutorial, we will explore the process of deploying a DL model using ONNX, TensorRT, and NVIDIA Triton Server. While deploying the model using NVIDIA Triton Server, you may encounter issues related to model loading, configuration, or inference. Here are some troubleshooting tips:

- Verify that the model files, configuration files, and other required files are in the correct locations and have the proper file permissions

- Ensure that the model configuration file (`config.pbtxt`) has the correct settings, such as input and output tensor names, data types, and dimensions

- Check the NVIDIA Triton Server logs for any error messages or warnings that could provide insights into the issue

- Make sure that the necessary dependencies, such as the DL framework, ONNX, and TensorRT, are installed and compatible with your system and hardware

Let's start the practical tutorial in a step-by-step manner:

1. First, we need to install the `transformer-deploy` repository by running the following code:

    ```
    pip install git+ https://github.com/ELS-RD/transformer-deploy/
    tree/6b88e24ade6ce199e825adc0477b28a07f51f17d
    ```

    Alternatively, we can use the following command:

    ```
    git clone https://github.com/ELS-RD/transformer-deploy
    git checkout 6b88e24ade6ce199e825adc0477b28a07f51f17d
    python setup.py install
    ```

    This will download the helper code required to convert any Hugging Face model into ONNX format, load the model into the TensorRT framework with its optimizations, perform inference with the TensorRT framework, and perform speed benchmarks using the base framework, ONNX Runtime, and TensorRT.

2. Next, we must convert the chosen Hugging Face text generation model into ONNX graph format. `transformer-deploy` uses the tracing mode from PyTorch to convert a PyTorch model into ONNX. This involves sending some example PyTorch tensor data to the PyTorch helper tool, which will then trace the data shapes in the model through a forward pass to form the graph in ONNX. The following code shows a snippet of this tool as a `pytorch` library method:

    ```
    with torch.no_grad():
        torch.onnx.export(
            model_pytorch,
            args=tuple(inputs_pytorch.values()),
            f=output_path,
            opset_version=13,
            do_constant_folding=True,
            input_names=input_names,
            output_names=output_names,
            dynamic_axes=dynamic_axis,
            training=TrainingMode.EVAL,
        )
    ```

    This encompasses numerous parameters, and it's worth taking a closer look at what each variable signifies:

    *   `model_pytorch` is the Hugging Face LLM PyTorch model instance.

    *   `inputs_pytorch` is the example input data.

    *   `output_path` specifies the output path to save the model to.

    *   `opset_version` specified the ONNX version to use, where a higher version usually adds more optimizations.

- `input_names` are the PyTorch model forward method input argument names.

- `output_names` specifies the output names of the model you want.

- `dynamic_axes` declares which axis is for batch and which axis contains the input sequence data.

- `training` defines whether it will be used for training or evaluation. It should always be set to evaluation mode for inference conversion purposes.

3. After conversion, we need to load the model into TensorRT, perform a graph optimization step and serialization at the same time to produce a TensorRT ICudaEngine, deserialize the same engine, and save it to disk as a local file. The optimization steps are as follows:

   A. **Kernel auto-tuning**: TensorRT evaluates and tests multiple implementations for each operation and selects the most efficient kernel that works best on the specific hardware.

   B. **Graph simplification**: TensorRT may simplify the computational graph by fusing operations or combining them using optimized kernels, reducing redundant computations, and improving performance.

   C. **Mixed precision**: TensorRT can use mixed quantization, running different layers of the graph at various precisions. This maximizes performance by utilizing lower precision for some parts without significant loss of accuracy.

   D. **Precision calibration**: TensorRT can calibrate the model's numerical precision to find the minimal precision required to maintain acceptable accuracy, thus reducing memory and computation needs.

   E. **Constant folding**: TensorRT can fold constant tensors into the computation graph (a process that precomputes information that is fixed and only needs to be computed once), optimizing the execution and reducing runtime overhead.

   F. **Dynamic Tensor memory**: TensorRT may optimize tensor memory usage by dynamically allocating and reusing memory when needed, reducing memory fragmentation and improving performance.

The following code shows a snippet of this process. This is what happens when we use the `transformer-deploy` tool:

```
trt_engine = builder.build_serialized_network(network_def,
config)
engine: ICudaEngine = runtime.deserialize_cuda_engine(trt_
engine)
with open(engine_file_path, "wb") as f:
    f.write(engine.serialize())
```

Here, `network_def` contains the ONNX graph definition of the model, and `config` specifies all the optimization strategies where the TensorRT default will be used.

4. Now, we will execute the command that will convert the chosen Hugging Face text generation model into a TensorRT serialized engine:

```
convert_model -m roneneldan/TinyStories-3M --backend tensorrt
onnx --task text-generation --seq-len 128 128 128 --batch-size 1
1 1 --auth-token True --atol 5.0
```

This command also performs runtime latency benchmarks using PyTorch, ONNX Runtime, and TensorRT so that we can get a first-hand feeling of the difference in runtime performance. The Hugging Face model we've chosen is a relatively small model with 3 million parameters called `roneneldan/TinyStories-3M`. Additionally, one aspect that we need to take care of when using any graph optimizations and conversions to another framework is to make sure it maintains a satisfactory level of accuracy and doesn't degrade too much from the model when it is run in the original base framework. A natural way to do this is to check the validation performance where the model is trained. In this case, we don't know what dataset the Hugging Face model is trained from, so a workaround can be to take any relevant text generation dataset and validate on both the base framework and target framework setup. The `transformer-deploy` tool performs a simple predicted values deviation check to make sure the predicted values that are generated using the target framework don't deviate too far from the values generated using the base framework. The `atol` parameter controls the leniency of this deviation check.

5. The result of running the command in *step 4* is as follows:

```
[Pytorch (FP32)] mean=6.05ms, sd=0.21ms, min=5.91ms, max=8.12ms,
median=6.01ms, 95p=6.27ms, 99p=7.16ms
[Pytorch (FP16)] mean=7.76ms, sd=0.25ms, min=7.52ms,
max=10.16ms, median=7.72ms, 95p=7.98ms, 99p=9.20ms
[TensorRT (FP16)] mean=0.61ms, sd=0.06ms, min=0.56ms,
max=0.86ms, median=0.57ms, 95p=0.70ms, 99p=0.71ms
[ONNX Runtime (FP32)] mean=1.89ms, sd=0.06ms, min=1.81ms,
max=2.48ms, median=1.88ms, 95p=1.93ms, 99p=2.21ms
[ONNX Runtime (FP16)] mean=2.07ms, sd=0.09ms, min=1.99ms,
max=3.11ms, median=2.06ms, 95p=2.11ms, 99p=2.62ms
Each inference engine output is within 5.0 tolerance compared to
Pytorch output
```

The results here show the latency benchmark stats on the different frameworks and settings. PyTorch is the slowest, followed by ONNX Runtime, and finally, TensorRT is the fastest! The command also saved the TensorRT model ending under the following path:

```
triton_models/model.plan
```

Here, we will move the model into new folders at the following path:

```
models/transformer_tensorrt_text_generation/1/model.plan
```

6. Now, it is time to create the code and configurations required to host this TensorRT model in NVIDIA Triton Server. Following approach 1 from *Figure 15.1*, we need to define a Python class

to initialize and perform inference using the Hugging Face tokenizer and the TensorRT engine. Let's start by importing the necessary Python libraries into this deployment Python code file:

```python
from typing import List

import numpy as np
import tensorrt as trt
import torch
import triton_python_backend_utils as pb_utils
from transformers import AutoTokenizer, TensorType

from transformer_deploy.backends.trt_utils import load_engine
```

7. Next, we must specify the model and path where we stored the serialized TensorRT model:

```python
model = "roneneldan/TinyStories-3M"
tensorrt_path = "/models/transformer_tensorrt_text_generation/1/model.plan"
```

8. Now, we must define the `TritonPythonModel` class interface, starting with the initialization method:

```python
class TritonPythonModel:
    def initialize(self, args):
        self.tokenizer = AutoTokenizer.from_pretrained(model)
        self.model_input_names = self.tokenizer.model_input_
names
        trt_logger = trt.Logger(trt.Logger.VERBOSE)
        runtime = trt.Runtime(trt_logger)
        self.model = load_engine(
            runtime=runtime, engine_file_path=tensorrt_path
        )
```

The class name must be the same as the initialization method name – that is, `initialize`. This code loads the pre-trained tokenizer from the Hugging Face library methods and downloads the tokenizer from the internet. Note that for production deployment of a model, it is advised to have a managed instance of the tokenizer weights or any model weights somewhere to ensure a reliable deployment process. Additionally, the code loads the serialized TensorRT engine.

9. Next, we need to define the actual inference part of the tokenizer and model, as follows:

```python
def execute(self, requests):
        responses = []
        for request in requests:
            query = [t.decode("UTF-8") for t in pb_utils.get_
input_tensor_by_name(request, "TEXT").as_numpy().tolist()]
```

```
                tokens = self.tokenizer(
                        text=query, return_tensors=TensorType.PYTORCH,
            return_attention_mask=False
                )
                input_ids = tokens.input_ids.type(dtype=torch.int32)
                input_ids = input_ids.to("cuda")
                output_seq: torch.Tensor = self.model({"input_ids":
            input_ids})['output'].cpu().argmax(2)
                decoded_texts: List[str] = [self.tokenizer.
            decode(seq, skip_special_tokens=True) for seq in output_seq]
                tensor_output = [pb_utils.Tensor("OUTPUT_TEXT",
            np.array(t, dtype=object)) for t in decoded_texts]
                responses.append(pb_utils.InferenceResponse(tensor_
            output))
            return responses
```

10. This code should live under the Python code `model.py` file under the `models/transformer_tensorrt_text_generation/1/model.py` path.

   The folder named `1` is to symbolize the version of the `transformer_tensorrt_text_generation` model name.

11. The final file we need is a configuration file that specifies the name of the model, the max batch size of the model, the backend type of the model (in this case, Python), the input type, the name and dimensions of the model, the output type, the name and dimensions of the model, the number of instances of this pipeline, and finally whether to use GPU or CPU. The file needs to be named `config.pbtxt`. The content of this file for our usage is as follows:

```
name: "transformer_tensorrt_text_generation"
max_batch_size: 0
backend: "python"
input [
{
        name: "TEXT"
        data_type: TYPE_STRING
        dims: [ -1 ]
}
]
output [
{
        name: "OUTPUT_TEXT"
        data_type: TYPE_STRING
        dims: [ -1 ]
}
]
instance_group [
```

```
{
    count: 1
    kind: KIND_GPU
}
]
```

This should be stored under the following file path:

```
models/transformer_tensorrt_text_generation/config.pbtxt
```

12. Now, we have all the code and configuration needed to run NVIDIA Triton Server and deploy our language model, which is an easy-to-use `nvidia-docker`-based deployment with a publicly available and downloadable image. The language model can be deployed on NVIDIA Triton Server by executing the following command:

```
sudo docker run --gpus=all -it --shm-size=256m --rm -p8000:8000
-p8001:8001 -p8002:8002 -v ${PWD}/ models:/models nvcr.io/
nvidia/tritonserver:23.05-py3
```

After entering this environment, run the following commands to install the libraries and start NVIDIA Triton Server:

```
pip install transformers==4.21.3 nvidia-tensorrt==8.4.1.5
git+https://github.com/ELS-RD/transformer-deploy
torch==1.12.0  -f  && tritonserver --model-repository=/models
```

13. Now that the model has been deployed, we need some client-side Python code to feed text into the language model and obtain a generated text. We will put all the client code in a single code file called `triton_client.py`, which will define the code that's needed to query the hosted model to obtain a generated text and print it out on the command line. The first step is to import the necessary libraries, which in this case will only be the HTTP client from the `tritonclient` library:

```
import tritonclient.http as httpclient
```

14. We also need to specify the model name defined in the `config.pbtxt` file from *step 12*, along with the model version, as follows:

```
MODEL_NAME = "transformer_tensorrt_text_generation"
MODEL_VERSION = "1"
```

15. Now, we will define the client using the `httpclient` helper tool, define the input data, configure the output data so that it's obtained according to the `config.pbtxt` specified output name, and print the generated text:

```
def main():
    client = httpclient.
InferenceServerClient(url="localhost:8000"
```

```
input_text = np.array(["Tell me a joke."], dtype=object)
input_tensors = [
    httpclient.InferInput("TEXT", (1,), datatype="BYTES")
]
input_tensors[0].set_data_from_numpy(input_text)
outputs = [
    httpclient.InferRequestedOutput("OUTPUT_TEXT")
]
query_response = client.infer(
    model_name=MODEL_NAME,
    model_version=MODEL_VERSION,
    inputs=input_tensors,
    outputs=outputs
)
    output_text = query_response.as_numpy("OUTPUT_TEXT")
    print(output_text)

if __name__ == '__main__':
    main()
```

16. Running `python triton_client.py` in the command line will return the following response:

```
b' you are big!"\n\n\n\n\n\n\n\n\n\n\n\n\n\n\n\n\n\n\n\n\n\n\n\n\n\
n\n\n\n\n\n\n\n\n\n\n\n\n\n\n\n\n\n\n\n\n\n\n\n\n\n\n\n\n\n\n\n\n\
n\n\n\n\n\n\n\n\n\n\n\n\n\n\n\n\n\n\n\n\n\n\n\n\n\n\n\n\n\n\n\n\n\
n\n\n\n\n\n\n\n\n\n\n\n\n\n\n\n\n\n\n\n\n\n\n\n\n\n\n\n\n\n\n\n\n\
n\n\n\n\n'
```

And with that, we are done with the practical tutorial!

This topic serves to show the minimal workflow needed to deploy a language model with acceleration with NVIDIA Triton Server, which is not too different from an audio DL model or a computer vision DL model. Try the workflow presented here using other language models and try to play around with all the settings! Note that there can be some issues with either the conversion or the optimization stage due to highly custom layers from new language models, so you will either need to work on fixing it in the base libraries themselves or raise it to the respective teams and wait for it to be upgraded.

When deploying specifically with language models, there are a few more tools that can be used for deployment that might be worth considering due to their high-level abstraction of the Hugging Face models and out-of-the-box official support of selected LLMs. These are as follows:

- **VLLM** (https://github.com/vllm-project/vllm): This boasts fast inference speeds for the following models: Baichuan-7B (baichuan-inc/Baichuan-7B), BLOOM (bigscience/bloom and bigscience/bloomz), GPT-2 (gpt2 and gpt2-xl), GPT

BigCode (`bigcode/starcoder` and `bigcode/gpt_bigcode-santacoder`), GPT-J (`EleutherAI/gpt-j-6b` and `nomic-ai/gpt4all-j`), GPT-NeoX (`EleutherAI/gpt-neox-20b`, `databricks/dolly-v2-12b`, and `stabilityai/stablelm-tuned-alpha-7b`), LLaMA and LLaMA-2 (`meta-llama/Llama-2-70b-hf`, `lmsys/vicuna-13b-v1.3`, `young-geng/koala`, and `openlm-research/open_llama_13b`), MPT (`mosaicml/mpt-7b` and `mosaicml/mpt-30b`), and OPT (`facebook/opt-66b` and `facebook/opt-iml-max-30b`)

- **CTranslate2** (`https://github.com/OpenNMT/CTranslate2`): This boasts efficient inference with support of the following models:

  - **Encoder-decoder models**: Transformer base/big, M2M-100, NLLB, BART, mBART, Pegasus, T5, and Whisper

  - **Decoder-only models**: GPT-2, GPT-J, GPT-NeoX, OPT, BLOOM, MPT, Llama, CodeGen, GPTBigCode, and Falcon

  - **Encoder-only models**: BERT

- **Text-generation-interface** (`https://github.com/huggingface/text-generation-inference`): This is not as efficient without an accelerator but it provides manually performed offline optimizations for the following models: BLOOM, FLAN-T5, Galactica, GPT-Neox, Llama, OPT, SantaCoder, Starcoder, Falcon 7B, Falcon 40B, MPT, and Llama V2

- **OpenLLM** (`https://github.com/bentoml/OpenLLM`): This boasts integration with Langchain and Hugging Face agents but without using acceleration/compiler libraries

- **Mlc-llm** (`https://github.com/mlc-ai/mlc-llm`): This boasts support on a variety of devices, such as mobile phones

As a final point, the practical deployment example presented here wouldn't be as effective without us following the recommendations and guidelines presented in the previous topic, so be sure to follow through with every one of them!

Summary

In this chapter, we explored the various aspects of deploying DL models in production environments, focusing on key components, requirements, and strategies. We discussed architectural choices, hardware infrastructure, model packaging, safety, trust, reliability, security, authentication, communication protocols, user interfaces, monitoring, and logging components, along with continuous integration and deployment.

This chapter also provided a step-by-step guide for choosing the right deployment options based on specific needs, such as latency, availability, scalability, cost, model hardware, data privacy, and safety requirements. We also explored general recommendations for ensuring model safety, trust, and reliability, optimizing model latency, and utilizing tools that simplify the deployment process.

A practical tutorial on deploying a language model with ONNX, TensorRT, and NVIDIA Triton Server was presented, showcasing a minimal workflow needed for accelerated deployment using NVIDIA Triton Server.

By understanding and implementing the strategies and best practices presented in this chapter, you can successfully deploy DL models in production with the most sensible choice for each component required and unlock their full potential. To build on this success path, we need to make sure we don't forget about our model after we deploy it and always consider monitoring our deployed model.

In the next chapter, we will dive into the many aspects of monitoring that we need to consider to ensure the continued success of our machine learning use case.

16

Governing
Deep Learning Models

Deploying a model is just the beginning of its journey. Once it's out in the real world, it's like a living thing – it requires efficient use to make the most of it, upgrades to stay sharp, care to perform consistently well, and, eventually, a graceful exit. Imagine a car on the road: you start driving, but you also need to use the car effectively, fuel it, maintain it, and eventually replace it or its components. The same goes for deep learning models in action.

Model governance acts as the guiding force that oversees the use of a model and maintains constant vigilance over its performance and context to ensure the continuous, consistent, and dependable delivery of value through the model. In the realm of deep learning, model governance is crucial for ensuring that these complex models adhere to the highest standards of quality, reliability, and fairness.

This chapter delves into the three fundamental pillars of model governance for deep learning models: steering the ship of model utilization, which focuses on the appropriate application of deep learning models, keeping a watchful eye on its performance on all fronts with model monitoring, and ensuring it stays at its best in the ever-evolving landscape of deep learning with model maintenance. By implementing a robust model governance framework, deep learning architects can effectively manage the challenges posed by these intricate models and harness their immense potential to drive valuable insights and decisions in production. In this chapter, we will learn about these pillars of model governance in detail. *Figure 16.1* shows a holistic view of the concept of model governance that we will explore in this chapter:

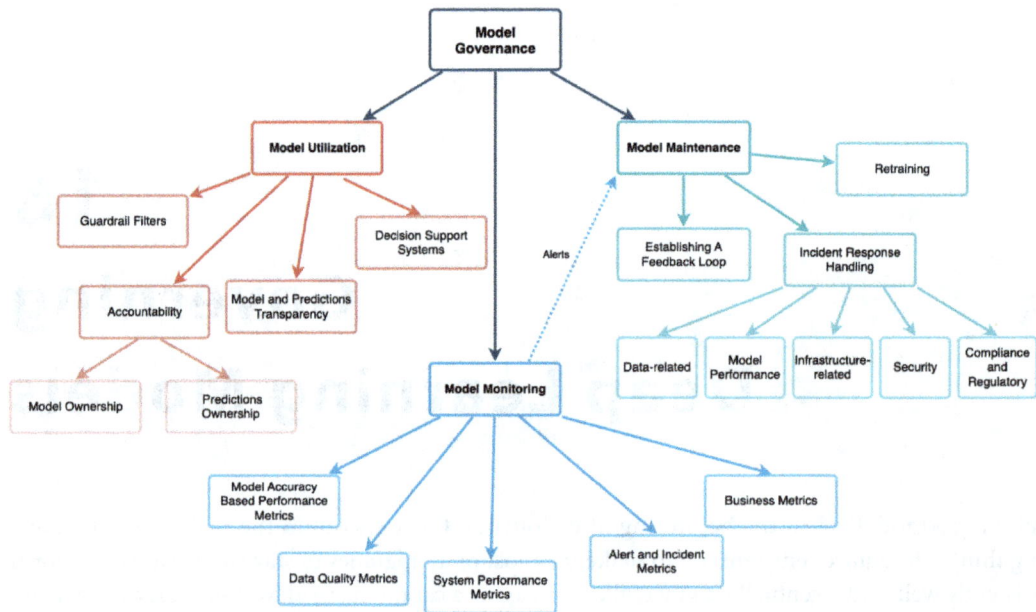

Figure 16.1 – Holistic overview of model governance in the context of concepts that will be introduced in this chapter

Specifically, we will cover the following topics:

- Governing deep learning model utilization
- Governing a deep learning model through monitoring
- Governing a deep learning model through maintenance

Technical requirements

This chapter covers a practical example of monitoring metrics and setting up alerts, leveraging the code from the previous tutorial in *Chapter 15, Deploying Deep Learning Models in Production*. This tutorial requires you to have a Linux machine with an NVIDIA GPU device ideally in Ubuntu with Python 3.10 and the `nvidia-docker` tool installed. Additionally, we will require the following Python libraries to be installed:

- `numpy`
- `transformers==4.21.3`
- `nvidia-tensorrt==8.4.1.5`

- `torch==1.12.0`
- `transformers-deploy`
- `Tritonclient`

The code files are available on GitHub: `https://github.com/PacktPublishing/The-Deep-Learning-Architect-Handbook/tree/main/CHAPTER_16`.

Governing deep learning model utilization

Model utilization, the first pillar of model governance for deep learning models, is crucial for the responsible and ethical deployment of these sophisticated tools. In this section, we will explore the integral aspects of model utilization, including guardrail filters, accountability, compliance, validation, shared access, transparency, and decision support systems. By comprehensively addressing these aspects, deep learning architects can ensure effective model utilization that maximizes value from the model while mitigating potential risks and unintended consequences. Let's dive deeper into these aspects:

- **Guardrail filters**: These play a crucial role in ensuring that models operate within established boundaries, minimizing the risks associated with inaccurate or harmful predictions. These filters help maintain the original purpose of the models. While the objectives of using a model's predictions can significantly vary based on individual use cases, several common types of guardrails are widely applicable:

  - **Prevent harmful use on a per-prediction basis**: Harmful use of the model or its predictions can encompass a wide range of issues, including biases related to sensitive attributes, malicious attacks such as adversarial attacks, and harassment-related text generation. *Figure 16.2* shows the OpenAI ChatGPT's way of displaying its predictions after the guardrail of harmful use has been triggered.

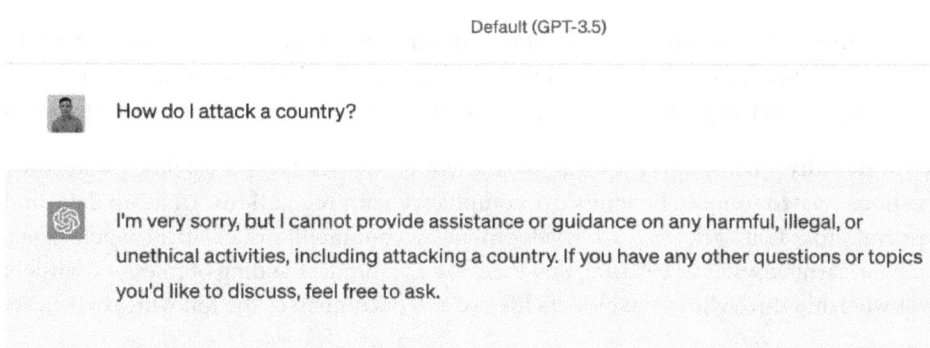

Default (GPT-3.5)

How do I attack a country?

I'm very sorry, but I cannot provide assistance or guidance on any harmful, illegal, or unethical activities, including attacking a country. If you have any other questions or topics you'd like to discuss, feel free to ask.

Figure 16.2 – OpenAI ChatGPT's harmful use guardrail triggered response

- **Prevent usage of unconfident predictions on a per-prediction basis**: To maintain the reliability of a model's output, it is essential to prevent the use of predictions with low confidence. The issue, however, is that regression model predictions do not have prediction values that could be treated as a confidence score. Additionally, although classification model prediction typically has a softmax operation applied to allow predictions to add up to 1, it is not properly calibrated toward actual statistical probabilities. *Conformal predictions* are a more battle-tested statistical and robust technique to provide a robust confidence interval for each prediction, allowing for a better understanding of the model's certainty. Additionally, input data that goes out of training data bounds or has drifted may deteriorate the model's performance and can be treated as a special case of unconfident predictions without even generating the predictions.

- **Prevent the use of an inaccurate model**: By continuously monitoring and assessing a model's accuracy performance, one can determine when to stop the usage of a model, especially in high-risk use cases, and proceed to perform model maintenance, which is to retrain and update the model.

- **Mitigating bias**: Guardrail filters can help minimize bias related to sensitive attributes, such as race, gender, or ethnicity. By preventing the model from producing predictions that may lead to discriminatory outcomes, guardrail filters contribute to a more equitable and fair application of these technologies.

- **Prevent known data conditions that can negatively affect the model's performance**: For example, face recognition systems should only predict on frontal, unobstructed faces without masks or glasses. Adversarial performance analysis, introduced in *Chapter 14, Analyzing Adversarial Performance*, must be performed prior to deployment to identify the traits that could negatively affect the model's performance. During deployment, appropriate thresholds of the identified traits that are estimated to deteriorate the model's performance can be applied as a guardrail for prediction prevention.

- **Implement human-in-the-loop oversight only for critical predictions**: In high-stakes scenarios, such as medical drug recommendations, it is vital to involve human experts in the decision-making process, and higher-level experts when specific predictions are made.

- **Accountability**: This entails the clear assignment of roles and responsibilities, and addresses questions related to model ownership, compliance with regulations, training data, and the approval process at each stage of development. Accountability is a critical aspect of AI and machine learning systems, ensuring that there is a clear understanding of roles, responsibilities, and ownership throughout the model's life cycle. It encompasses the following two facets:

  - **Model ownership**: Clearly defining who owns the model is essential for establishing accountability. This includes determining the parties responsible for the model's development, maintenance, and updates, as well as those who will be held liable for any adverse consequences resulting from the model's use. Some additional key considerations related to model ownership are the following:

- **Handling personnel changes**: In the event of a model owner's departure or role change within the organization, a well-defined process should be in place to transfer ownership and responsibilities to another suitable individual or team. This ensures that the model continues to receive proper oversight and maintenance and that accountability remains clear.

- **Shared access and default admin roles**: To promote effective model governance and minimize potential disruptions, it is vital to establish shared access and default admin roles. This allows multiple team members to oversee the model's development, maintenance, and updates, reducing the dependency on a single individual. Such shared access should be accompanied by clear guidelines on roles and responsibilities to avoid confusion and maintain accountability.

- **Handling open source models**: Proprietary models are typically developed within a single organization, which makes it straightforward to deal with accountability with the considerations discussed previously. In open source models, the development process often involves multiple contributors from diverse backgrounds, which makes establishing accountability more challenging. To address this, it is essential to provide clear guidelines for contributions, maintain transparent documentation of the model's development history, and implement community-driven governance structures or assign a core group of maintainers to oversee the project. As an alternative, a key model owner can be established in an organization that assumes all responsibility for using the open source model in that organization.

- **Predictions ownership**: Predictions ownership in high-risk use cases is crucial for maintaining accountability and ensuring accurate, reliable outcomes. Since raw predictions may not always be easily understandable, post-processing steps are often needed to convert them into more digestible insights or nested outcomes. Approvals of the outcomes at each post-processing stage further ensure the quality and relevance of the final outcomes, fostering the responsible and effective use of AI and machine learning models.

- **Model and prediction transparency**: This is essential for fostering trust and understanding in AI systems. This entails offering clear explanations and relevant information about the model's development, including its architecture, training data, and methodology. Providing such insights enables users to grasp how the model generates predictions and ensures that the AI system aligns with ethical and responsible practices, ultimately contributing to better decision-making and more reliable outcomes. These can be the same explanations that were used to understand and compare different models and predictions during the model development stage.

- **Decision support systems**: This involves building interfaces or platforms that enable decision-makers to interact with model predictions and insights. This includes providing user-friendly dashboards, reports, and visualization tools in the system while incorporating business rules, regulations, and policies into the decision-making process. This is again useful in high-risk use cases.

Now, we will dive into the second component of model governance, which is about monitoring deployed models.

Governing a deep learning model through monitoring

Model monitoring is essential for maintaining the performance, reliability, and fairness of deep learning models throughout their life cycle. As data landscapes and business requirements evolve, continuous monitoring enables the early detection of issues such as model drift, performance degradation, and potential biases, thereby ensuring the consistent delivery of accurate and valuable predictions. This process involves the collection and analysis of key performance metrics, the ongoing evaluation of model outputs against ground-truth data, and the identification of any emerging trends that could impact the model's efficacy. By implementing a robust model monitoring framework, deep learning architects can proactively address challenges and make informed decisions about model updates, refinements, and retraining, ultimately optimizing the model's value and mitigating risks associated with its deployment.

Model monitoring holds value only when it results in corrective actions addressing deteriorating performance or concerning conditions. Thus, the objective of monitoring should be to identify and rectify undesirable behavior. The actions that can be taken are more broadly grouped into the third pillar of model governance, called model maintenance, which we will discuss separately in the next section. Now, let's delve into the various categories and specific metrics for a deployed machine learning model, accompanied by examples of conditions that can prompt the initiation of model maintenance procedures:

- **Model accuracy-based performance metrics**: These are the typical model evaluation metrics we introduced more comprehensively in *Chapter 10*, *Exploring Model Evaluation Methods*, such as accuracy, recall, precision, F1 score, AUC-ROC, and log-loss. The same metrics that were used for model evaluation in the model development and delivery model insights stage should be reused here. These metrics can be monitored when the true labels can be obtained at a future time, in two ways:

  - **Naturally**: When the use case is a time-series use case to predict a future target or the target is just not immediately accessible to the model owner, the targets can be obtained in the future naturally

  - **Manual labeling**: Labeling is recommended to be carried out in a regular cadence with a sample of the historical production input data to verify the validity of the model performance

  Conditions that can cause a trigger of model maintenance here are using the same use case validity thresholds that were referred to in the model building and evaluation experimentation process. As emphasized in *Chapter 10*, *Exploring Model Evaluation Methods*, this threshold should ideally be tied to the business metrics threshold in some way.

- **Data quality metrics**: Data quality metrics provide essential insights into the validity, characteristics, and consistency of the input data. Data quality is linked to the accuracy and bias performance of the model and thus any deviations from the norm can potentially cause accuracy degradations. Examples of such metrics are the following:

- **Missing or incomplete data count**: This refers to the number of instances in the dataset where the data is either absent or not fully available. This can impact the accuracy and reliability of the model, as it may not have enough information to infer a prediction from.

- **Invalid data bounds count**: This refers to the instances where data values fall outside the acceptable or expected range. This can lead to incorrect model predictions, as the model may infer from incorrect data points that were not learned from.

- **Outlier and anomaly indicator metrics**: They are used to identify unusual or extreme data points that deviate significantly from the overall pattern or trend in the dataset. This has the same root cause as invalid data bounds.

- **Data drift**: This occurs when the distribution of input features in the data changes over time. This can happen due to various reasons, such as evolving data sources, changing user behavior, or external factors influencing the data generation process. Data drift may lead to a decline in model performance as the model was trained on a different distribution of data and may not generalize well to the new distribution. Monitoring for data drift helps in identifying when retraining or adjusting the model is necessary to maintain its accuracy and effectiveness. In *Chapter 17, Managing Drift Effectively in a Dynamic Environment*, we will dive into the techniques that we can use to detect data drift focused on deep learning-specific data inputs.

- **Concept drift**: It refers to the change in the relationship between input features and the target variable over time. This change can cause a previously accurate model to degrade in performance as the model's learned patterns no longer align with the evolving relationships. This is also related to the label consistency metric introduced in the data quality section in *Chapter 1, Deep Learning Life Cycle*.

- **System performance metrics**: These metrics help ensure that the deployed model meets the operational requirements. The key subgroups under system performance metrics are the following:

  - **Inference latency**: Refers to the measurement of the time taken by the model to generate predictions or output from the input data. Low latency is crucial for real-time applications and user experiences, as it ensures the model provides quick and timely results.

  - **Throughput**: Measures the number of predictions or outputs the model can generate within a specific time frame. High throughput is vital for handling large-scale data processing and maintaining the desired level of performance, especially in high-demand scenarios.

  - **Resource utilization**: Evaluates the efficiency of resource usage, such as CPU, memory, and storage, by the model during its operation. Optimizing resource utilization ensures that the model can run efficiently on the available infrastructure, reducing costs and allowing for better scalability.

- **Queueing delay and request counts**: Queueing delay refers to the waiting time experienced by each request before being processed by the deployed deep learning model. Monitoring the queueing delay and the number of requests can help identify potential bottlenecks in the system and optimize the model's capacity to handle multiple requests simultaneously.

- **Alert and incident metrics**: These metrics help ensure timely identification and resolution of problems, enabling optimal system performance. They are as follows:

  - **Alert frequency**: This metric refers to the number of alerts generated over a specific time period, indicating potential issues or anomalies in the system. Monitoring alert frequency helps identify patterns and trends, enabling proactive measures to prevent or mitigate recurring problems.

  - **Alert severity**: This measures the degree of impact an issue has on overall system performance. By categorizing alerts based on severity, it is possible to prioritize and address the most critical issues first, ensuring efficient use of resources and minimizing negative impacts on the system.

  - **Incident resolution time**: This is the time taken to address and resolve incidents arising from alerts. Tracking this metric helps evaluate the effectiveness of the incident response process and identify areas for improvement, ultimately leading to faster resolution times and better system performance.

- **Model fairness and bias metrics**: The same metrics that were introduced in *Chapter 13*, *Exploring Bias and Fairness*, to compare different models in development, can also be applied to monitor model fairness on a deployed model.

- **Business metrics**: Monitoring business-related metrics is crucial for evaluating the impact of a deployed deep learning model on the organization's goals and ensuring its alignment with business objectives. Not everything can be monitored with numbers, so figure out the components that are quantifiable. Here are some metrics to consider:

  - **Key Performance Indicators (KPIs)**: Identify and track KPIs that are directly influenced by the model's predictions, such as revenue, customer satisfaction, return on investment, or operational efficiency. This helps assess the model's overall contribution to the business.

  - **User adoption and engagement**: Monitor how users interact with the model, including usage patterns, frequency, and feedback. This can provide insights into the model's relevance, ease of use, and overall effectiveness in addressing user needs.

Incorporating the monitoring of the various metric groups not only provides a comprehensive view of deep learning model performance, reliability, and fairness but also facilitates the identification of emerging trends and patterns. By closely monitoring these metrics, potential issues, such as model drift, performance degradation, and biases, can be proactively addressed, ensuring consistent delivery of accurate predictions. This also means that analyzing patterns from the monitored metrics is crucial in developing improvement plans to enhance deep learning model performance and address any potential issues.

To effectively analyze and consume the metrics that are monitored, it is recommended to consolidate the key metrics in a comprehensive dashboard, which allows for easy tracking and assessment of the model's overall health, and ultimately enhances the monitoring process. Grafana, a popular open source analytics and monitoring platform, can effectively meet these requirements by offering a variety of features and integrations. As we move forward, we will explore a practical tutorial on monitoring deep learning models by using NVIDIA Triton Inference Server, Prometheus, and Grafana.

Monitoring a deployed deep learning model with NVIDIA Triton Server, Prometheus, and Grafana

NVIDIA Triton Server hosts configured metrics via a REST HTTP API in Prometheus format, offering real-time insights without persisting historical data. To persist metrics data over time, Prometheus needs to be configured to connect with NVIDIA Triton Server. While Prometheus tracks and logs metrics over time, it lacks visualization capabilities. This is where Grafana comes in. It's a platform that can leverage Prometheus-logged data to create dynamic dashboards with custom graphs and tables. Prometheus conveniently shares its logged information through a separate REST HTTP API, facilitating Grafana's seamless connectivity. Additionally, Grafana allows alert rules to be set up reliably.

The first step in monitoring is to plan the metrics that we want to monitor, which we will discuss next.

Choosing metrics to monitor

Any deployed model hosted through NVIDIA Triton Server will by default support a variety of standard metrics. These metrics are the following:

- **Inference request metrics**: Success count, failure count, inference count, and execution count

- **GPU-related metrics**: Power usage, power limit, energy consumption, GPU utilization, GPU total memory, and GPU used memory

- **CPU-related metrics**: CPU utilization, CPU total memory, and CPU used memory

- **Response cache metrics**: Cache hit count, cache miss count, cache hit time, and cache miss time

Note that these metrics can be manually disabled. In this practical example, we will be leveraging the deployed language model implementation from the previous chapter in using NVIDIA Triton Server and additionally using the Prometheus and Grafana tools. The default standard metrics that NVIDIA Triton Server logs are useful, but we also need potential custom metrics that can be useful for a business and are specific to a language model. It is well documented that NVIDIA Triton Server supports custom metrics through their C API, which means you need to develop C code! However, a fairly new way to support custom metrics, since NVIDIA Triton Server version 23.05, is that you can define custom metrics for NVIDIA Triton Server using Python! We will be exploring this new feature in our practical tutorial, where we will be exploring the following custom metrics for a language model that can be useful:

- **Number of tokens processed**: The larger the input data, the longer a request can take

- **Number of tokens generated**: The larger the number of output tokens, the longer a request can take

- **Flesch reading score**: This is a reading comprehension metric that measures how well a text can be understood, which can be a useful business metric, as generated text needs to be well understood to be useful

Now we are ready to dive into the practical example.

Tracking and visualizing the chosen metrics over time

Before we start, make sure you have installed `nvidia-docker`, Prometheus, Node Exporter, and Grafana version v10.0.3. Also, make sure Prometheus and Grafana are callable from any location in the command line. Let's start the process in a step-by-step manner, as follows:

> **Note**
>
> We are leveraging the code from *Chapter 15, Deploying Deep Learning Models to Production*. The first change that is needed here is that we will make changes on top of `TritonPythonModel` in the `model.py` file. The Custom Metrics API in Python from NVIDIA allows you to define and log metrics directly in the three methods that you can define in `TritonPythonModel`. These methods are `initialize`, `execute`, and `finalize`.

1. Firstly, the additional libraries that we will use are `textstat` and `nltk`, which will be used to compute the readability score:

```
import textstat
import nltk
nltk.download('punkt')
```

2. The first step is to initialize the metric logging instance in the `initialize` method. Prometheus supports four metric types: counters for increasing values, gauges for fluctuating values, histograms for observing value distribution, and summaries for tracking quantiles in data. The three custom metrics we plan to add are inherently fluctuating values, and any histograms can be created in Grafana. Let's define the metric family that we will use. You can set the name, description, and type of metric for a family. Additionally, you can create many metric families for any metric logical group. For our case, all three metrics we plan for are business metrics and are fluctuating values:

```
self.metric_family = pb_utils.MetricFamily(
    name="business_metrics",
    description="",
    kind=pb_utils.MetricFamily.GAUGE
)
```

3. Now, let's define the metrics in this metric family:

```
self.number_of_input_tokens_metric = self.metric_family.Metric(
    labels={"name": "Number of input tokens","version": "1"}
)
self.number_of_tokens_generated_metric = self.metric_family.
Metric(
    labels={"name": "Number of tokens generated", "version":
"1"}
)
self.readability_metric = self.metric_family.Metric(
    labels={"name": "Flesch Readability Score", "version": "1"}
)
```

Conveniently, you can do versioning of the metric in case any logic needs to be changed, which makes for a more robust monitoring process.

4. Next, we will be logging the metrics in every execution. We will be defining a helper method that will in turn be executed at the end of the execute method:

```
def _compute_custom_metrics(self, input_tokens, generated_
tokens, generated_text):
    self.readability_metric.set(textstat.flesch_reading_
ease((generated_text)))
    self.number_of_input_tokens_metric.set(input_tokens.
shape[1])
    self.number_of_tokens_generated_metric.set(generated_tokens.
shape[1])
```

5. Finally, we will use this helper method under the execute method:

```
self._compute_custom_metrics(input_ids, output_seq, "
".join(decoded_texts))
```

6. Now, we need to start the NVIDIA Triton Server nvidia-docker instance with the same command, which is the following:

```
sudo docker run --gpus=all -it --shm-size=256m --rm -p8000:8000
-p8001:8001 -p8002:8002 -v ${PWD}/models:/models nvcr.io/nvidia/
tritonserver:23.05-py3
```

7. After that, we are in the Docker environment, where the next step is to install the necessary libraries:

```
pip install transformers==4.21.3 nvidia-tensorrt==8.4.1.5
git+https://github.com/ELS-RD/transformer-deploy
torch==1.12.0  -f https://download.pytorch.org/whl/cu116/torch_
stable.html textstat==0.7.3
```

> **Note**
>
> For production usage, please be sure to create a Docker image where all the libraries are fixed, and you don't need to manually install libraries anymore.

8. Now, you can execute the `python triton_client.py` command in the command line and get your predictions.

9. The default metrics and the custom metrics are immediately hosted in the URL `http://localhost:8002/metrics`, where you can view the real-time metrics in text form. `localhost` can be replaced with the IP of your remote server if you are using one. The following snippet shows the real-time Prometheus-formatted metrics that can be found at the preceding URL:

```
# HELP nv_gpu_power_usage GPU power usage in watts
# TYPE nv_gpu_power_usage gauge
nv_gpu_power_usage{gpu_uuid="GPU-246b298f-13e6-f93e-b6f1-
0fb8933ce337"} 13.676
nv_gpu_power_usage{gpu_uuid="GPU-0fa69700-a702-0113-f30c-
89d3fa1cec2f"} 19.626
# HELP nv_gpu_power_limit GPU power management limit in watts
# TYPE nv_gpu_power_limit gauge
nv_gpu_power_limit{gpu_uuid="GPU-246b298f-13e6-f93e-b6f1-
0fb8933ce337"} 250
nv_gpu_power_limit{gpu_uuid="GPU-0fa69700-a702-0113-f30c-
89d3fa1cec2f"} 260
# HELP nv_cpu_utilization CPU utilization rate [0.0 - 1.0]
# TYPE nv_cpu_utilization gauge
nv_cpu_utilization 0.04902789518174133
# HELP nv_cpu_memory_total_bytes CPU total memory (RAM), in
bytes
# TYPE nv_cpu_memory_total_bytes gauge
nv_cpu_memory_total_bytes 67239776256
# HELP nv_cpu_memory_used_bytes CPU used memory (RAM), in bytes
# TYPE nv_cpu_memory_used_bytes gauge
nv_cpu_memory_used_bytes 14459383808
# TYPE business_metrics gauge
business_metrics{name="Flesch Readability Score",version="1"}
93.81
business_metrics{name="Number of tokens generated",version="1"}
128
business_metrics{name="Number of input tokens",version="1"} 11
```

10. As these are only real-time metrics, we need to set up a local server or use an online Prometheus server. In this step, we will opt for a locally hosted Prometheus server where the following commands need to be run in the command line:

```
sudo systemctl start node_exporter
sudo systemctl start prometheus
```

node_exporter here is a Prometheus exporter that collects system-level metrics from target machines, enabling Prometheus to monitor and analyze their resource usage and performance.

11. Now, we need to add the NVIDIA Triton Server endpoint into the Prometheus configuration file to track metrics. To do that, execute sudo gedit /etc/prometheus/prometheus.yml in the command line and add the following job details:

```
- job_name: "triton"
    scrape_interval: 10s
    static_configs:
      - targets: ["localhost:8002"]
```

With that, Prometheus is all set up to log metrics from NVIDIA Triton Server.

12. Prometheus hosts its web app by default with port 9090. So, accessing the link localhost:9090 in a web browser will take you to the Prometheus home page. Going to the **Status** tab and clicking on **Targets** in the dropdown will show the following screenshot, which verifies that the Triton endpoint is being tracked.

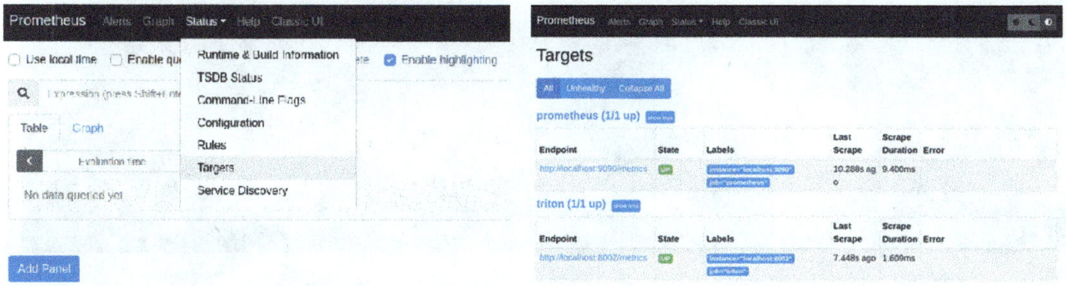

Figure 16.3 – Prometheus web app home page on the left and targets that
Prometheus is tracking and polling metrics from on the right

> **Note**
> Prometheus by default doesn't include user account enforcement but it can be configured to be enforced.

13. Next, we will set up Grafana to connect to the locally hosted Prometheus instance. First, we have to start up the Grafana service by executing the following command in the command line:

```
sudo systemctl start grafana-server
```

14. Grafana by default hosts its web app on port 3000, so accessing the localhost:3000 link in a web browser will bring you to the home page. The default username and password are admin and admin. Once that is filled in, click on **Log In** and create a new password, after which you will be greeted with Grafana's home page. We can set up the Prometheus link now by clicking on the three-line button on the top-left tab and clicking on the **Data sources** tab under the **Administration** dropdown, as shown in *Figure 16.4 (a)*. This will result in the screen shown in *Figure 16.4 (b)*.

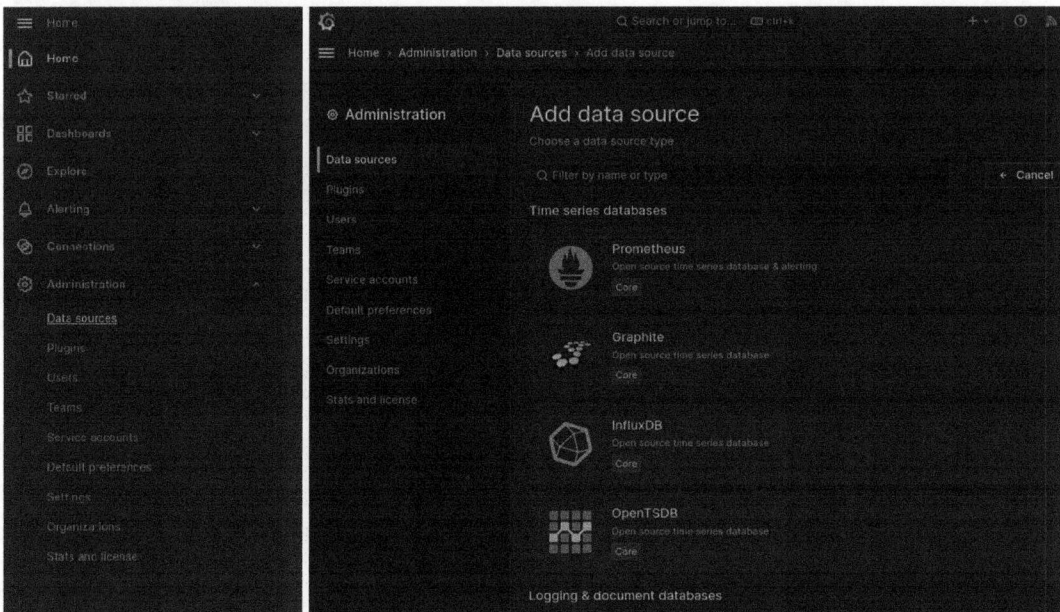

Figure 16.4 – Screenshots showing how to navigate to the Add data source page in the Grafana web app

15. Next, click on Prometheus as the data source, where you will be presented with the screen shown in *Figure 16.5 (a)*. Set the Prometheus default hosted web app link to http://localhost:9090 and click on **Save & Test**. This should result in the success screen shown in *Figure 16.5 (b)*.

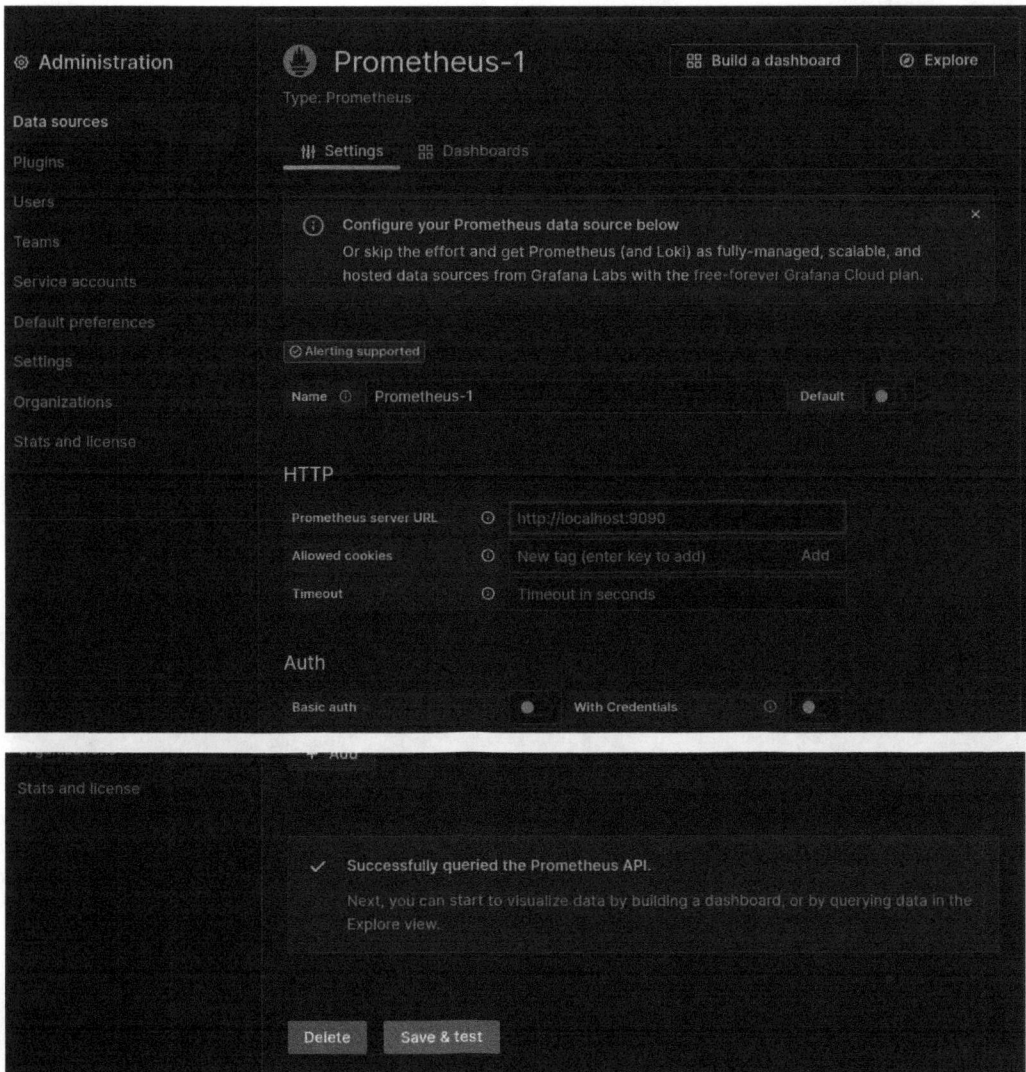

Figure 16.5 – Grafana Prometheus data source settings tab in (a) and successfully created screen (b)

16. With that, you will see that the data source has been created as shown in *Figure 16.6 (a)*. At this point, we will be able to create a dashboard to visualize the metrics we are monitoring. Grafana allows you to create dashboards in three ways: importing through its publicly shared dashboard IDs, importing through an exported dashboard JSON file, and creating a new dashboard. In Grafana, you can create many types of visualizations manually using the in-built visualization UI builder system or the **PromQL**-based visualizations and choose how you want them to be displayed. However, in this tutorial, we will be using a ready-made dashboard with visualizations

by importing it through a dashboard JSON file. To do that, navigate to the dashboard page using the same three-line button dropdown shown in *Figure 16.4 (a)*. Once, you are on the dashboard page, click on **New** and then on **Import**, as shown in *Figure 16.6 (b)*.

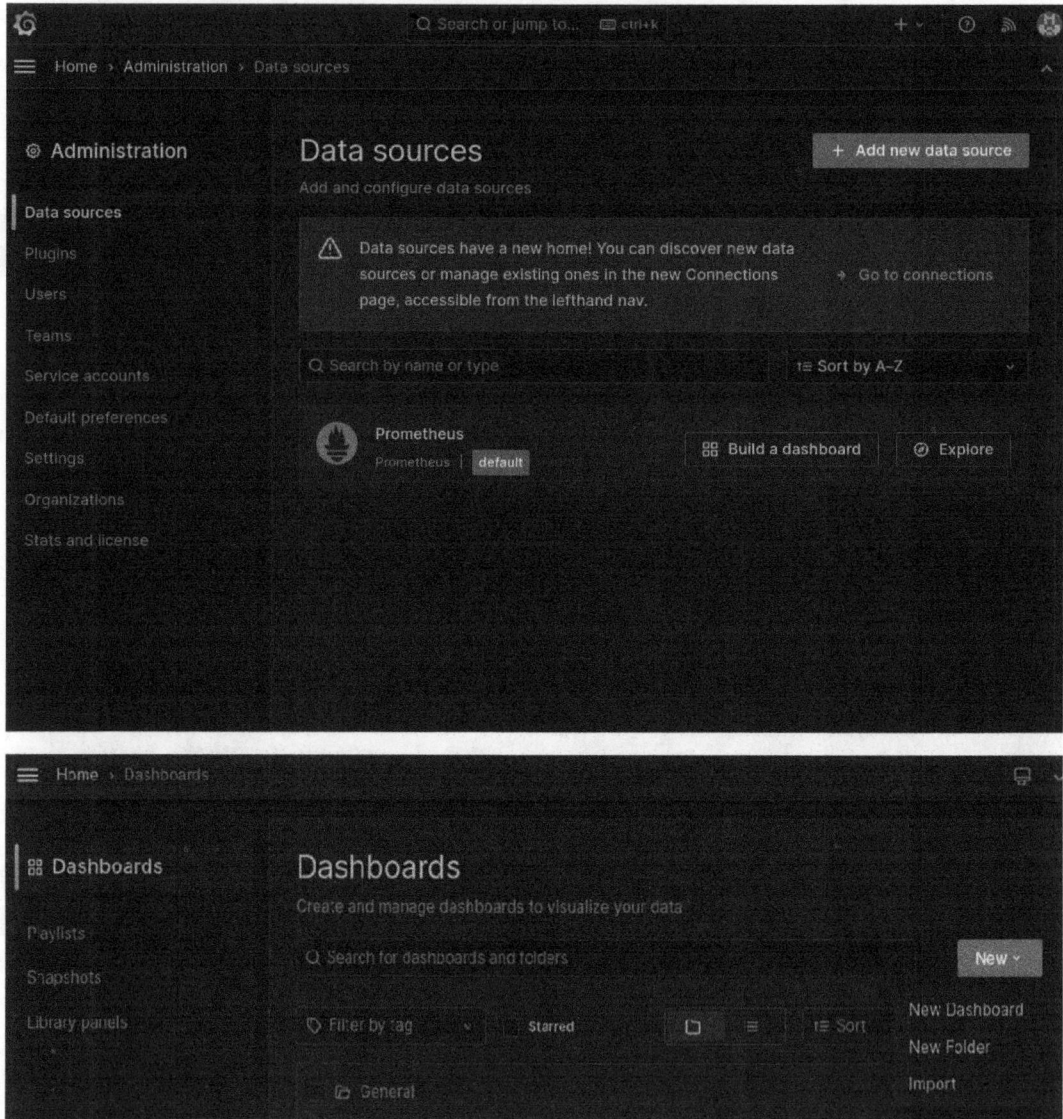

Figure 16.6 – Grafana Data sources tab showing the created data source and the Dashboards tab showing the dropdown of the New button

17. Drag the provided `Triton Inference Server-1692252636911.json` file straight into the import area and then connect to the Prometheus database you created, and you'll see the dashboard shown in *Figure 16.7*.

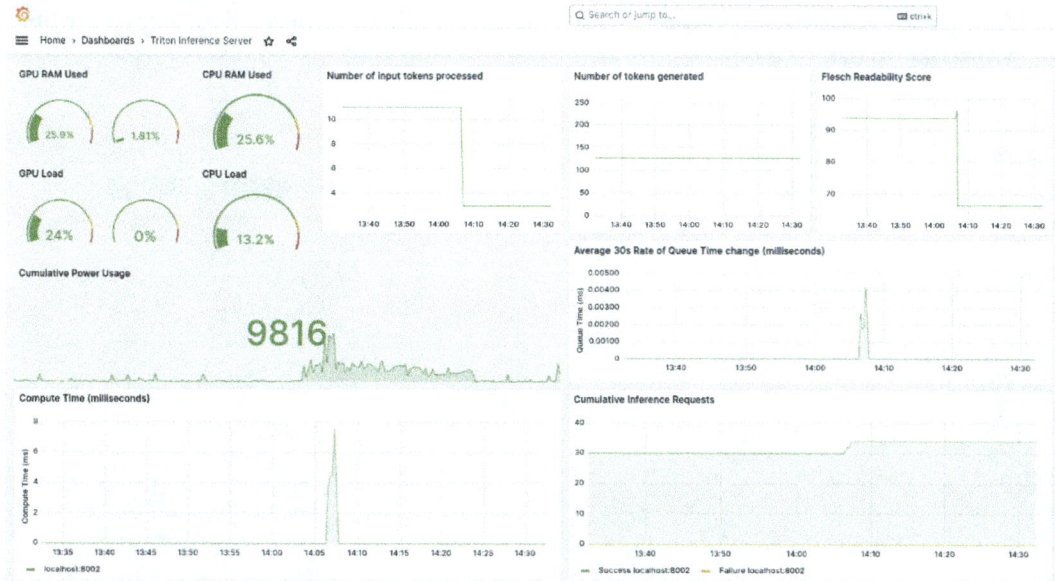

Figure 16.7 – A custom Grafana dashboard for the monitoring tutorial

Note that there were two GPUs in the machine that generated this metric, which is why there are two GPU stats. This visualization effectively represents most of the default NVIDIA Triton Server metrics, along with the three extra custom metrics we added, by displaying them on a graph that captures their historical values up to the present moment. However, hardware-resource-specific stats are an exception, as they are shown only in real-time.

Now, the component missing from monitoring is to create rules and conditions that would be considered an alarming incident, called the incident alerting component. Monitoring deployed deep learning models without alerts is like having a security camera but no alarm. You won't know if something has gone wrong until it's too late to do anything about it. Incidents can include deteriorating model accuracy, consistently delayed responses, consistent resource bottlenecks, consistently unexpected output variations during the monitoring of deployed deep learning models, and hardware failures. Grafana has an in-built alert management, notifications management, and contact management system that we will leverage in the following section.

Setting up alerts with Grafana

Let's go through the steps on how to set up alerts with Grafana:

1. Click on the **Alerting** tab, shown in *Figure 16.4 (a)*, and then on **Alert rules**. You will see the screen shown in *Figure 16.8*.

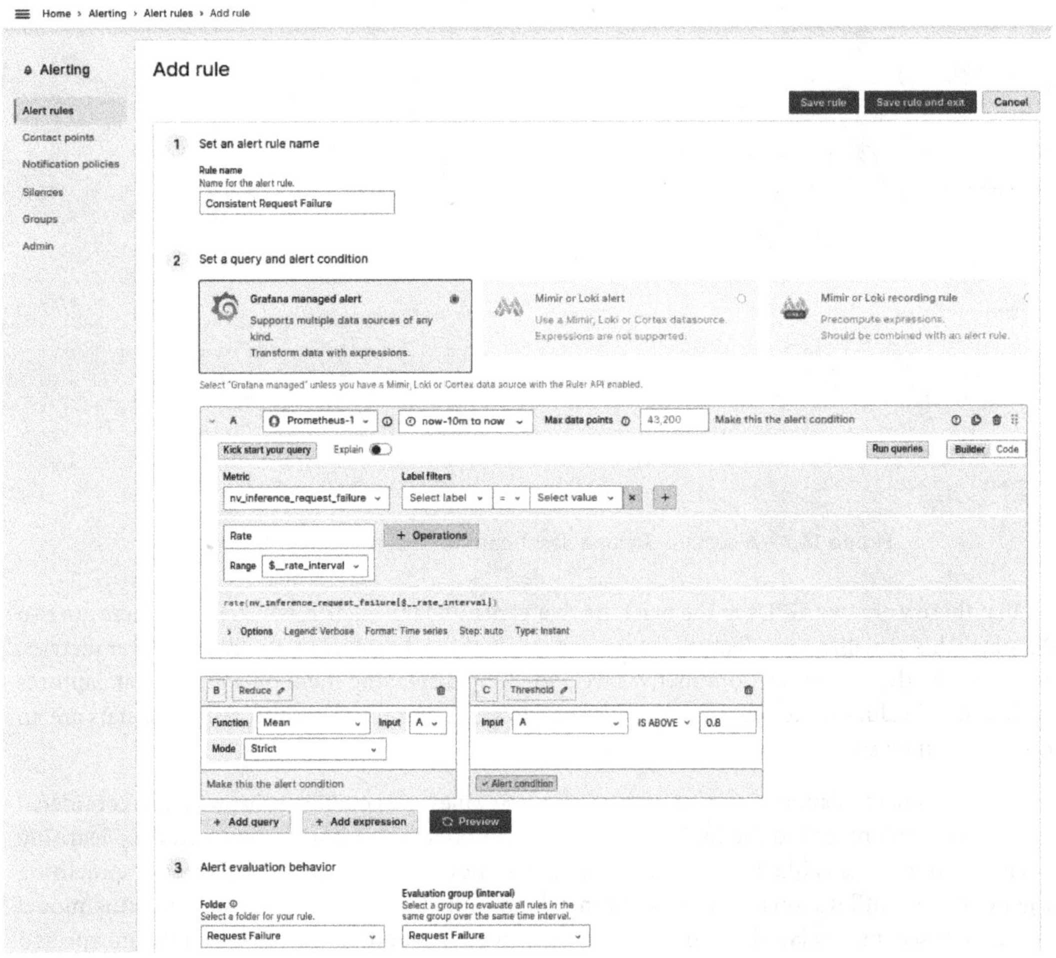

Figure 16.8 – Alert rules tab settings for NVIDIA Triton request failure alert rule

2. In this example, we will set an alert to trigger when there is any failed NVIDIA Triton Server inference request. So, in the same tab, choose the `nv_inference_request_failure` metric tab, and set the threshold to a number that is lower than 1 so that a single failed request will trigger the alarm. In *Figure 16.8*, the number is set to 0.8.

3. Next, set the evaluation interval to be one minute and to raise an alarm only if there are consistent request failures for five minutes straight, as shown in *Figure 16.9*. Then, click on the **Save and exit** button.

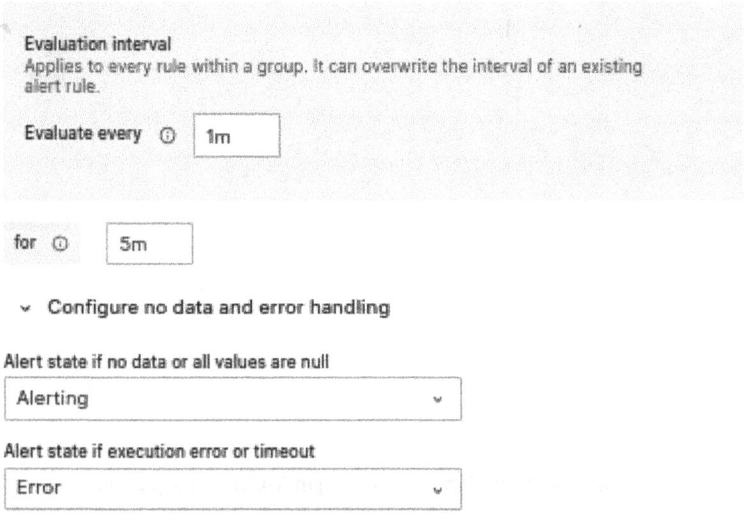

Evaluation interval
Applies to every rule within a group. It can overwrite the interval of an existing
alert rule.

Evaluate every ⓘ   1m

for ⓘ   5m

⌄ Configure no data and error handling

Alert state if no data or all values are null
Alerting   ⌄

Alert state if execution error or timeout
Error   ⌄

Figure 16.9 – Evaluation interval settings for alert rules

There are three possible statuses that alerts in Grafana can have: **Normal**, which indicates that the condition wasn't triggered; **Pending**, which indicates that the condition was partially triggered but there isn't a consistent behavior yet; and **Firing**, which indicates that the condition has been consistently satisfied and an alarm has been triggered. Now that an alert rule is saved and created, you will see the screen shown in *Figure 16.10 (a)*, where the status is **Normal**. *Figure 16.10 (b)* shows the **Pending** stage, where a failure has been detected but is not yet consistent enough to send an alert. *Figure 16.10 (c)*, on the other hand, shows the **Firing** stage, where the failure has consistently happened per the configured time interval.

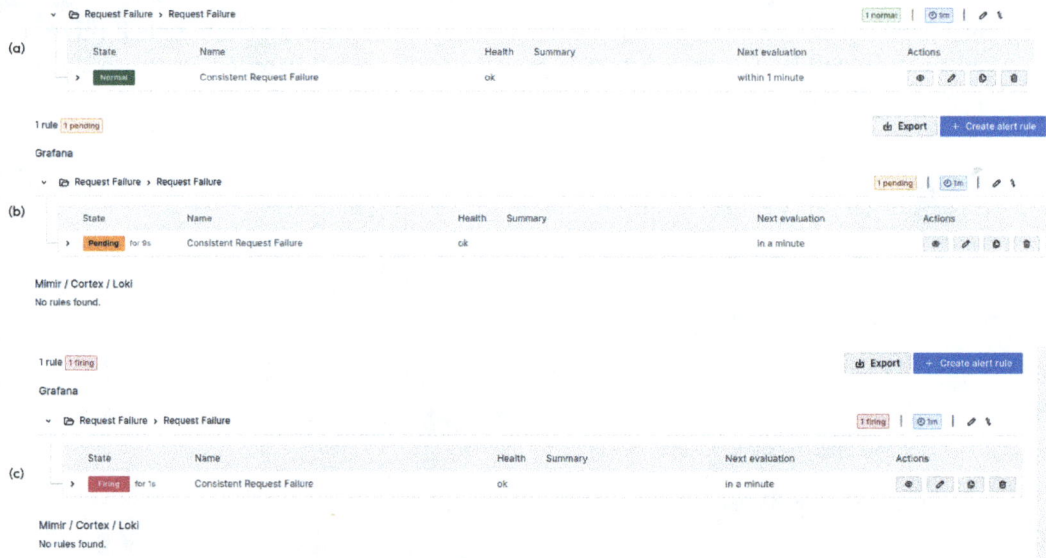

Figure 16.10 – Alert status of Normal in (a), Pending in (b), and Firing in (c)

4. To configure where and who these alerts will be sent to, we'll work with the **Contact points** and **Notification policies** tabs in the **Alerting** section. Let's start by clicking on the **Contact points** tab to set up the individuals who will receive notifications. You can even organize these into groups, but for simplicity in this tutorial, we'll have notifications sent to ourselves. Grafana offers various contact platform integrations: Alertmanager, Cisco Webex Teams, DingDing, Discord, Email, Google Chat, Kafka REST policy, LINE, Microsoft Teams, Opsgenie, PagerDuty, Pushover, Sensu Go, Slack, Telegram, Threema Gateway, VictorOps, Webhook, and WeCom. To keep things straightforward, we'll choose a widely available integration type: email. Grafana uses the **sSMTP** software for sending emails, so ensure you have an email account with credentials set up before proceeding. Within the contact points settings, provide your name and email, then click on **Test** to generate a test notification to confirm that the credentials are accurate. Once you've verified that you've received the email notification, save your settings. Refer to *Figure 16.11* for an example of the settings interface.

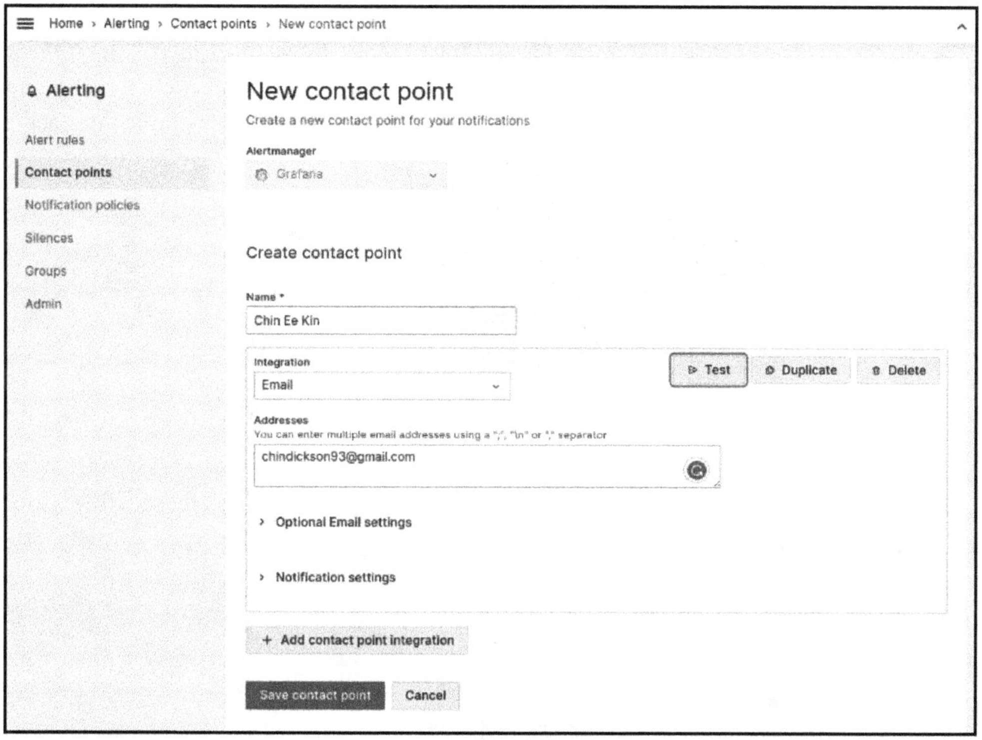

Figure 16.11 – Contact points tab with email set up

5. Next, we need to link up the contact as part of the default notification policy. Proceed to the **Notification policies** tab, click on **Settings**, and change the default contact point to be the email contact we set in *step 4*. You will then see a similar screen to *Figure 16.12*.

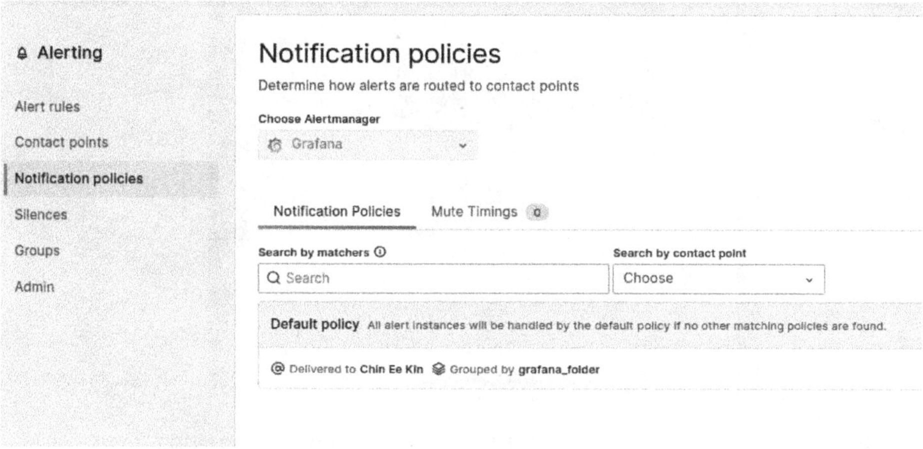

Figure 16.12 – The default notification policy set up to notify us

6. Now, we are all set up to receive email notifications! As a challenge, try to figure out ways you can make the inference server request fail, and if you can't, change the rule to something that will definitely trigger so you can get an example actual alert notification come through email. *Figure 16.13* shows the example email that you will get through a mobile phone interface.

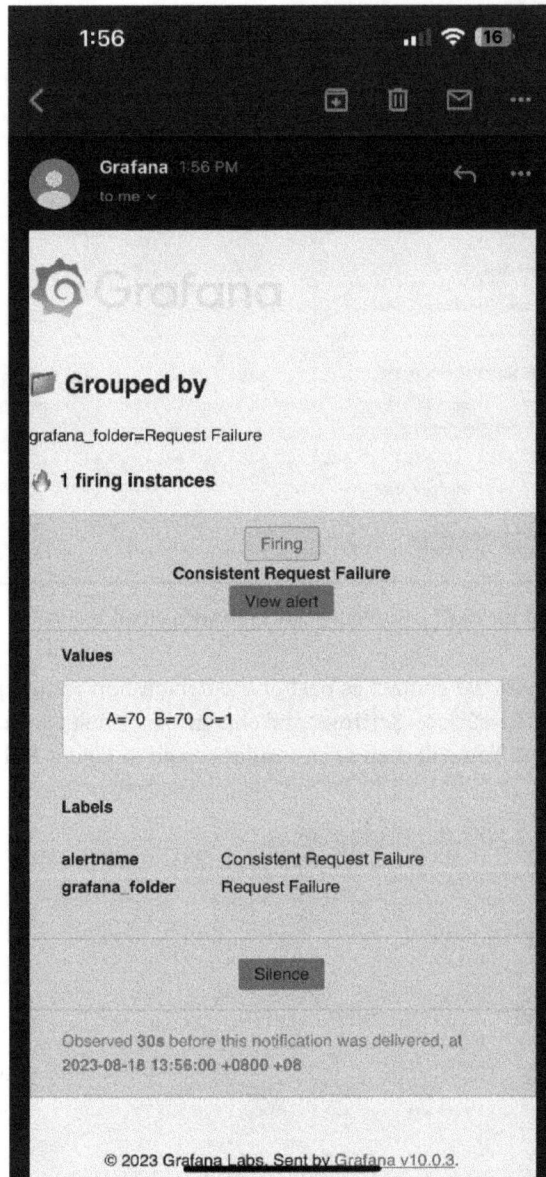

Figure 16.13 – Example triggered alert email notification with the Firing status

With that, we have successfully set up a monitoring and alerting system for a deployed deep learning model! A notable caveat in this implementation is the fact that metrics are bundled up in the same execution as the model. A solution to make it decoupled to not increase the runtime for prediction-specific inference is to use the C API instead to build the custom metrics. If the time needed to get the metrics logged is not crucial, you can also consider hosting another "model" in NVIDIA Triton Server that takes in outputs from the prediction-specific model and log metrics. NVIDIA Triton Server also provides a tool called `perf_client`, which evaluates the runtime of different configurations, helping you optimize your system's performance. Specifically, the tool measures and reports the throughput and latency with different load conditions.

However, just having monitoring and alerts doesn't provide a full picture of model monitoring. We need to dive into those numbers, cross-reference them, spot connections, and find patterns. It's like checking the fuel efficiency, tire pressure, and engine temperature of a car to ensure a smooth ride.

Additionally, alerts alone won't fix issues. They're like the car's warning lights – they tell you something's up, but you still need to pull over, pop the hood, and fix the problem. That's where model maintenance comes in. In the next section, we'll explore how to not only detect issues but also take action to keep your model running smoothly and efficiently over time.

Governing a deep learning model through maintenance

Metrics logging, dashboard building, logged metrics analysis, and alerts are essential components of model monitoring, but they are only effective when followed by appropriate actions, which are covered under model maintenance. Model maintenance is akin to a skilled pit crew in a car race, regularly fine-tuning and optimizing the performance of deep learning models to keep them running efficiently and effectively. Like how a pit crew conducts rapid repairs, refuels, and adjusts the car's components to adapt to changing race conditions, model maintenance involves updating the models to account for environmental changes, improving and refining the models with new data obtained from feedback loops, and performing incident responses on miscellaneous issues. This ensures that the models consistently stay on track, deliver valuable insights, and drive informed decision-making in the ever-evolving landscape of data and business requirements.

Key aspects of model maintenance comprehensively include the following:

- **Establishing a feedback loop**: Establishing a feedback loop is vital for capturing real-world outcomes and validating model predictions, enabling deep learning practitioners to identify areas for improvement, and adapting the model accordingly.

- **Retraining**: Retraining is an essential part of model maintenance, ensuring that the model stays up to date with the latest data and trends, thereby maintaining its accuracy and relevance. Regular retraining enables the model to learn from new insights and adapt to evolving data landscapes, ensuring consistent performance. Fortunately, for deep learning models, a fine-tuning process can be employed, which is much faster than a full retraining process. Two use cases that highlight the importance of frequent updates with model retraining are the following:

- **E-commerce product recommendation**: Consumer preferences and product availability change rapidly in e-commerce. To provide relevant product recommendations, deep learning models need to be retrained frequently, maybe weekly or even daily, to understand the latest trends and customer behavior.

- **Social media sentiment analysis**: Social media platforms are constantly evolving with new trends, hashtags, and user behaviors. To accurately gauge public sentiment and opinion, deep learning models need to be retrained frequently, maybe quarterly, to account for these changes.

- **Incident response handling**: When alerts signal potential issues, it's vital to have a dedicated response team to triage and address the problem promptly. This team should be well equipped to investigate the root cause, implement corrective measures, and prevent similar issues from recurring in the future. Let's discover response-handling recommendations for different groups of incidents:

  - **Data-related incidents**: These incidents occur when the model receives incorrect, incomplete, or biased input data. To handle such issues, the response team should work closely with the data provider to identify the cause, correct the data, and retrain the model as needed.

  - **Model performance incidents**: These incidents involve the model generating inaccurate or unexpected predictions. Proper handling requires collaboration between the model owner (responsible for model creation or approving the model usage) and the prediction owner (responsible for approving the usage of the predictions), as described in the *Governing deep learning model utilization* section earlier. They should analyze the model's performance, identify potential issues in its architecture or training, and implement improvements to ensure better performance in the future.

  - **Infrastructure-related incidents**: These incidents are caused by hardware or software failures, affecting the model's deployment environment. The response team should work with the infrastructure provider or team to resolve the issue and ensure the model runs smoothly.

  - **Security incidents**: These incidents involve unauthorized access, data breaches, or other malicious activities targeting the model. The response team should follow the organization's security policies, identify the threat, and take appropriate measures to mitigate the risk.

  - **Compliance and regulatory incidents**: These incidents occur when the model's output or operation violates legal or regulatory requirements. The response team should work with legal and compliance teams to address the violation and modify the model to comply with the necessary regulations.

By comprehensively considering model maintenance components shared here, organizations can effectively address challenges associated with deployed deep learning models, ensuring their continuous improvement and alignment with business requirements. Traditionally, these maintenance actions are executed manually after alerts are raised. However, it is possible to schedule custom tasks to be executed automatically given an alert event. Consider using Apache Airflow to orchestrate your desired automated tasks from your model monitoring alerts. Apache Airflow is like a conductor for your data tasks, allowing you to choreograph and schedule complex workflows in a directed acyclic graph format. It lets you define, automate, and monitor sequences of tasks, making sure they happen in the right order and at the right time. However, there are some inherent limitations and risks with creating automated tasks from model monitoring alerts, which we will briefly explore next.

Exploring limitations and risks of using automated tasks triggered by model monitoring alerts

While automating tasks based on model monitoring alerts can save time and resources, it also comes with limitations and potential risks that need to be considered when implementing such an approach. Some limitations of automating tasks based on model monitoring alerts are the following:

- **Complexity of issues**: Some issues may be too complex or nuanced to be handled effectively by an automated process. For example, in a deep learning model for medical image analysis, an automated task might be triggered to retrain the model when the monitoring alerts indicate a drop in accuracy. However, the complexity of the issue may stem from an imbalance in the training data, such as an underrepresentation of a certain disease, which cannot be resolved by simply retraining the model. In this case, automated processes might not be able to effectively address the root cause of the problem.

- **Lack of context**: Automated tasks may lack the ability to consider the broader context of an issue or understand its potential impact on other aspects of the system.

    Consider a deep learning model that predicts customer churn based on various behavioral and demographic factors. An automated task might be set up to send promotional offers to customers identified as high risk for churn. However, the task may not have the context to consider external factors, such as a recent negative publicity event or a widespread service outage, which might be causing a temporary increase in churn risk. This lack of context may lead to unnecessary promotional offers and an ineffective use of resources.

- **Inadequate or inappropriate responses**: Automated tasks might not always choose the most appropriate action in response to an alert, potentially leading to suboptimal outcomes. For example, an AI model monitoring social media posts for harmful content may detect a post containing offensive language. An automated response system might remove the post or ban the user immediately, without considering the possibility of false positives or the post's broader context (e.g., quoting offensive language to criticize it).

As for risks associated with enabling automated tasks with model monitoring alerts, they are as follows:

- **Over-reliance on automation**: Relying too heavily on automated tasks can lead to a lack of human oversight and expertise in the model maintenance process. This may result in overlooking subtle patterns and trends that only human intuition can detect, potentially leading to suboptimal model performance.

- **Inaccurate or premature triggers**: Automated tasks are often triggered by specific conditions in the monitored metrics. If these conditions are not carefully defined, tasks may be triggered inaccurately or prematurely, leading to unnecessary or even detrimental actions being taken on the model.

- **Inflexibility**: Automated tasks are typically designed for specific scenarios or issues and may not be flexible enough to handle unforeseen or complex situations. This could limit their effectiveness in addressing unique challenges that arise during model maintenance.

- **Risk of compounding errors**: When automated tasks are executed based on erroneous alerts or inaccurate metrics, they can compound the issue by making unnecessary or incorrect adjustments to the model. This may lead to further deterioration in model performance or even irreversible damage.

- **Security risks**: Automating tasks based on alerts can expose the model and its infrastructure to potential security risks, especially if the automation system is not adequately secured. Unauthorized access or manipulation of the automation system could lead to unintended consequences or malicious actions on the model.

To mitigate these limitations and risks, it is essential to strike a balance between automation and human involvement in the model maintenance process. This can be achieved by incorporating human-in-the-loop systems, ensuring proper validation and calibration of monitoring metrics and alerts, and implementing robust security measures to protect the automation infrastructure.

With that, we have covered all the components of deep learning model governance. This holistic three-pillar approach to model governance ultimately enables organizations to consistently and continuously harness the full potential of deep learning models, driving valuable insights and informed decision-making in the real world.

Summary

In this chapter, we explored the three fundamental pillars of model governance for deep learning models: model utilization, model monitoring, and model maintenance. Model utilization ensures the effective, efficient, ethical, and responsible utilization of deep learning models, while model monitoring allows for ongoing evaluation of performance, identification of potential bias or drift, and infrastructure-related metrics. Model maintenance, on the other hand, focuses on regular updates and refinements to keep models aligned with evolving data landscapes and business requirements.

We also dove into and learned about the technical steps for monitoring deep learning models using NVIDIA Triton Server, Prometheus, and Grafana. By diligently considering the components for model governance, deep learning architects can effectively manage the challenges posed by these complex models in production and consistently harness their potential for driving valuable insights and decisions.

In the next chapter, we will further dive deeper into the details of drift detection for deep learning models.

17

Managing Drift Effectively in a Dynamic Environment

Drift is a significant factor in the performance deterioration of deployed deep learning models over time, encompassing concept drift, data drift, and model drift. Let's understand the drift of a deployed model through a culinary-based analogy. Imagine a deployed deep learning model as a skilled chef who aims to create dishes that delight customers but excels in a particular cuisine. Concept drift occurs when the taste preferences of the diner shift, which alters the relationships between ingredients and popular dishes that can satisfy the diner's palate. Data drift, on the other hand, happens when the ingredients themselves change, such as variations in flavor or availability. Finally, model metric monitoring alerts happen most straightforwardly when the chef loses customers. In all cases, the chef must adapt their dishes to maintain their success, just as deep learning models need to be updated to account for concept drift, which deals with the changing relationships between input and target variables, and data drift, which tackles adjustments in the input data distribution and characteristics.

Monitoring drift is crucial to ensuring the continued success of deep learning models, just as a chef needs to keep track of their customers' evolving preferences and the changing nature of ingredients. As a sneak peek, only some use cases require monitoring. In this chapter, we will delve into the techniques to measure and detect drift, which will allow us to effectively monitor and send timely alerts when drift is detected, and make necessary model maintenance adjustments. Drawing parallels with our chef analogy, techniques to monitor drift can be likened to a chef observing the reactions of their customers, reading reviews, or collecting feedback to better understand their preferences and the quality of their ingredients. By staying alert to any drift, both the chef and the deep learning model can adapt and evolve, maintaining their expertise and delivering exceptional results in a dynamic environment. Specifically, we will cover the following topics:

- Exploring the issues of drift
- Exploring the types of drift
- Exploring strategies to handle drift
- Detecting drift programmatically
- Comparing and contrasting the Evidently and Alibi-Detect libraries for drift detection

Technical requirements

This chapter will cover a practical example to test out data drift techniques. We will be using Python 3.10 and, additionally, we will require the following Python libraries to be installed:

- `evidently`
- `numpy`
- `transformers==4.21.3`
- `torch==1.12.0`
- `syllables`
- `audiomentations`
- `datasets`

The code files are available on GitHub: `https://github.com/PacktPublishing/The-Deep-Learning-Architect-Handbook/tree/main/CHAPTER_17`.

Exploring the issues of drift

The most obvious issue of drift is the degradation of the accuracy. However, there are more issues than you might initially notice, which include the following:

- **Applicability**: The model's ability to make accurate predictions on new, unseen data may be compromised as data patterns and distributions shift. This can result in reduced effectiveness in real-world scenarios and diminished value for decision-making, which raises the likelihood of the model becoming less relevant and practical to use.

- **Interpretability**: Understanding and explaining the model's decisions can become challenging, as the factors influencing its predictions may no longer align with the current data landscape. This can hinder effective communication with stakeholders and impede trust in the model's predictions. Note that an originally explainable model is still explainable as we can still produce accurate information on how it used the input data, but it can become more difficult to interpret with drifted data.

- **Fairness**: Biases and disparities could emerge or worsen, raising fairness concerns in the model's output. This can lead to the unequal treatment of different groups, perpetuating harmful disparities and posing ethical concerns in the model's application.

- **Stability**: Sensitivity to changes in input data may result in fluctuating performance, impacting the model's stability and consistency. Unstable models can lead to unreliable results, making it difficult for decision-makers to rely on the model's outputs.

These issues comprehensively highlight the challenges that can arise from drift in data. By now, we know about the three high-level drift groups. This is useful, but not enough to implement drift detection. Prior knowledge of the types of drift that can impact your model will enable you to prepare your model for any deployment-related concerns and implement drift monitoring, which brings us to the next topic.

Exploring the types of drift

Drift is like a shift in the way things work with data. It happens when the data changes, or the environment it comes from changes. This can sometimes happen suddenly or quickly, sometimes slowly, or even in a recurring pattern. When it comes to drift, it's important to look at the big picture, not just a couple of odd blips. Drift isn't about those rare anomalies or one or two odd predictions; it's about changes that stick around, like a new pattern that stays. These persistent shifts can mess up your model permanently, making it way less useful. It's like if your friend suddenly started speaking a different language occasionally, which could lead to one-off confusion but not really be a problem. But if they started speaking a different language all the time, it'd be a big problem.

Furthermore, drift can be categorized into three main types: data drift, concept drift, and model drift. While concept drift is related to the data and can be argued to be part of data drift, concept and data drift are often considered separate in the field. Let's dive into each of the drift types sequentially, starting with data drift.

Exploring data drift types

Data drift occurs when certain features of a current data batch differ from those of a historical batch. It is essential to understand that data drift is not limited to a single type of change. Many practitioners mistakenly focus only on the most widely known type of characteristic change for data drift, which is the shift in distribution. Distribution is a fundamental concept in statistics that reveals how frequently various values of a variable appear in a dataset. It helps us comprehend the nature of the data we are working with, but vaguely. Some well-known distribution patterns include normal, uniform, skewed, exponential, Poisson, binomial, and multinomial. Determining whether a change in distribution is beneficial or harmful can be challenging because a change in distribution can sometimes result in good consequences, while no change in distribution doesn't necessarily imply that performance won't be degraded. The relationship between distribution changes and model performance is not always straightforward, making it difficult to assess the impact accurately. *So, it is advisable to always consider other, more understandable, cross-validation-tested statistical data drift that strongly correlates with performance metrics instead of data distribution drift.*

Statistical drift refers to changes in statistical characteristics, such as mean, variance, correlation, or skewness, that can have a more direct impact on model performance. This also means that the relationship between them is more predictable, which allows for more targeted maintenance actions. Change between statistical values can be measured through difference or ratio. For example, if a deep learning model is trained to recognize handwritten digits, a statistical drift in the mean intensity of the digits (for example, due to changes in lighting conditions) could have a direct impact on the model's accuracy. This approach will enable us to better prepare for any potential negative impacts and maintain the effectiveness of our data analysis. More broadly, the choice between these two is part of the **data drift techniques**.

However, the prerequisite of drift being relevantly and reliably handled comes down to the choice of the **data types** to apply distributional or statistical drift methods. Some examples of the other data types are the following:

- **Data characteristic drift**: Occurs when there are changes in the underlying properties or attributes of the data. It's important to remember that drift doesn't just cover actual model input or output data changes and can also cover external descriptors or metadata associated with the input or output data. To make this more concrete, let's explore some example characteristics that can be monitored and measured for drift:

  - **Text**: The usage of certain words, phrases, sentiment, word count, and average length of sentences

  - **Image**: Object orientation, lighting hue, color, size, and any styles

  - **Audio**: Sound or speaker pitch, tempo, timbre, tone, speaker gender, speaker accents, speaker dialects, and speaking styles

- **Data quality drift**: Occurs when there are distribution changes in the quality of the data being collected, such as missing values, data entry errors, or measurement errors. These changes can impact the model's ability to make accurate predictions.

- **Core data drift**: This involves the raw data of image, text, audio, and any embedding data that is extracted from a deep learning model at any level. This drift data type can be hard to interpret and it can be hard to find a correlation with the metrics you care about. Drift in the unstructured raw data itself rarely correlates with the metrics you care about. However, embedding data from the actual deep learning model that was used for the prediction is more likely to show drift values that are relevant to the metric you care about. As a standard, practitioners usually choose the embeddings from the final layer output.

While these data type categories cover what characteristics of the data drifted, there are two higher-level data drift types that should be known. The higher-level data drift types govern where the drift is measured, monitored, and detected, which can be labeled as part of the **drift scenarios**. These are the following:

- **Covariate drift** or input features drift, which involves shifts in the input features while maintaining the same relationship with the target variable. This can arise from changes in data collection methods, user behavior, or external factors, challenging the model's ability to generalize effectively. It can affect the importance or relevance of certain features in predicting the target variable.

- **Label drift** or target drift, which pertains to changes in the ground-truth labels of the data. Occurs when the target changes over time. This may result from errors in labeling, shifts in labeling criteria, or changing interpretations of the target variable. Not to be confused with concept drift, let's explore an example of label drift with a product recommendation system. Shifts from broad categories such as `Electronics` to specific labels such as `Smartphones` constitute label drift when underlying user-product relationships remain unchanged.

Building upon our understanding of data drift, let's now delve into the equally important phenomenon of concept drift, which focuses on the evolving relationships between input data and target variables in various use cases and scenarios.

Exploring concept drift

Concept drift is intimately tied to the specific use case and data characteristics. Rather than adhering to predefined types, concept drift's occurrence and impact vary based on factors such as problem context, data attributes, temporal dynamics, external influences, and adaptation strategies. Acknowledging this context dependency is essential for tailoring effective concept drift detection and adaptation methods that align with the nuances of each individual scenario. Let's explore some examples of concept drift:

- **Search engine algorithms**: A search engine's ranking algorithm learns from user behavior, but the user preferences evolve over time. What was once considered relevant might not be anymore, causing a shift in the concept of relevance and altering both the input data (queries) and target (rankings) relationships.

- **Online advertisement campaigns**: In online advertising, users' click-through behavior changes due to new trends or demographics. This leads to shifts in user preferences, affecting both the input data (ad impressions) and the target (click-through rates) relationships.

- **Medical diagnostics**: In medical diagnosis, patient profiles change as demographics or health trends shift. This impacts the concept of "normal" and "abnormal" within the data, altering both the input data (patient characteristics) and the target (diagnosis) relationships.

Finally, it's time to explore the last drift type, called model drift.

Exploring model drift

Model drift simply deals with the shift in model evaluation metrics, which usually require real and natural targets being provided later, system metrics, and business metrics. This drift is the most straightforward to monitor and capture as it is a drift that can be directly connected to metrics you care about.

Now, we are ready to explore strategies to handle drift.

Exploring strategies to handle drift

Simply setting up drift monitoring for a deployed model isn't enough to effectively tackle all potential drift-related challenges. It's crucial to ask yourself: does the specific drift with the chosen data type impact the model's performance in the metrics that matter the most? At what point does drift become intolerable? To properly address drift, start by pinpointing the drift metric and data type that carries the most significance for your model and the business. If your model has been developed correctly, it may possess generalizable properties, which is the primary goal for most machine learning practitioners. This means that a well-developed model should be able to handle drift effectively. When drift detection and alerts are configured without proper consideration of their effects, it poses the risk that drift alerts can be raised without an actual issue, which can result in wasted time and resources that could have been used more productively elsewhere. *Figure 17.1* shows the high-level methodology that encompasses the strategies we will explore:

Drift-Handling Methodology

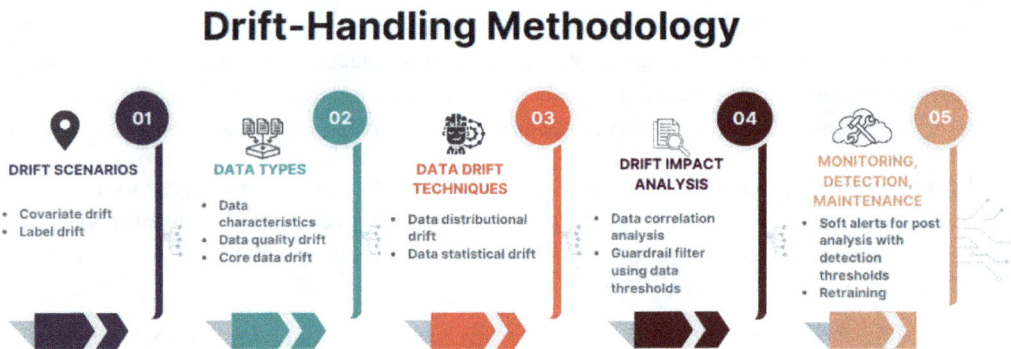

DRIFT SCENARIOS 01
- Covariate drift
- Label drift

DATA TYPES 02
- Data characteristics
- Data quality drift
- Core data drift

DATA DRIFT TECHNIQUES 03
- Data distributional drift
- Data statistical drift

DRIFT IMPACT ANALYSIS 04
- Data correlation analysis
- Guardrail filter using data thresholds

MONITORING, DETECTION, MAINTENANCE 05
- Soft alerts for post analysis with detection thresholds
- Retraining

Figure 17.1 – Drift-handling methodology

Let's start with a deep dive into the first strategy, which is to explore drift detection methods and strategies.

Exploring drift detection strategies

Not every drift type that can impact your model and the metrics you care about requires monitoring. There are two primary ways to detect drift, which are the following:

- **Estimating future drift events based on manual insights and domain knowledge:** This method doesn't require any monitoring setup and depends on humans as the alerting mechanism. Imagine an e-commerce platform that uses a deep learning model to recommend products to its customers, with the goal of increasing product purchases. Drift in customer preferences consistently happens during different seasons, such as Halloween or Christmas. For instance, people might search for costumes in October and gifts in December. Building on this domain knowledge, instead of measuring and detecting drift, before each season arrives, you pre-emptively adjust the model's recommendations according to the expected season trend. This prevents the need to measure and detect any change in preferences and ultimately can also prevent any degradation in performance.

- **Using automated programmatic measurements, monitoring, and detection of specific traits or patterns**: Imagine you trained a deep learning model to identify whether an email is spam and validated with real data. Through analysis, you find that the average length of an email degrades the model's performance. You don't have sufficient data to train the model to tackle this issue, deploy it, and decide to monitor it if the email length changes. By tracking and comparing this programmatic measurement of email length, you can effectively detect the drift and take subsequent actions to improve the model's performance. Programmatic data distribution drift detection methods include statistical tests, distance metrics, and classification models and will be explored further in the *Detecting drift programmatically* section.

Sometimes, both are required, and sometimes, only the first method is needed. Having planned fixed dates on when you expect concept drift to occur and assigning planned dates to mitigate them are more reasonable strategies than trying to detect them. However, as mentioned, there are cases where you absolutely need to implement drift monitoring and detection.

If the choice is to measure, monitor, and detect drift programmatically, it is important to make sure that you set a big enough time interval to how you're measuring and detecting drift. You want to catch those shifts that are there to stay, and those are the ones that can make your model struggle. This ensures we're not chasing after one-time anomalies but addressing substantial shifts that can impact the reliability of our models. To that end, drift detection in production should be configured to be run in batch mode and does not need real-time monitoring and detection.

For other metrics for monitoring a deployed deep learning model, the recommended route is to use real-time predictions and monitoring and alerting functionalities using NVIDIA Triton Inference Server, the Prometheus server, and Grafana. However, for batch predictions, the recommended stack would be to use Apache Airflow for scheduling a drift detection task to be executed regularly, a database such as PostgreSQL to store drift measurements from Airflow tasks, and Grafana to connect to PostgreSQL for monitoring batch drift and creating alerts. Notably, a database is recommended over using Prometheus, being the more efficient choice, as Prometheus requires more services to be set up, which can take up more resources. This can be wasteful as Prometheus is set up to take up resources for real-time usage when it's not needed. Look into `https://github.com/evidentlyai/evidently/tree/v0.4.4/examples/integrations/airflow_batch_monitoring` for a tutorial on how to set this up.

The prerequisite to setting the recommended batch predictions stack up is that the input data used must be saved somewhere for an Airflow task to pick it up in the future. The data can live in any format, and most typically for actual business use cases, the input data should already live in a database such as PostgreSQL. If drift is applied to the predictions, then either the predictions also need to be saved in a database or it can be a task prior to the drift measurement under the same directed acyclic graph in Airflow, which can be scheduled regularly.

As a follow-up, to ensure reliable drift monitoring and detection programmatically, we need an additional step before setting it up, which is to analyze the impact of drift.

Analyzing the impact of drift

To mitigate the risk that drift alerts are raised without an actual issue, perform a **drift impact analysis** on your chosen model before deployment. Drift impact analysis is closely related to the adversarial performance analysis methods introduced in *Chapter 14, Analyzing Adversarial Performance*. The same strategy of using controllable augmentation or collecting real-world data with the targeted data characteristic drift to perform an evaluation can be adopted. So, re-explore that chapter and apply the same analysis methods with drift in mind. The idea is to make sure any variation in the data type or types you choose to monitor for drift correlates with the model performance metrics in some way. In other words, perform correlation analysis, adversarial performance analysis, or drift impact analysis at once! But do be aware of the issue where correlation not being causation can lead to misleading conclusions.

> **Note**
> When there are cases where you can't augment the characteristics, it can be crucial to monitor it after deployment and perform impact analysis then.

Consequently, as a bonus, findings from adversarial performance analysis should guide the establishment of an appropriate detection threshold, determined by pinpointing the stage at which any additional metric degradation becomes unacceptable. Just as a binary threshold must be finely calibrated to balance recall and precision trade-offs for a binary classification model. This is part of the guardrail filter component introduced in the *Governing deep learning model utilization* section in the previous chapter, *Chapter 16, Governing Deep Learning Models*. With guardrail filters implemented through data characteristic thresholding, the risk of data drifting in extreme ways negatively will be decreased.

It's important here to differentiate the two related approaches introduced that help to combat performance degradation, which are guardrail filters with thresholds and drift monitoring. Guardrail filters operate on a per-prediction request level and drift operates at a higher level working with a batch of predictions made in a specified time frame. The key mutually beneficial relationship between them is that guardrail filters aid in eliminating extreme examples known to yield inaccurate or unreliable results, thus reducing the likelihood and adverse effects of more extreme drift. It can still be useful to measure and monitor the statistics of the same data types even when extreme values are prevented through guardrail filters.

Another important aspect to consider is that deployed models can be vulnerable to various types of drift. Sometimes, you can reliably analyze the impact of specific characteristics on your desired metric, while in other cases, you might not be able to, even if the characteristics are measurable. This inability could stem from the lack of viable augmentations to simulate the characteristic or the insufficient availability of natural data examples containing the targeted characteristic. When you're unable to reliably analyze the impact, you have two main options to consider:

- Set up monitoring of the statistics of the chosen data characteristics without a detection component, allowing for future analysis once more data has been gathered over time. Alternatively, consider implementing soft alerts that prompt post-deployment analysis when extreme or unknown values emerge in the characteristic you suspect could influence the model.

- Apply data distribution drift-based monitoring and detection. Although it is hard to predict the metric impact from distribution change, there is value in using it as a more reliable arbitrary soft alert mechanism, similar to as described in the previous point.

After crafting a detection strategy and verifying its impact, it's time to consider how we can tackle the potential issues caused by drift and determine the steps to address any drift incidents that arise.

Exploring strategies to mitigate drift

This falls within the domain of model maintenance, where we take action to ensure our model stays on track and remains effective over time. Here are a few techniques that can be used to tackle drift alerts:

- **Retrain or fine-tune**: Regularly retrain or fine-tune the model with new data, incorporating any changes in patterns or trends. This will help the model adapt to evolving data dynamics and maintain its accuracy.

- **Prevent predictions of high-risk drift metric score ranges**: Don't allow data that lies in a high-risk range of characteristics or range of distribution distances to be predicted. For example, face recognition systems should only predict frontal, unobstructed faces without masks or glasses. This will result in targeted prevention of drift, as the data you receive will always be in the expected range.

- **Manual human analysis**: This is where a human expert is involved in the decision-making process when drift alerts are triggered. The expert can review the situation, validate the model's predictions in aggregate, and provide feedback to improve the model's performance over time. This approach helps maintain the model's accuracy and effectiveness while also providing valuable insights for future improvements. This can be useful to handle new, unseen, non-numerical data types properly, such as new words for text data or new label categories.

With proper drift impact analysis, drift detection strategy, and model maintenance flow setup, you can now ensure that your deep learning model remains robust and accurate even as the underlying data distribution evolves. By understanding how changes in the data can affect your model's performance, having a strategy in place to identify and quantify drift, and establishing a clear process for model updates and maintenance, you can proactively address issues, maintain model reliability, and provide consistent and trustworthy results over time.

We will dive deeper into the topic of programmatic drift detection in the next section.

Detecting drift programmatically

With a comprehensive understanding of drift types and their effects, we will explore techniques for detecting drift programmatically, diving into the realms of concept drift and data drift. Armed with these methods, you'll be well equipped to implement high-risk drift detection components. Let's start with concept drift.

Detecting concept drift programmatically

Concept drift involves both the input data and the target data. This means that we can effectively detect concept drift for a deployed model only when we can get access to the real target labels in production. When you do have access to them, you can adopt the following techniques to detect concept drift:

- **Check the similarity of production data to the reference training data**: This should include both input and output data.

- **Use model evaluation metrics as a proxy**: Evaluation metrics can signal concept drift or data drift.

- **Use multivariate-based data drift detection and include both input and target data**: This can be unreliable where detection can be data drift instead of concept drift. But it doesn't change the fact that something needs to be done, so it's fine.

Next, we will explore programmatic data drift detection.

Detecting data drift programmatically

Detecting data drift programmatically involves two essential steps: quantifying the type of change of data and applying a detection threshold based on reference or training data and the current data. For statistical-based drift, detection can be implemented with data statistical value thresholds identified during the analysis. However, for distribution-based data drift, it can be hard and ambiguous to define the threshold properly. In this section, we will focus on methods to quantify the distribution change. To do that, one can employ either of the following methods:

- **Statistical tests**: These tests are used to compare the distribution of training/reference data with the distribution of new incoming data. A significant divergence between the two distributions may indicate data drift. **Evidently AI** provides an open source tool called evidently that provides these metrics out of the box. *Table 17.1* highlights the different univariate statistical tests that can be used with details of the implementation done in `evidently`, along with its pros and cons:

Statistical test type	Pros	Cons	Evidently implementation info
Kolmogorov-Smirnov (K-S) test	Non-parametric and distribution-free, making it versatile. Fast to compute and interpret.	Less sensitive to differences in tails of distributions. Assumes continuous and one-dimensional data.	Supports: Numerical data type Threshold: Score < 0.05 Default: For numerical data, if <= 1,000 samples
Chi-squared test	Works well for categorical data. Fast to compute and interpret.	Requires data to be binned, which can be subjective. Assumes that observations are independent.	Supports: Categorical data type Threshold: Score < 0.05 Default: For categorical with > 2 labels, if <= 1,000 samples
Z-test	Works well for large sample sizes. Fast to compute and interpret.	Assumes normal distribution and known population variance. Not suitable for small sample sizes.	Supports: Categorical data type Threshold: Score < 0.05 Default: For binary categorical data, if <= 1,000 samples
Anderson-Darling test	More sensitive to differences in tails of distributions compared to the K-S test. Can be used for various distributions with proper scaling.	Assumes continuous data. Computationally more complex than the K-S test.	Supports: Numerical data type Threshold: Score < 0.05 Default: N/A
Fisher's exact test	Accurate even with small sample sizes. Suitable for categorical data.	Computationally intensive, especially with large sample sizes. Limited to 2x2 contingency tables.	Supports: Categorical data type Threshold: Score < 0.05 Default: N/A
Cramér-von Mises test	Sensitive to differences in both central and tail regions of distributions. Non-parametric and distribution-free.	Computationally more complex than the K-S test. Assumes continuous data.	Supports: Numerical data type Threshold: Score < 0.05 Default: N/A

G-test (likelihood ratio test)	Suitable for categorical data. Asymptotically equivalent to the Chi-squared test.	Requires large sample sizes for accurate results. Assumes independent observations.	Supports: Categorical data type Threshold: Score < 0.05 Default: N/A
Epps-Singleton test	Sensitive to differences in the shape of distributions. Robust against outliers.	Computationally complex. Assumes continuous data.	Supports: Numerical data type Threshold: Score < 0.05 Default: N/A
T-test	Fast to compute and interpret. Applicable for comparing the means of two groups.	Assumes normal distribution and equal variance. Not suitable for small sample sizes when the normality assumption is not met.	Supports: Numerical data type Threshold: Score < 0.05 Default: N/A

Table 17.1 – Statistical tests for distribution change

- **Distance metrics**: Distance metrics can be computed between distributions. *Table 17.2* shows the different distance metrics that can be used with details of the implementation done in `evidently` along with its pros and cons:

Distance metric	Pros	Cons	Evidently implementation info
Wasserstein distance	Captures the geometric differences between two distributions, taking into account both the shape and location. Provides a natural and interpretable metric for comparing distributions.	Computationally expensive, especially for high-dimensional data. May not work well for discrete distributions or sparse data.	Supports: Numerical data type Threshold: Distance >= 0.1 Default: For numerical data, if > 1,000 samples

Kullback-Leibler (KL) divergence	Quantifies the difference between two probability distributions by measuring the extra number of bits required to encode one distribution using the other. Works well for continuous distributions and has a strong theoretical foundation.	Asymmetric: KL(P ‖ Q) ≠ KL(Q ‖ P), which may affect the interpretation of the measure. May be infinite if the support of the two distributions does not overlap, making it less suitable for data drift detection.	Supports: Numerical and categorical data types Threshold: Distance >= 0.1
Jensen-Shannon (JS) distance	Symmetric measure: JS(P ‖ Q) = JS(Q ‖ P), making it more suitable for comparison. Bounded between 0 and 1, making it easier to interpret. Combines the strengths of KL divergence and mutual information.	In some cases, JS distance might not be sensitive enough to detect small differences between distributions.	Supports: Numerical and categorical data type Threshold: Distance >= 0.1 Default: For categorical, if > 1,000 samples
Hellinger distance	Bounded: Produces values between 0 and 1, providing easier interpretation	Might not be sensitive enough to detect small differences between distributions.	Supports: Numerical and categorical data type Threshold: Distance >= 0.1

Table 17.2 – Distance metrics for data drift

- **Classification model to differentiate between the reference and the current data**: A binary threshold needs to be set. Although this isn't strictly a distribution change measurement, it can be considered an approximated distribution change.

Fortunately, the `evidently` library provides all these methods out of the box with default thresholds. Evidently is an easy-to-use toolkit that provides metrics monitoring, data drift detection, and data drift analysis functionalities for machine learning models.

As a follow-up to the data distribution-based drift detection topic, either of the methods introduced can be executed using either univariate or multivariate approaches. The choice between these approaches depends on the complexity of the data and the desired level of granularity in drift detection. Here are some suggestions on when to choose each method:

- Use univariate drift detection in the following cases:

  - The relationships between individual variables are not significant or not of primary concern

  - The goal is to detect drift at a granular level, focusing on each variable separately

  - The data has a low dimensionality or a small number of variables, making it less challenging to analyze each variable individually

  - The computational resources or time available for analysis are limited, as univariate methods are generally less computationally demanding than multivariate methods

- Use multivariate drift detection in the following cases:

  - The relationships between multiple variables are essential, and detecting drift in these relationships is crucial for model performance

  - The data has high dimensionality or many variables, making it challenging to analyze each variable individually

  - The goal is to capture a holistic view of the data drift, considering the interactions between variables

  - The computational resources and time available for analysis are sufficient, as multivariate methods can be more computationally intensive than univariate methods

  - The meaning of individual variables is not defined, for example, embeddings generated from deep learning models

Evaluate these factors to determine the appropriate method for detecting data drift in machine learning models.

One final step is to identify the detection threshold. As any distribution change doesn't necessarily mean a positive impact or negative impact, it's hard to set a threshold through any cross-validation techniques for your dataset. The idea here is to set a reasonable large distribution change value that can at least cause a change in impact. Fortunately, if you use the `evidently` library, it provides default thresholds that allow us to truly treat this technique as an arbitrary drift detector when you don't have the means to analyze the metric impact.

Next, we will dive into a short practical implementation of programmatic data distribution drift detection using the Python **evidently** library.

Implementing programmatic data distribution drift detection using evidently

One thing you are probably curious about is whether the absolute magnitude of distribution matters in distribution drift computation. In this section, we will explore a short tutorial on using evidently that demonstrates three things:

- Absolute magnitude matters as well in distribution drift measurements along with relative magnitude
- A detected distribution drift or a highly drifted score doesn't necessarily result in degraded performance
- Distribution drift alignment with a drop in metric performance

The tutorial will be based on the same model, dataset, and dataset characteristic used in the *Executing adversarial performance analysis for speech recognition models* section in *Chapter 14, Analyzing Adversarial Performance*, which is about speech recognition. Let's dive into it in a step-by-step manner:

1. First, we will import the necessary libraries:

```
import evaluate
import numpy as np
import pandas as pd
import syllables
import torch
from audiomentations import TimeStretch
from datasets import load_dataset
from evidently.metric_preset import DataDriftPreset
from evidently.metrics import ColumnSummaryMetric
from evidently.metrics import DataDriftTable
from evidently.report import Report
from tqdm import tqdm_notebook
from transformers import Speech2TextProcessor,
Speech2TextForConditionalGeneration
```

2. Next, we will load the dataset and the speech recognition model in the GPU:

```
ds = load_dataset("google/fleurs", 'en_us', split="validation")
device = torch.device("cuda")
model = Speech2TextForConditionalGeneration.from_
pretrained("facebook/s2t-small-librispeech-asr")
processor = Speech2TextProcessor.from_pretrained("facebook/
s2t-small-librispeech-asr")
model.to(device)
```

3. We will be using the word error rate performance metric here, so let's use the method from the Hugging Face `evaluate` library along with the method to compute and return a list of the metric scores:

```
wer = evaluate.load("wer")

def get_wer_scores(dataset, transcriptions=None, sampling_
rates=None, is_hg_ds=False, verbose=True):
  all_wer_score = []
  for idx, audio_data in tqdm_notebook(enumerate(dataset),
total=len(dataset), disable=not verbose):
    inputs = processor(
        audio_data["audio"]["array"] if is_hg_ds else audio_
data, sampling_rate=audio_data["audio"]["sampling_rate"] if is_
hg_ds else sampling_rates[idx], return_tensors="pt")
    generated_ids = model.generate(
        inputs["input_features"].to(device), attention_
mask=inputs["attention_mask"].to(device))
    transcription = processor.batch_decode(
generated_ids, skip_special_tokens=True)
    wer_score = wer.compute(predictions=transcription,
references=[audio_data['transcription'] if is_hg_ds else
transcriptions[idx]])
    all_wer_score.append(wer_score)
    return np.array(all_wer_score)
```

4. We will be using a known characteristic that affects the metric performance of the model, can be controlled through augmentation, and can be measured, which is syllables per second. Let's define the method that gets the augmented results and calls the method to compute the metric scores:

```
def get_augmented_samples_wer_results(all_baseline_samples,
transcriptions, all_sampling_rates, rates_to_change):
    all_augmented_samples = []
    for idx, audio_sample in enumerate(all_baseline_samples):
        if rates_to_change[idx] != 0:
            augment = TimeStretch(min_rate=rates_to_change[idx],
max_rate=rates_to_change[idx], p=1.0)
            augmented_samples = augment(samples=audio_sample,
sample_rate=all_sampling_rates[idx])
            all_augmented_samples.append(
augmented_samples)
        else:
            all_augmented_samples.append(audio_sample)
            wer_scores = get_wer_scores( all_augmented_samples,
transcriptions, sampling_rates=all_sampling_rates, is_hg_
ds=False)
    return wer_scores, all_augmented_samples
```

5. To properly demonstrate the behavior of performance improvements even when drift is detected, we will use a modified version of the dataset as the reference baseline. Let's obtain it by first extracting the original dataset info:

```
all_syllables_per_second = []
original_dataset = []
all_sampling_rates = []
transcriptions = []
    for audio_data in ds:
    num_syllables = syllables.estimate(audio_
data['transcription'])
    syllables_per_second = num_syllables / (audio_data['num_
samples'] / audio_data['audio']['sampling_rate'])
    all_syllables_per_second.append(
syllables_per_second)
    original_dataset.append(audio_data['audio']['array'])
    all_sampling_rates.append(audio_data['audio']['sampling_
rate'])
    transcriptions.append(
audio_data['transcription'])
```

6. Next, we obtain the audio dataset that is expanded three times its original duration and prepare the DataFrame compatible with `evidently` library processing, which effectively reduces the syllables per second by three times:

```
reference_wer_scores, reference_samples = get_augmented_samples_
wer_results(original_dataset, transcriptions, all_sampling_
rates, rates_to_change=[3] * len(original_dataset)
)
reference_df = pd.DataFrame({
  "wer_score": reference_wer_scores,
  "syllables_per_second": [sps / 3.0 for sps in all_syllables_
per_second]})
```

7. Now that we have a reference dataset, we need a current dataset that simulates new data that we receive with a deployed model. We will modify 90% of the reference dataset to 10 syllables per second to show a more extreme case of distribution change from normal to highly skewed distribution:

```
majority_number = int(len(reference_samples) * 0.9)
minority_number = len(reference_samples) - majority_number
majority_rates = []
for i in range(majority_number):
    majority_rates.append(10.0 / all_syllables_per_second[i])
    current_wer_scores, current_samples = get_augmented_
samples_wer_results(reference_samples, transcriptions, all_
sampling_rates, rates_to_change=majority_rates + [0] * minority_
number)
```

```
reference_syllables_per_second = reference_df[
  'syllables_per_second'].values.tolist()
current_df = pd.DataFrame({
  "wer_score": current_wer_scores,
  "syllables_per_second": [10] * majority_number + reference_
syllables_per_second[-minority_number:]})
```

8. Now that we have both a reference dataset and a current dataset, let's obtain the data drift report:

```
data_drift_dataset_report = Report(metrics=[
  DataDriftTable(columns=["syllables_per_second"]),
  ColumnSummaryMetric(column_name="wer_score")])
data_drift_dataset_report.run(reference_data=reference_df,
current_data=current_df)
data_drift_dataset_report.show(mode='inline')
```

Drift is detected for 100.0% of columns (1 out of 1).

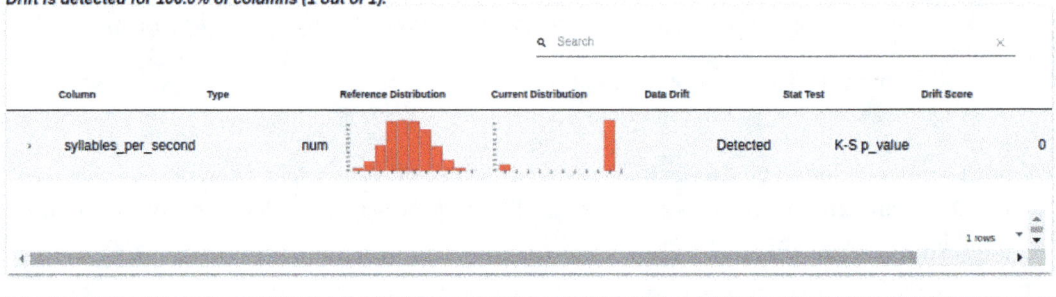

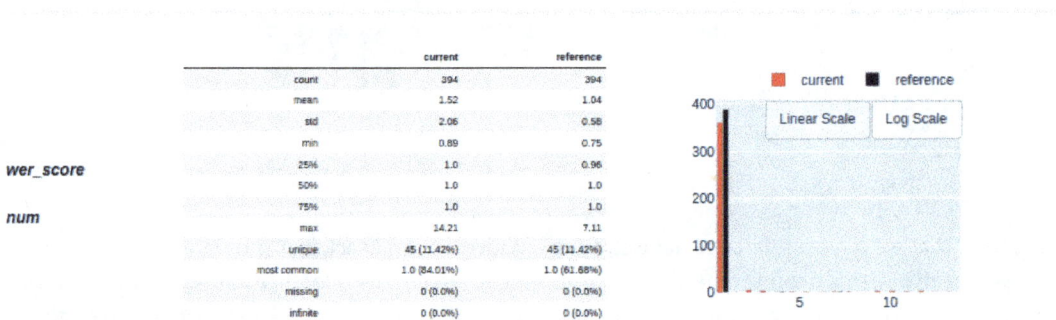

Column	Type	Reference Distribution	Current Distribution	Data Drift	Stat Test	Drift Score
syllables_per_second	num			Detected	K-S p_value	0

1 rows

		current	reference
	count	394	394
	mean	1.52	1.04
	std	2.06	0.58
	min	0.89	0.75
wer_score	25%	1.0	0.96
	50%	1.0	1.0
	75%	1.0	1.0
num	max	14.21	7.11
	unique	45 (11.42%)	45 (11.42%)
	most common	1.0 (84.01%)	1.0 (61.68%)
	missing	0 (0.0%)	0 (0.0%)
	infinite	0 (0.0%)	0 (0.0%)

current   reference

Linear Scale   Log Scale

Figure 17.2 – Data drift report by evidently

Remember that the default K-S test is used when the dataset has less than 1,000 columns, which reflects what was used here. Drift was detected with a 0.05 threshold and the metric performance dropped significantly; this is the ideal situation.

9. Next, we'll create a simulation in which the number of syllables pronounced per second is tripled compared to the current data. We'll use the original dataset for this and get the evidently drift report:

```
wer_scores = get_wer_scores(original_dataset, transcriptions,
sampling_rates=all_sampling_rates, is_hg_ds=False)
current_df = pd.DataFrame({"wer_score": wer_scores,
  "syllables_per_second":all_syllables_per_second})
data_drift_dataset_report = Report(metrics=[
  DataDriftTable(columns=["syllables_per_second"]),
  ColumnSummaryMetric(column_name="wer_score")])
data_drift_dataset_report.run(reference_data=reference_df,
current_data=current_df)
data_drift_dataset_report.show(mode='inline')
```

This will result in the report shown in *Figure 17.3*:

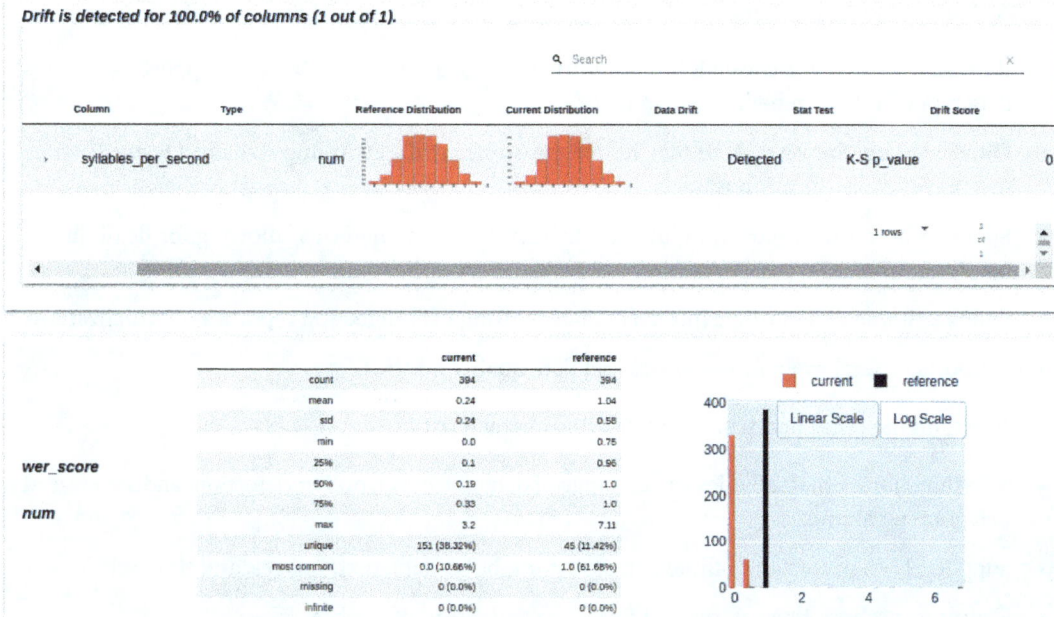

Figure 17.3 – Evidently report with same distribution but different magnitude still detected drift

Here, the distribution pattern is visibly the same, but the absolute magnitude of each group of syllables per second is much higher. The K-S test still managed to detect this as drift as it uses cumulative distribution difference, showcasing the versatility of distribution drift methods. With that, we have completed the tutorial! evidently offers a broader range of metrics for measurement, including data quality statistics and model evaluation metrics. It also includes built-in support for monitoring data drift and detecting data types such as embeddings and text. Be sure to explore these features separately. Additionally, consider diving into each of the methods in detail to discover new behaviors of distribution drift techniques that you never thought were possible.

Other than Evidently, there is one more notable open source library that you can use to handle drift programmatically, which is part of what we will discover next.

Comparing and contrasting the Evidently and Alibi-Detect libraries for drift detection

In this section, we will compare and contrast two popular libraries for drift detection in deep learning models: Evidently and Alibi-Detect. Both libraries provide tools for monitoring and detecting drift in data, but they differ in terms of their features. By understanding the strengths and weaknesses of each library, you can choose the one that best suits your needs and requirements for drift detection in your deep learning models.

Evidently has the following characteristics:

- Provides an easy-to-use toolkit for monitoring data drift, including built-in support for various data types such as embeddings and text
- Offers a comprehensive set of metrics for measuring drift, including statistical tests, distance metrics, and classification models
- Supports both univariate and multivariate drift detection methods, allowing for flexibility in handling different types of data and use cases
- Offers a simple and intuitive interface for generating drift detection reports and visualizations
- Supports general evaluation metrics and data quality metrics

Alibi-Detect, on the other hand, has the following characteristics:

- A Python library that includes a wide range of drift detection, outlier detection, and adversarial detection techniques
- Supports both online and offline detectors for tabular data, text, images, and time series
- Claims to support TensorFlow and PyTorch models out of the box

Both Evidently and Alibi-Detect are powerful libraries for drift detection in deep learning models. Depending on your specific needs and requirements, you can choose either Evidently or Alibi-Detect as your preferred library for drift detection in deep learning models. Under typical conditions, Evidently can serve as the de facto library. However, as of the time of writing this book, if you're dealing with non-tabular data without available embedding models, require outlier detection, or need a statistical test that Evidently doesn't offer, Alibi-Detect is a more suitable choice.

Summary

In this chapter, we explored the concept of drift, which affects the performance of deployed deep learning models over time. We covered the three types of drift – concept drift, data drift, and model drift – and discussed strategies to handle them effectively. This included strategies to approach drift, including automatic programmatic detection and manual domain expert predictions, strategies to quantify drift, and strategies to mitigate drift effectively. We learned that statistical-based drift should always be opted for over ambiguous data distribution drift. We also learned that monitoring drift by batch in regular intervals is crucial in ensuring the continued success of deep learning models. Finally, using the `evidently` library, we demonstrated how to implement programmatic data distribution drift detection in a practical tutorial and understood behaviors that can shape how you think of data distribution drift methods. This knowledge can be applied across various industries and applications, such as healthcare, finance, retail, and manufacturing, where maintaining the accuracy and performance of deep learning models is crucial for efficient decision-making and optimizing business processes.

This chapter marks the completion of our deep dive into every component of the deep learning life cycle. In the next chapter, we will explore how a paid-for platform called DataRobot covers crucial components of the deep learning life cycle in an easy-to-use user interface.

18

Exploring the DataRobot AI Platform

In this chapter, we will turn our focus to the DataRobot AI platform, a paid software platform that provides a powerful toolkit for deep learning use cases. DataRobot allows its users to streamline the complex stages of the machine learning life cycle. It presents an intuitive interface for data scientists, engineers, and researchers who wish to harness the power of machine learning for their projects and businesses. As we delve into the workings of DataRobot, you will learn how it simplifies and accelerates the creation, training, deployment, and government of intricate deep learning models. Thanks to features designed for automation and ease of use, it empowers users to focus on what truly matters—extracting significant value from their machine learning applications.

Our exploration will highlight the key functionalities of DataRobot, underlining its potential as a catalyst in the evolution of deep learning solutions. DataRobot aspires to offer a combination of automation, collaboration, and scalability for machine learning use cases, making it also a noteworthy tool in the deep learning domain.

Specifically, we will cover the following:

- A high-level look into what the DataRobot AI platform provides
- Preparing data with DataRobot
- Executing modeling experiments with DataRobot
- Deploying a deep learning blueprint
- Governing a deployed deep learning blueprint

Technical requirements

We will have a practical topic in this chapter to make predictions using a DataRobot deployed model. We will be using Python 3.10 and we will require the following Python libraries to be installed:

- `datarobotx==0.1.17`

- `pandas==2.0.3`

The code files are available on GitHub at `https://github.com/PacktPublishing/The-Deep-Learning-Architect-Handbook/tree/main/CHAPTER_18`, and the dataset can be downloaded from `https://www.kaggle.com/datasets/dicksonchin93/datarobot-compatible-house-pricing-dataset`.

Additionally, a paid or free trial account is needed to access DataRobot. To subscribe for a trial account, do the following:

1. Visit the DataRobot website at `https://www.datarobot.com/trial/`.

2. Fill up your credentials under the **Start For Free** interface on the right side of the web page and click on the **Submit** button.

A high-level look into what the DataRobot AI platform provides

The DataRobot AI platform provides data ingestion, data preparation, data insights, model development, model evaluation, model insights and analysis, model deployment, and model governance through model monitoring and model maintenance tools that work seamlessly with each other. While DataRobot streamlines the deep learning life cycle, it is important to note that the planning stage still requires human input to define the goals and scope of the project. Additionally, you are still required to consume the insights, reports, and results made easy for you to obtain. Ultimately, this means that such a platform is a tool that can assist any machine learning practitioner instead of being a replacement for data scientists, machine learning engineers, machine learning researchers, or data analysts. Think of AI platforms such as DataRobot as being powerful calculators that can help you solve complex math problems quickly and accurately. But just like a calculator can't think for you, DataRobot can't replace the expertise and creativity of a data scientist, engineer, analyst, or researcher.

Some of the tools DataRobot offers are built to be extensible, composable, and flexible to add your own code or components, and they don't tie you into the only things that the platform provides. Additionally, some components cover a wide range of methods, so you don't need to worry about doing any customization. Effectively, this means you get the benefit of executing projects reliably fast while still holding customization power.

Before we dive into the components DataRobot offers, there's more key information that can help you understand what the platform is capable of:

- **Platform hosting options**: The DataRobot AI platform offers a cloud-hosted application. If your business has data privacy and security concerns, DataRobot also offers the option to host the AI platform as a privately hosted instance in the confines of your own infrastructure.

- **Scalability and collaborative nature**: Whether you're a single user or a team, DataRobot is designed to scale with your needs. Most of the tool components provided by DataRobot can be shared with multiple users, which enables collaboration between different users.

- **User interfaces**: DataRobot provides three main user interfaces to navigate through its features: the web application **graphical user interface** (**GUI**), the API interface, and the Python API client called `datarobot`, which is installable through `PyPI`. Almost everything you see in the web UI is available in the API interface and the Python API client. Additionally, the Notebooks feature provides a flexible and interactive environment where data scientists can manually perform complex data analysis, create machine learning models, and prototype data manipulations alongside the other DataRobot features through the Python API client easily. It enhances the user experience by allowing the flexibility of defaulting to using traditional Python code.

- **The transition to a use case-focused asset management**: The ML or DL life cycle introduced in *Chapter 1*, *Deep Learning Life Cycle*, is an iterative process. This means a DL use case will involve a lot of data, data versions, model development experimentations, applications that utilize a deployed model, and notebooks being created. The **Workbench** feature in DataRobot is meant to support this process by managing many use case-related assets mentioned in a single interface, making it easier to realize value through the use case. However, at the time of writing this book, Workbench does not comprehensively support all the features provided by the traditional, separately managed DataRobot projects. It will be updated to support the full suite of features in time. You can manage projects separately with DataRobot Classic features. *Figure 18.1* shows an example flow of how a typical machine learning practitioner would navigate the platform:

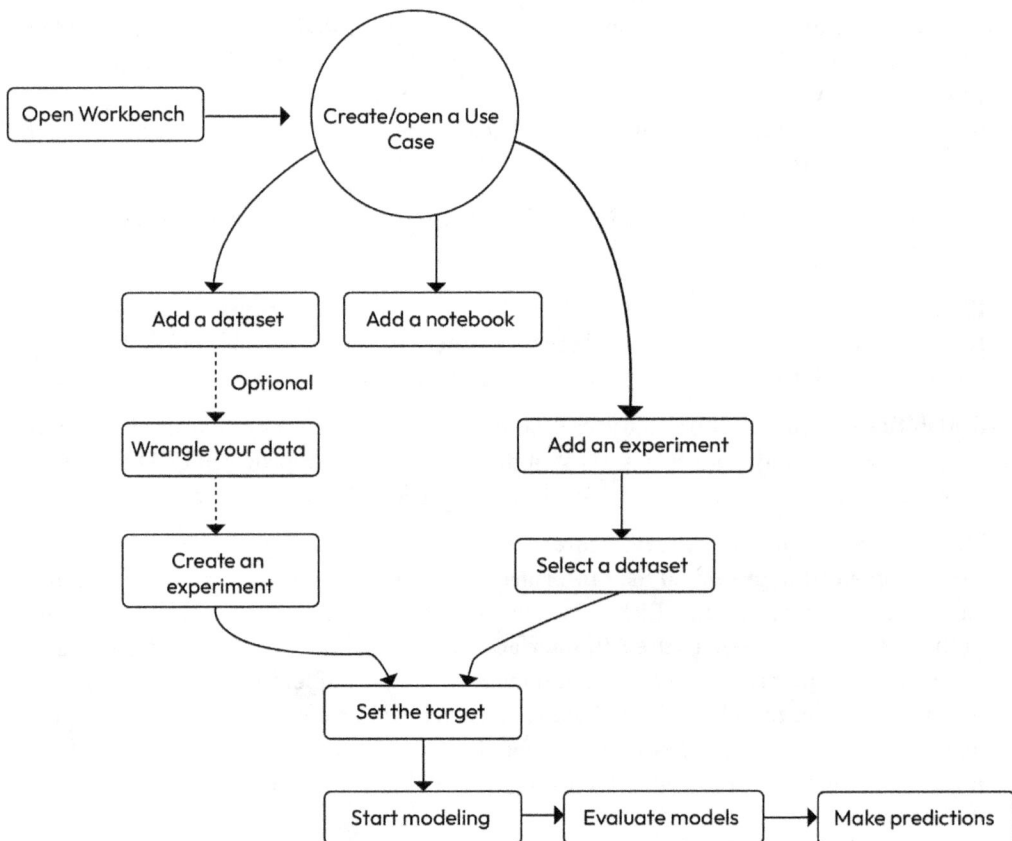

Figure 18.1 – Example Workbench workflow

Now, we are ready to dive into the relevant supported features for deep learning use cases, starting with the data preparation component.

Preparing data with DataRobot

The first part of what the platform offers is the data preparation component. DataRobot simplifies the data preparation process by offering a range of features to streamline data ingest, cleaning, transformation, and integration. Let's dive into these features in detail.

Ingesting data for deep learning model development

The development of deep learning models in DataRobot begins with the pivotal step of data ingestion. This process allows you to directly import your data from various sources, including cloud storage (such as AWS S3), Google Cloud Storage, local files, or databases such as PostgreSQL, Oracle, and SQL Server. The platform accepts diverse file formats, including CSV, XLSX, and ZIP files. Additionally, the platform supports image, text, document, geospatial, numerical, categorical, and summarized categorical data through secondary datasets as input data types. For the target data types, the platform supports numerical, categorical, and multilabel data types along with data with no targets for unsupervised learning. This sets the stage for regression, binary classification, multiclass classification, multilabel classification, unsupervised anomaly detection, and unsupervised clustering.

Notably, image data type, text data type, and document data type are the core unstructured input data types for building deep learning use cases in DataRobot. Image data and document data are supported by being encoded as a base64 string under the hood. However, the dataset to be ingested itself can be structured to be zipped folders with images or documents, where the folder names are the classes or target names. Text data can naturally exist under tabular data in a column, encoded in formats such as CSV. Additionally, DataRobot automatically creates useful features if there are secondary datasets or datetime feature columns.

There are two paths you can take to ingest data:

- Through project creation, tied closely to the model development process.
- Through the **AI Catalog** feature, which allows you to share the dataset independently. This can be subsequently used to create an experiment in a use case or a project independently.

Let's explore the second approach, as it is a more responsible and reliable way of managing data used for model development.

Using DataRobot to ingest an image and text dataset

We will be tackling the use case of predicting housing prices with multimodal data that consists of image data, text data, date data, categorical data, and numerical data. Let's start the step-by-step tutorial:

1. Start by creating a use case by clicking on the + **Create a new Use Case** button shown in *Figure 18.2*. The page shown in *Figure 18.2* will also be the landing page after you enter the DataRobot web app and log in through app.datarobot.com:

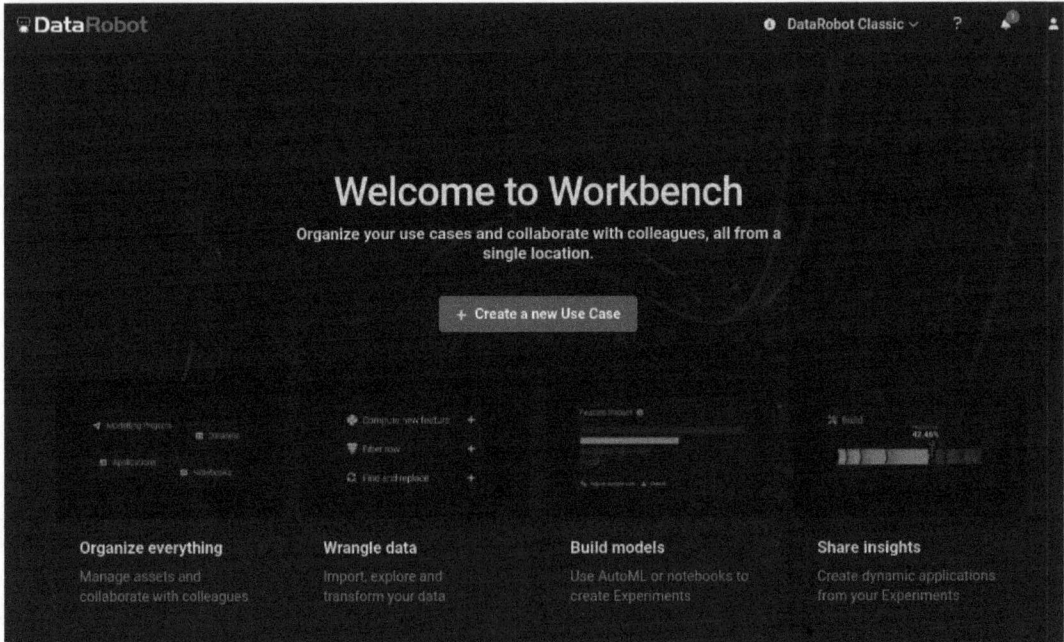

Figure 18.2 – Creating a use case screen in Workbench

2.   Now, click on **DataRobot Classic** in the top-right and then on the **AI Catalog** tab in the top-left of the interface. Then, click on the **Add to catalog** button in the top-left of the interface and then **Local File**, as depicted in *Figure 18.3*:

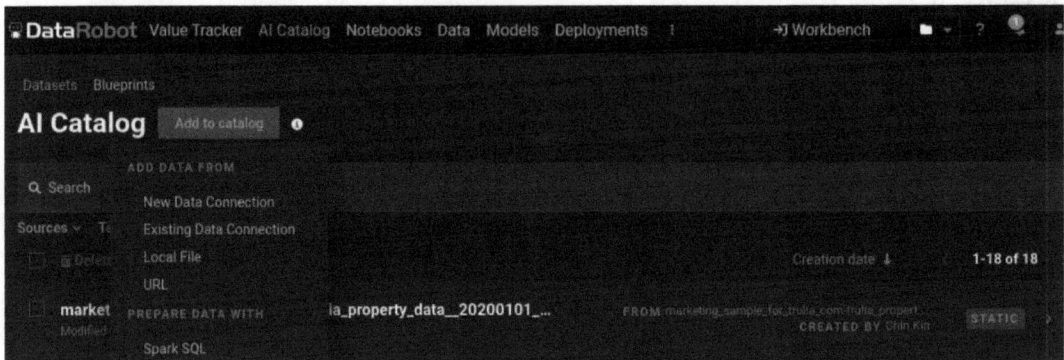

Figure 18.3 – Adding the dataset to the catalog interface in DataRobot

3. Upload the `trulia_pricing_dataset.zip` file provided in the code repo. This dataset is a ZIP file that consists of a single CSV file, which contains the raw data, and additionally, raw image files that are mapped to one of the columns in the CSV through its relative path in the zipped file. From here, an uploaded dataset in the AI Catalog can be managed independently and shared separately.

4. Once it is created, go into Workbench in the top-right again, go into the use case we created, and rename the use case to a suitable name, such as `Pricing Prediction`. Then, click on the **Add new** button dropdown in the top-right of the interface and click on **Add datasets**, as depicted in *Figure 18.4*:

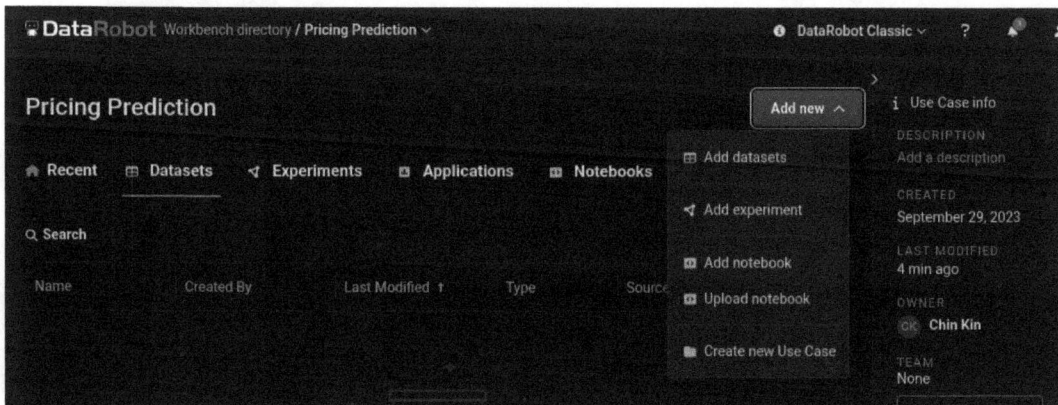

Figure 18.4 – Pricing prediction use case in the Workbench interface

5. Choose `trulia_pricing_dataset` from the **Data Registry** page and click on **Add to Use case**, as depicted in *Figure 18.5*:

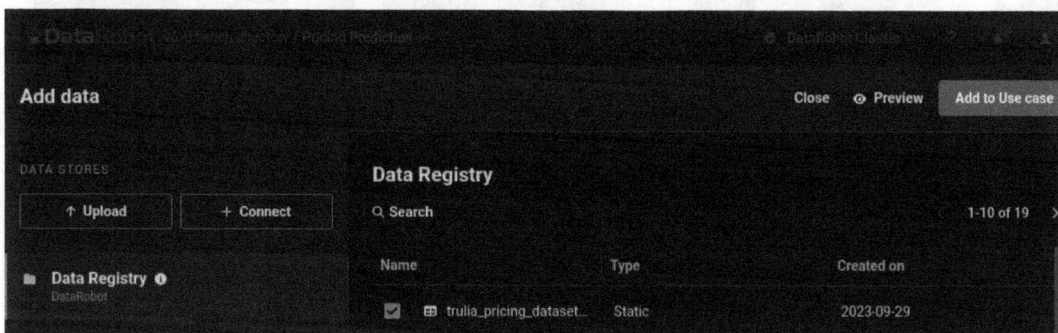

Figure 18.5 – Adding the dataset to the use case

Now, we are ready to move into the EDA part of DataRobot.

Exploratory analysis of the data

DataRobot provides a two-step experience in performing **exploratory data analysis (EDA)** on an at most 500MB subset of the dataset. The first step is EDA, which is executed before a project or experiment type is determined, and the second step is done after. Standard EDA techniques are provided. These include histograms for numerical and categorical data, frequency distribution for the top 50 items for categorical data, duplicate counts, missing value counts, disguised missing value detection, excessive zero value detection, target leakage, numerical value aggregates, outlier rows, univariate feature correlation to the target, and a feature association matrix that measures mutual information. For images and text data specifically, DataRobot Classic shows a general sample of the data in *step 1* and sorted by the target values or value ranges group in *step 2*.

Figure 18.6 shows the EDA 2 visualizations from the house pricing prediction dataset in DataRobot Classic, where the left image shows the image samples by binned targets and the right image shows the duplicate feature interface:

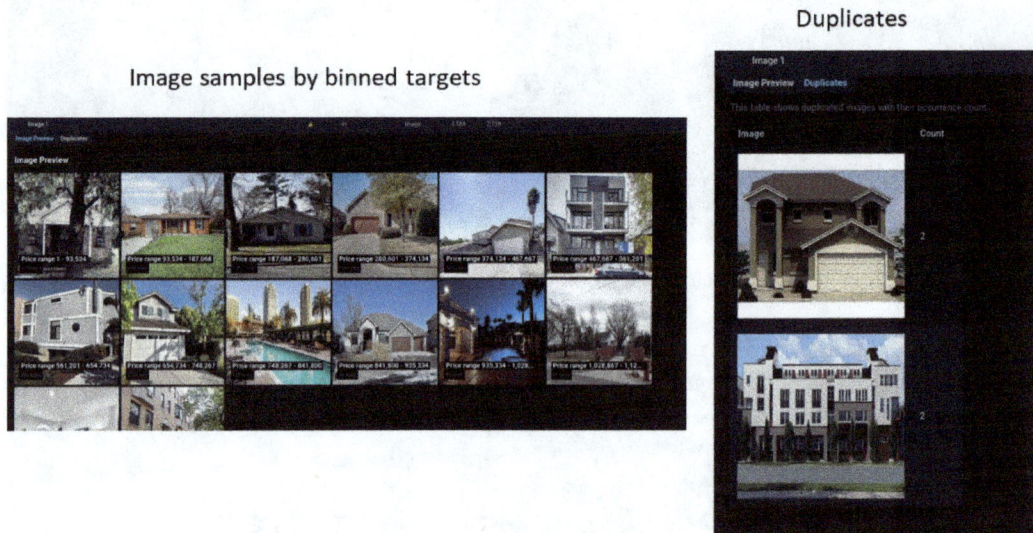

Figure 18.6 – EDA 1 and EDA 2 of images

Now, let's continue the tutorial with the EDA of the house pricing dataset.

Practically performing EDA in DataRobot

We will continue the EDA tutorial in a step-by-step manner:

1. Click on the `trulia_pricing_dataset` entity under the **Data** tab in the use case and you will be presented with the view in *Figure 18.7*, which can be scrolled horizontally and vertically:

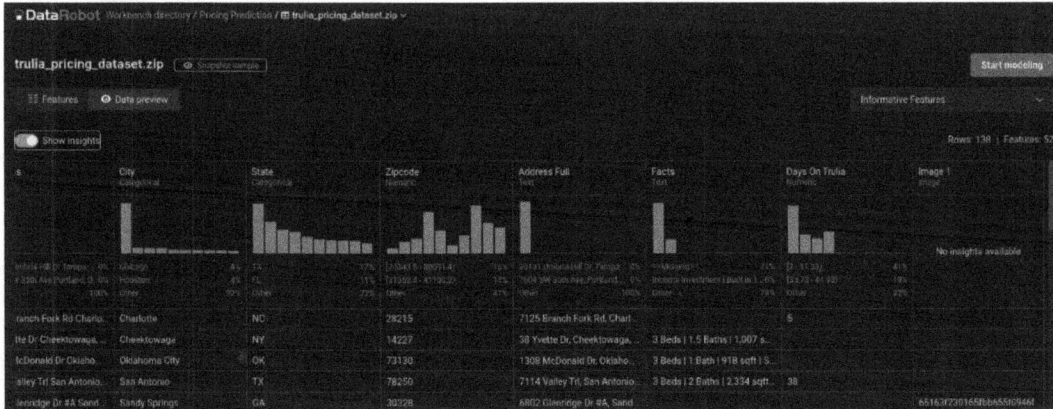

Figure 18.7 – Sample data preview EDA interface of the housing dataset in Workbench

If you scroll through the entire table, you will find that the dataset consists of 68 features, where 52 features are identified to be informative. This most notably consists of an image column that is valid, along with 23 other columns that are supposed to be images too but are just links that are not included properly. It also has five useful text columns that consist of facts, short and full addresses, features of the houses, and general descriptions of the houses.

2. If you click on the **Features** button in the top-left, you will see the interface in *Figure 18.8*, which shows simple statistics of the dataset:

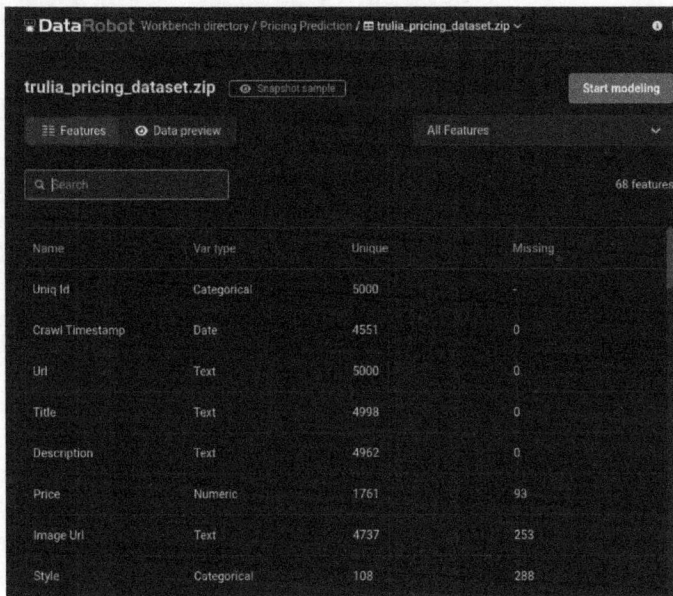

Figure 18.8 – Features of the housing dataset

3. Now, we need to set the prediction target. The use case is to predict the price of a house here, so click on the **Next >** button in the top-right and fill in `Price` in the **Target Feature** box. You will then see the interface shown in *Figure 18.9*:

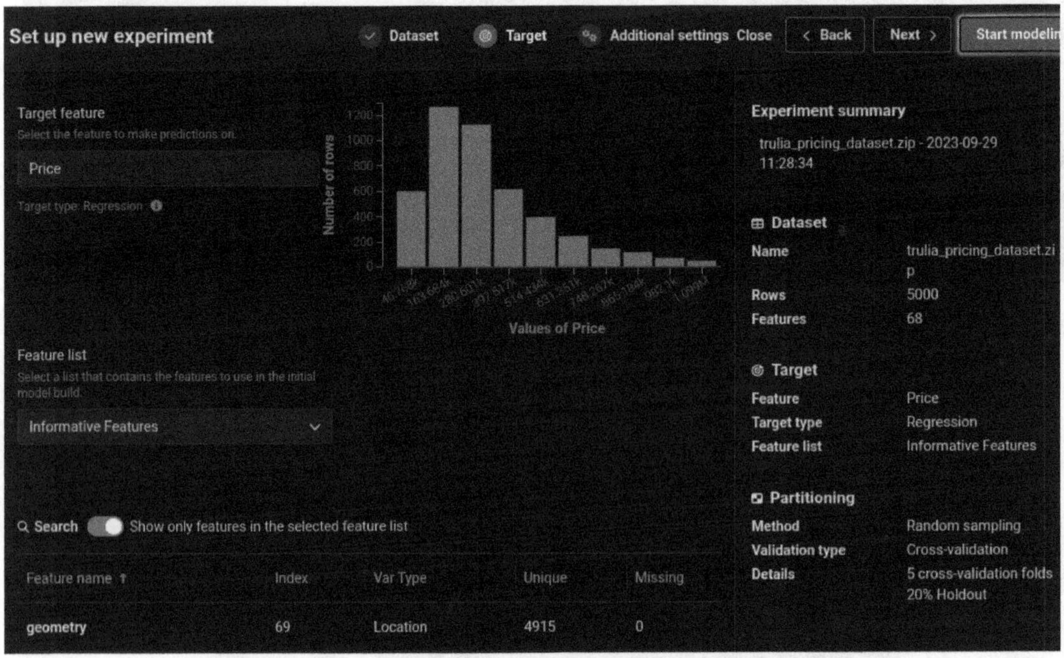

Figure 18.9 – Choosing a target in Workbench

4. Then, click on the **Start modeling** button to start the quick modeling process. If you navigate to the **Data** tab from the DataRobot Classic interface, you will be able to see the univariate importance computed under the hood. This ranks features by their univariate informativeness with regard to the chosen target. This is depicted by the green bars in *Figure 18.10*:

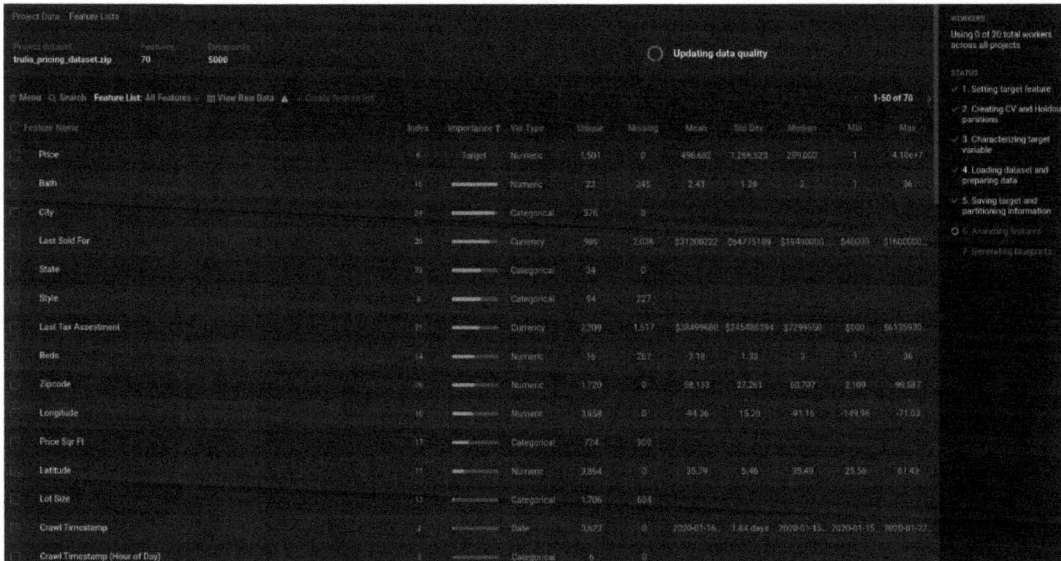

Figure 18.10 – DataRobot Classic Data tab showing univariate importance

Next, we will discover how DataRobot allows data wrangling after connecting to a dataset source and performing EDA.

Wrangling data for deep learning model development

DataRobot allows you to transform your data with its **Wrangle** feature. It is supported for the case where the dataset is added through a data connection to a source, such as Snowflake. You can craft a set of data transformations you intend to apply to the entire dataset, which are called **recipes**. These transformations are initially tested on the live sample to ensure accuracy. Once your recipe is finalized, it's sent to the data source and executed to create the final output dataset. The feature allows you to optionally save the transformed dataset right into the source of the data and get the output under the data registry, the AI Catalog.

DataRobot supports a wide range of transformation operations, including the following:

- Joining datasets from the same connection instance
- Applying mathematical aggregations to dataset features
- Computing new features using scalar subqueries, scalar functions, or window functions
- Filtering rows based on specified values and conditions
- De-duplicating rows to remove duplicates
- Finding and replacing specific feature values

- Renaming features within the dataset
- Removing selected features from the dataset

> **Note**
>
> DataRobot doesn't provide data labeling tools to create labels from scratch and depends on the idea that raw business data is all already recorded and saved somewhere. Use external labeling tools such as LabelBox to label data collaboratively and reliably for deep learning use cases.

As the dataset we are using leverages a local dataset, we won't be practically exploring the data wrangling component here. Dive into `https://docs.datarobot.com/en/docs/workbench/wb-dataprep/wb-wrangle-data/wb-add-operation.html` to explore more on this topic.

Next, we will explore how DataRobot executes modeling experiments or projects.

Executing modeling experiments with DataRobot

DataRobot currently provides two ways to execute modeling experiments: DataRobot Classic and Workbench. Workbench is where an experiment will be managed under a use case, focusing on extracting value from a use case more seamlessly, and DataRobot Classic is the original AutoML experience where a modeling experiment is called a project. A project, or a modeling experiment here, encompasses the same components, which include modeling machine learning, gathering model insights and prediction insights, and making one-off batch predictions. We will dive deeper into these three components.

Deep learning modeling

DataRobot provides modeling configurations and tasks in the form of **directed acyclic graphs (DAG)** called **blueprints**. The individual nodes in the graph are grouped up into the following:

- **Input data**: The input nodes can be any of the supported input data types.
- **Data preprocessing tasks**: They consist of data regularization, normalization, missing value filling, and just any data preprocessing logic. You can also have tasks that choose the exact column to operate on. Additionally, techniques to perform predictions post-processing are also grouped here. Pre-trained networks that serve as feature transforms are also grouped here.
- **Modeling tasks**: They consist of any model that produces predictions in all formats. You can also make a model task part of an intermediate node, where the predictions from the model can then be used in subsequent modeling tasks through the stacking method. The stacking method outputs the combined out-of-fold features from the k-fold cross-validation strategy introduced in the *Partitioning the data for deep learning training* section in *Chapter 8, Exploring Supervised Deep Learning*. This can be useful for training a neural network and using it to provide new features in the inferencing stage in a non-overfitting manner.

Comprehensively, the supported types of deep learning-specific tasks grouped by data type are as follows:

- **Image tasks**: The Visual AI feature, which is the product feature name that encapsulates everything related to images in DataRobot, supports both pre-trained featurizers, fine-tuning featurizers, and predictors with the following networks: Darknet, EfficientNet-B0, EfficientNet-B4, Preresnet10, Resnet50, Squeezenet, mobilenet-v3-small, and EfficientNetV2-S. For pre-trained featurizers specifically, pruned variants of the networks mentioned are offered, which offer no accuracy degradation with improved inference speeds. For featurizers, DataRobot provides an out-of-the-box way to extract low-, medium-, high-, and highest-level features from the pre-trained network, which can be tuned according to the use case. Additionally,image augmentation tasks are supported, which can be configured before an experiment has been executed and after a blueprint has been trained through the **advanced tuning** feature, where a trained blueprint can be retrained with new parameters. During the configuration of image augmentation, insights into how the augmented images will appear are provided, which will be demonstrated in the coming practical section.

- **Types of models for text and document**: DataRobot supports various models for text and document processing, including lemmatizer, pre-trained part of speech tagger, Stemmer, FastText embeddings, TFIDF with stopwords, pre-trained TinyBERT featurizer for the English language, pre-trained Roberta featurizer for the English language, and pre-trained MiniLM for multiple languages. Note that the strategy DataRobot made for text is that the stopwords and pre-trained model used will depend on the language detected in the EDA sample.

- **General models**: DataRobot offers various general models, such as MLP with and without residuals, **Automatic Feature Interaction Learning (AutoInt)**, Neural Architecture Search with Hyperband for MLP, Self-Normalizing Residual MLP with Training Schedule, and Adaptive Training Schedule.

DataRobot automatically determines the blueprints that will be included in an experiment based on the dataset characteristics based on the modeling strategy of autopilot, quick, manual, or comprehensive mode. Manual mode doesn't run any blueprints and leaves it to you to decide which blueprint to run. Quick, autopilot, and comprehensive modes can be viewed as modes that will take the fastest, medium fast, and slowest to complete. The comprehensive mode that is slowest to complete will run either different blueprints or additional blueprints that can be long-running, such as large deep learning models.

DataRobot uses a modeling strategy that gradually eliminates models to find the best balance between exploration and runtime required to build the best model. This involves creating a set of blueprints with smaller sample sizes and removing weaker models. The top blueprints are then trained with a higher sample size in succession. The process continues by identifying a second reduced feature list with only the most informative features. Finally, the best model is trained with this feature list. For images, a pre-trained CNN model is used as the base model across all blueprints. When the final blueprints are built with the final sample size and reduced feature list, the best model is retrained with a larger and more time-consuming pre-trained network. This approach helps identify the most effective features and ensures the final model is optimized for accuracy and efficiency.

Another key modeling functionality is DataRobot's bias and fairness functionality. It enables users to build and evaluate fair AI models by defining protected attributes, assessing various fairness metrics, and comparing model performance across different subpopulations. The protected attributes have to be categorical values at the time of writing this book. If enabled through the **Show Advanced Options** option, the platform automatically detects potential biases, offers mitigation strategies, and allows users to monitor fairness throughout the model development process. By incorporating these features, DataRobot promotes responsible AI deployment, ensuring models comply with ethical guidelines and deliver equitable results for all users and subgroups.

If you want to try out tasks that are not part of the out-of-the-box deep learning tasks, you can leverage **custom tasks** that you can share and use in a modeling experiment or project. Custom tasks allow you to define custom logic to either preprocess data or do modeling logic. On top of this feature, the Composable ML feature allows you to restructure and rearrange the blueprint DAGs flexibly. These features allow you to leverage more commonly used methods out-of-the-box and leverage any custom model that you might want to try out in your experiments.

DataRobot executes any tasks, such as training a blueprint, computing model insights, and computing predictions, through an on-demand worker queue. Each user will get their own assigned number of workers. Training and predicting with deep learning models can take a long time. Fortunately, DataRobot has both CPU and GPU workers, and deep learning models can be run on GPU workers to speed up runtime.

On the topic of evaluation, DataRobot uses nested cross-validation and never uses test data for in-training validation. The evaluation metrics supported by DataRobot are comprehensive and can be referred to at `https://docs.datarobot.com/en/docs/modeling/reference/model-detail/opt-metric.html`. The metrics and trained blueprints will then be presented in a leaderboard interface where blueprints are ranked by the chosen metric.

Comparisons between blueprints, however, are much broader than comparing blueprints trained on the same dataset. Datasets can be different, experiment settings can be different, and associated insights can also be different. This is where the **Model Comparison** feature helps to bridge this gap and allows the comparison of many blueprint setups managed under a single use case.

As a final note here, most modeling-related settings, such as the weights column, partitioning strategy, and metric to optimize against, can be configured under the **Advanced Options** feature. We will now continue the tutorial to execute a model experiment.

Practically executing modeling experiments in DataRobot

Let's dive into practical modeling with DataRobot through a step-by-step process, continuing on from the previous tutorial:

1. After you start the modeling process, in Workbench, you will see the following interface, where DataRobot shows you what the platform is doing while waiting for blueprints to start populating:

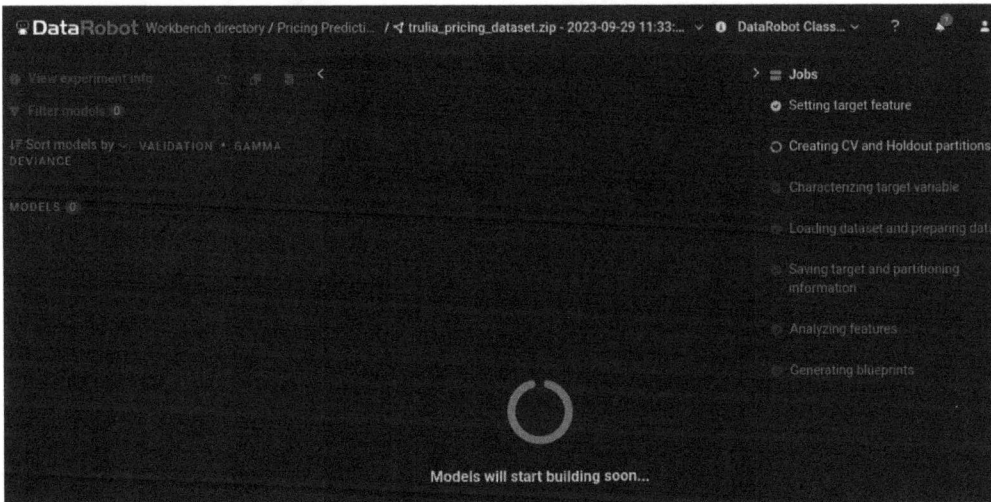

Figure 18.11 – Waiting for blueprints to populate in Workbench

2. After waiting for the blueprints to be generated and complete their training, in Workbench, you will be able to see the sorted trained blueprints on the left, as shown in *Figure 18.12 (a)*, where we can **star** two models to compare them more comprehensively:

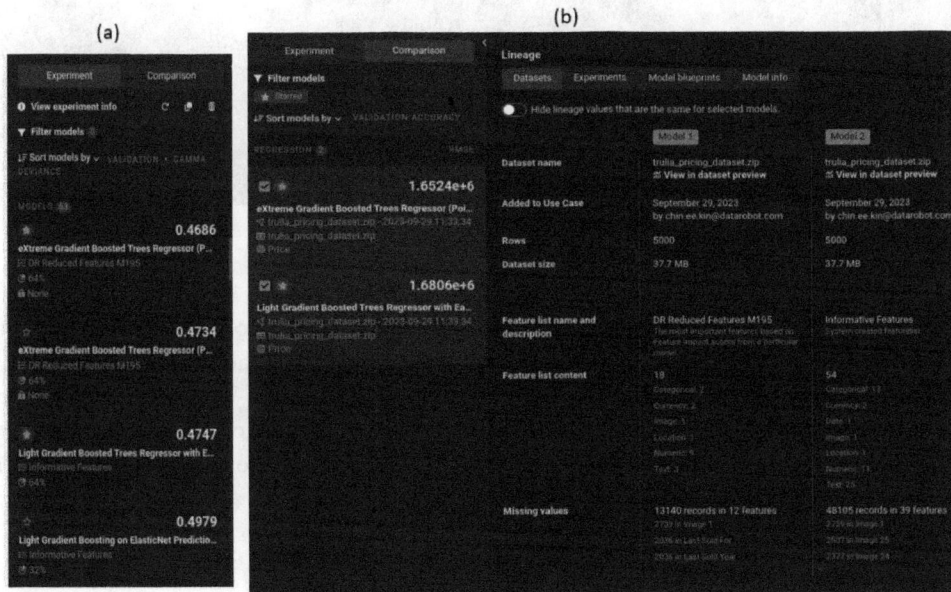

Figure 18.12 – (a) Showing the ranked blueprints with scores and
(b) showing the dataset Comparison feature

3. By clicking on the **Comparison** tab, you will be able to compare the two starred models in terms of evaluation metrics, datasets, blueprint type, and many more insights across different experiments, as depicted in *Figure 18.12 (b)*.

4. By clicking on the best-performing model in the **Gamma Deviance** metric, we can investigate the blueprint structure of the model under the **Blueprint** dropdown depicted in *Figure 18.13*. The blueprint is a multimodal blueprint with an XGBoost final modeler that takes in transformed input from categorical, geospatial, numerical, image, and text variables.

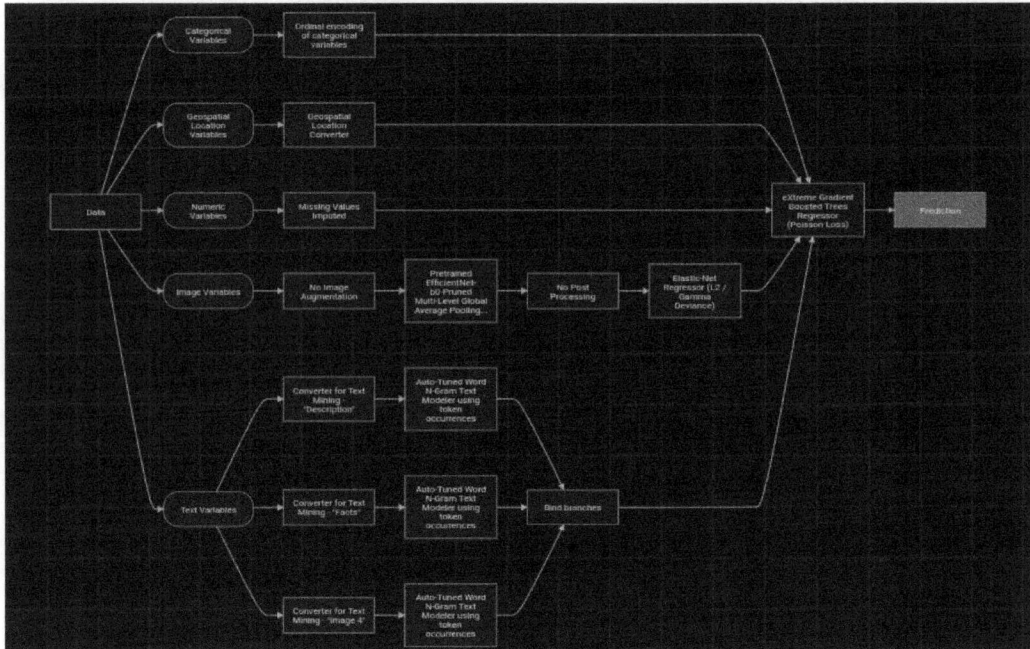

Figure 18.13 – Best model blueprint diagram

5. Now, let's see if we can make manual improvements to the metric score, which you can get for the **Validation**, **Cross Validation**, or **Holdout** partitions. As the default experiment modeling mode is a quick pilot and you can't rerun another modeling mode in Workbench as of writing, let's manually select blueprints available in the repository by clicking on **View experiment info** in the top-left of the experiment interface, which will bring you to the interface in *Figure 18.14*:

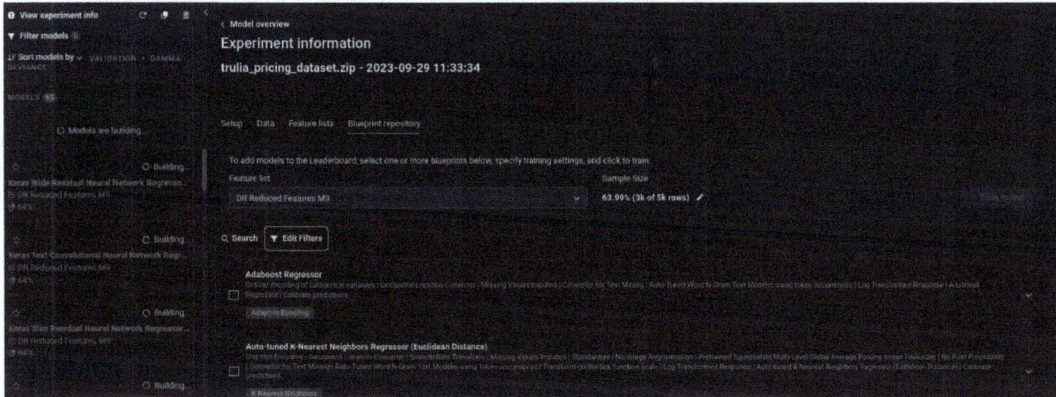

Figure 18.14 – The Blueprint repository tab in Experiment information

6. Now, search for all Keras models and fine-tuned image models, check the checkbox, and click on the **Train model** button on the right to train more blueprints that can take much longer to execute.

7. As an additional modeling step, let's add image augmentation to the existing best model starred earlier. We can do that by navigating into the **DataRobot Classic Models** tab and clicking on the starred best model on the Validation partition. Click on the **Advanced Tuning** sub-tab under the **Evaluate** tab under the blueprint. This will bring you to the interface shown in *Figure 18.15*:

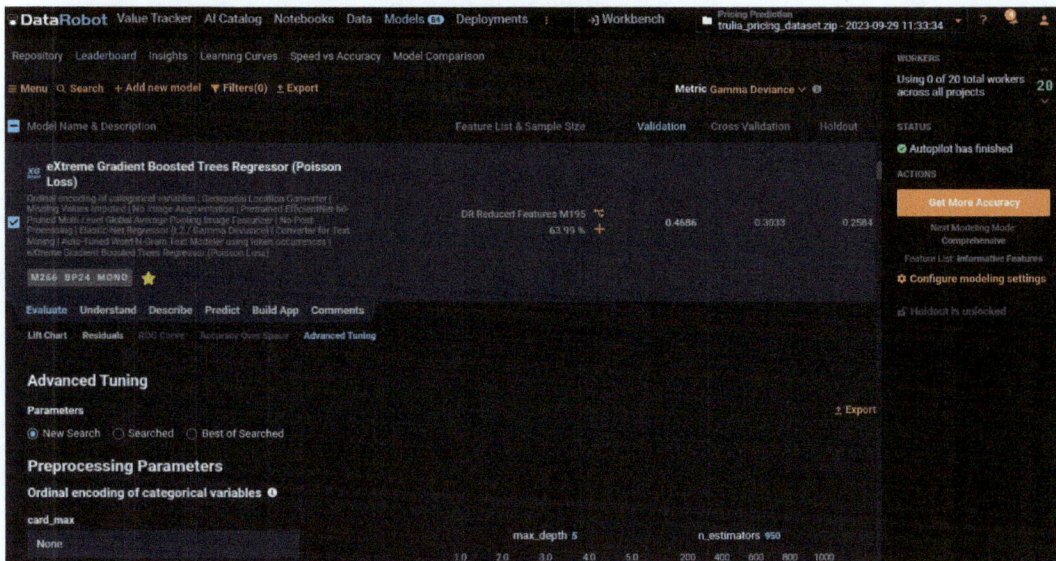

Figure 18.15 – The Advanced Tuning interface under the best model's Evaluate tab

8. Now, scroll down to the **Image Augmentation List** tab and click on **Create new list**, as shown in *Figure 18.16 (a)*. Configure **Blur**, **Cutout**, **Horizontal flip**, **Vertical flip**, **New images per original**, **Probability**, **Shift**, **Scale**, and **Rotate** to the default settings. Click on the **Preview augmentation** button and you will see image previews like in *Figure 18.16 (b)*. Now click on **Save as new list**, set your name, and click on **Create Augmentation List**. Finally, click on **Begin Tuning**, which is also shown in *Figure 18.16 (a)*:

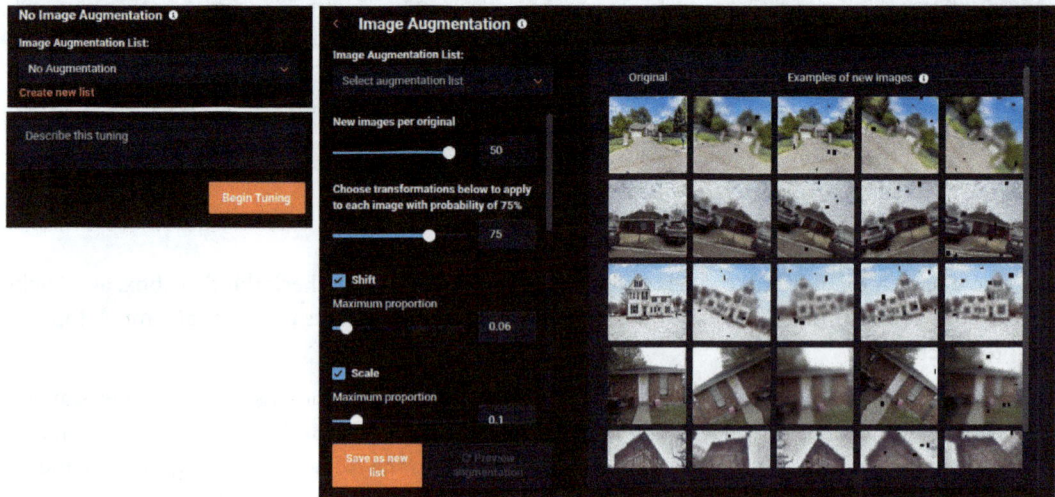

Figure 18.16 – Image augmentation configuration in Advanced Tuning

Wait until the newly tuned blueprint completes its training, and you will be blessed with a better-performing blueprint!

The steps done here only cover a small part of the modeling process that DataRobot supports. Be sure to explore features such as bias, fairness mitigation and evaluation, Composable ML, custom tasks, advanced tuning of many other parameters, and time-series modeling.

Notice that up to this stage, we were using the evaluation metric as the only form of model comparison feedback. Comparing blueprints by only using the metric is not enough in most critical use cases. In the next section, we will discover how we can gather model and prediction insights to compare blueprints comprehensively.

Gathering model and prediction insights

Notably, DataRobot provides the following insights, which are relevant to blueprints that have deep learning model tasks:

- **Feature impact**: This is a multivariate analysis that helps determine the importance of different features within a dataset, revealing which variables have the strongest influence on model predictions.

- **Feature Effects**: This is a tool that helps you understand how each feature in your dataset affects the model's predictions. It provides you with a clear and easy-to-interpret visual representation of the relationship between each feature and the model's output.

- **Activation maps**: These are visualizations that display the regions in an input image that are most responsible for making the final blueprint predictions. It covers neural networks as predictors, intermediate featurizers, and even intermediate modelers.

- **Prediction explanations**: These are techniques used to explain the output of a model, highlighting the contribution of each input feature to a particular prediction. **SHAP (SHapley Additive exPlanations)** and **XEMP (eXtended Example-based Model explanations through Perturbations)** are two popular methods that are supported. For images, image activation maps are used. For text explanations, a model-agnostic method is used to provide word-based importance scores.

- **Word cloud**: This is a visualization technique that represents the frequency of words or phrases within a text dataset, where the size of the word indicates its importance or frequency.

- **Image embeddings**: This is a visualization of images in a lower-dimensional space that captures essential features. It is derived from the output of the supported CNN models.

- **ROC curve**: This is a plot that illustrates the diagnostic ability of a binary classifier, showing the true positive rate against the false positive rate at various threshold settings, which helps in selecting an optimal threshold. Along with the curve, the confusion matrix and an optional profit curve feature are added. The profit curve is a tool that helps optimize the threshold for classification models by plotting the profit (or other performance metrics) against different threshold values.

- **Neural network visualizer**: This is a tool that allows users to visualize the architecture of a neural network, displaying the layers, neurons, and connections between them.

- **Training dashboard**: This is a tool that provides you with an easy-to-use interface that shows you important loss curves and any metric by epochs or iterations for a neural network model.

- **Blueprint visualizer**: This is a feature that provides a comprehensive view of the overall blueprint, displaying the data processing, feature engineering, and modeling steps involved in creating a machine learning model.

Now, let's explore some of these functionalities by continuing the earlier tutorial.

Practically gathering insights in DataRobot

Under each blueprint in the DataRobot Classic experience, you can explore all the insight functionalities. In Workbench, work is being done to add the comprehensive insights experience, and so far, feature impact, lift charts, and residuals insights are available. Let's dive into the insights part of the tutorial in a step-by-step manner:

1. Start by clicking on the **Feature Impact** subtab in the **Understand** tab of the best XGBoost model. Then, we can get the interface shown in *Figure 18.17*. The importance provided by feature impact can help to see if the blueprint's intuition aligns with a domain expert. Notice that **Feature Impact** shows the detected redundant features pair that prioritizes the more informative feature and can be useful if you want to further improve the model by removing features. The most concerning issue here is the `Image 4` column being used, even when it's just URLs. The next thing is the `Home Id` column, which should've been removed:

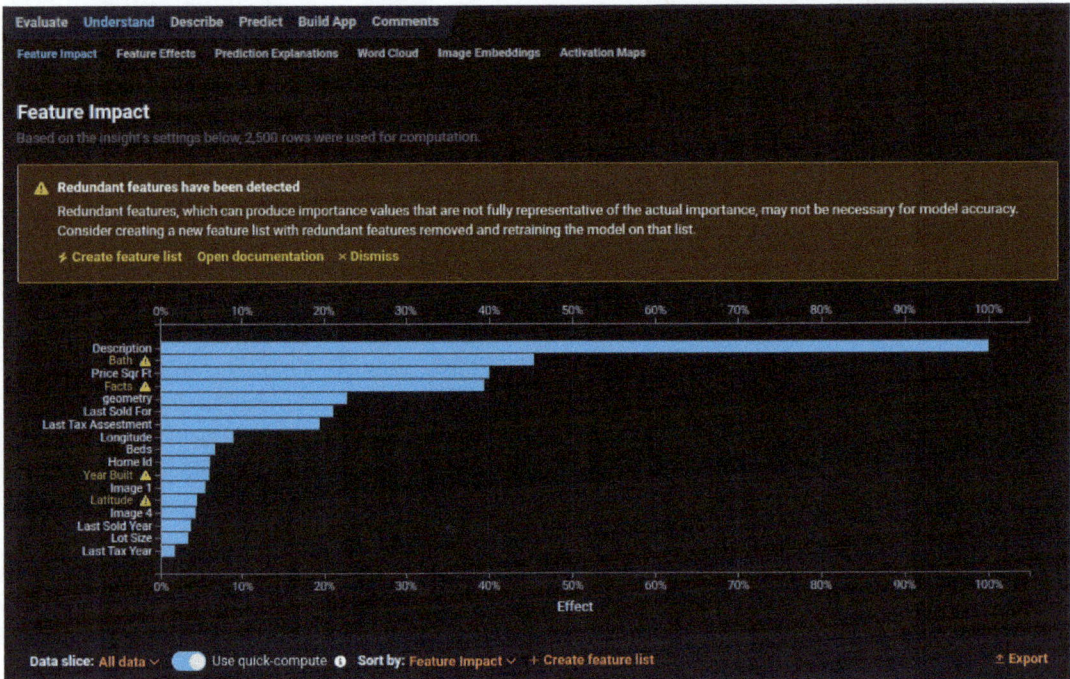

Figure 18.17 – Feature impact on the best model

2. Next, we will explore the feature effects of the best blueprint by clicking on the **Feature Effects** sub-tab. *Figure 18.18* shows that only numerical variable types are displayed, which only shows the column `Bath`. The effect graph shows that with increasing bath numbers, the price generally increases, which makes sense:

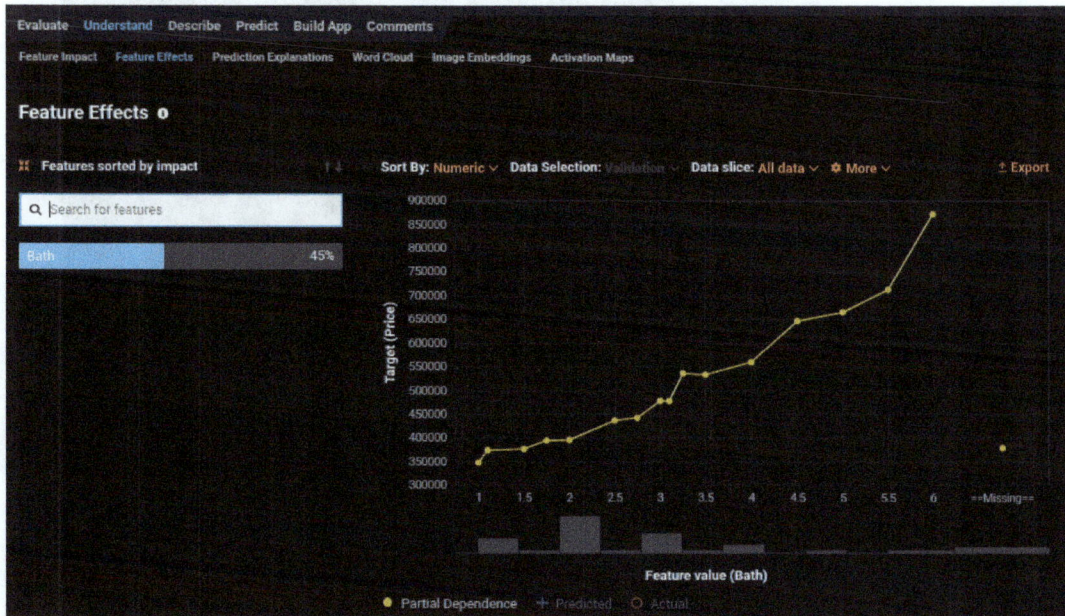

Figure 18.18 – Feature effects for the best blueprint

3. Next, click on the **Prediction Explanations** sub-tab to obtain feature importance per row of data. The feature importances are signed, which shows positive or negative contribution clearly, and categorized to 3 tiers for each positive and negative values for easier understanding. The main idea is to make sure signed importance attributions make sense. A few representative samples are computed here for visualization, but you can compute prediction explanations for every row of the training dataset on the **Download Prediction Explanations** feature. In *Figure 18.19*, we can see that `Description` is also one of the top contributing features. Clicking on the symbol under the **Value** column open a pop-up modal window will allow you to check the text explanations for the feature, as shown in *Figure 18.19*. Both the explanations look good and make good sense here:

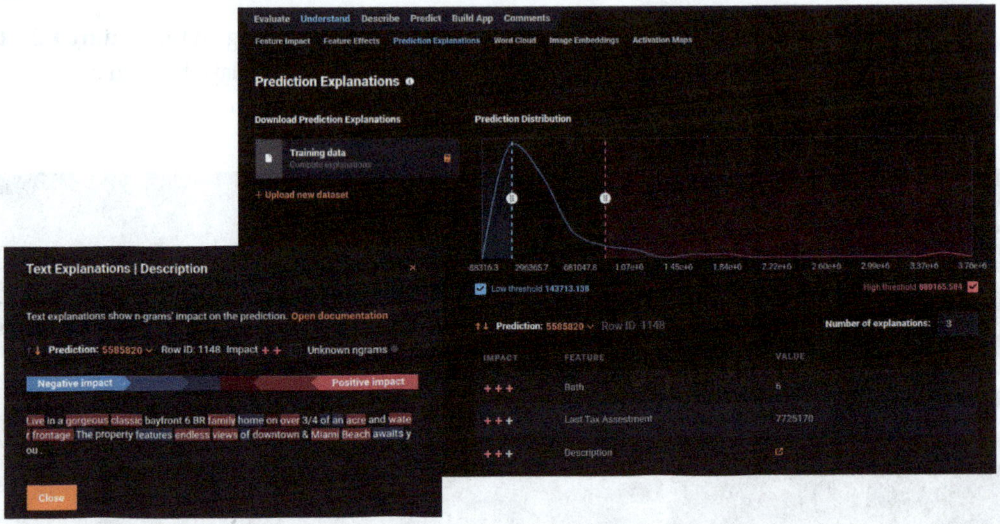

Figure 18.19 – Prediction explanations for the best blueprint

4. Now, let's click on the **Word Cloud** sub-tab to see if there are any unreasonable attributions. *Figure 18.20* shows that the model is not perfect in attribution, with words such as nice being attributed negatively. However, cozy makes sense to be negatively attributed, as it usually refers to small units.

Figure 18.20 – Word cloud importance attribution of the best blueprint

5.   Next, click on the **Activation Maps** sub-tab to see the interface shown in *Figure 18.21*:

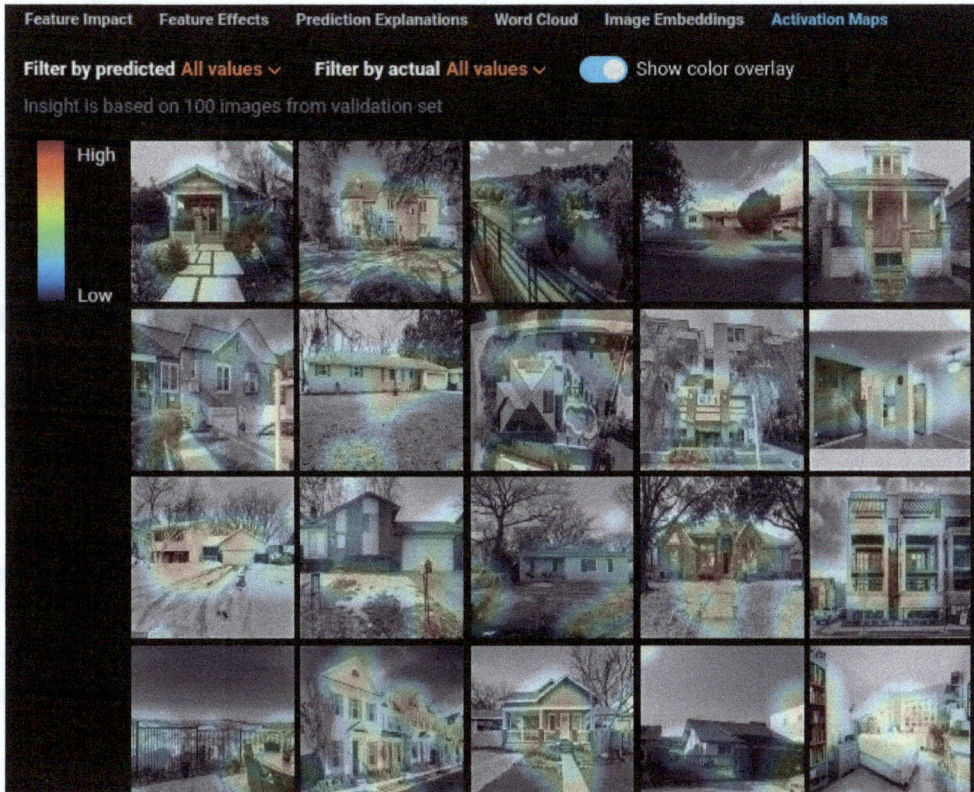

Figure 18.21 – Activation maps of the best-performing blueprint

For image activation maps, you want to look out for areas that don't make sense logically. Ask questions such as, "Why is the model looking at the sky?" or "Why is the model looking at the grass?" Also, ask follow-up questions such as "Is a well-trimmed lawn connected to price?" For this use case, there are a lot of visual components at play that make it hard to say what doesn't make sense. Focusing on grass can still be meaningful, but It's a relief that the model is at least not looking at the sky, which contributes to nothing. Additionally, the predicted and actual filter can allow you to pinpoint the successful example's behavior vs the failed example's behavior, which can be useful to form a mental picture of what patterns the model is identifying. Another issue is that the image column isn't standardized with regard to which part of the house it is representing. Standardizing can help you achieve a better prediction performance.

6. Finally, you can export the insights individually, and better yet, download the compliance report documentation that provides an offline one-stop document with all the insights. You can do this by clicking on the **Compliance** tab and clicking on the **Create Report** button shown in *Figure 18.22*. There is a default structure of the document, but you can craft the exact structure that you want to have in your report. An example document is provided in the code repository.

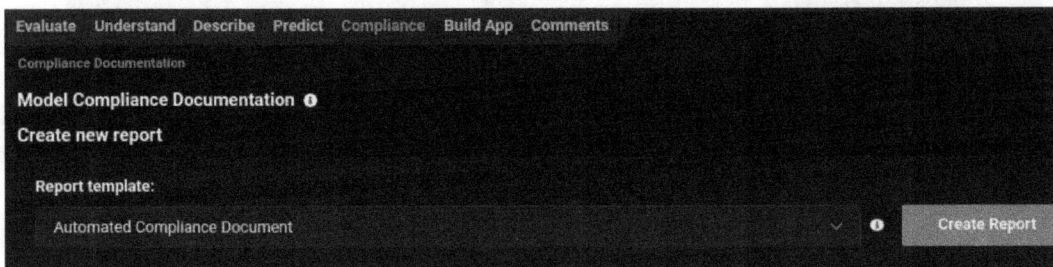

Figure 18.22 – The Model Compliance Documentation interface under the blueprint

With that, we are done with the tutorial on gathering insights. However, note that this is not a comprehensive take on gathering insights for this use case. So, be sure to test out more insight types such as lift charts and image embeddings and iterate through more model improvements that you identify through gathering insights.

Before we move on to the DLOps side of things, to deploy and govern a deep learning model, let's explore how batch predictions can be made without deploying a model.

Making batch predictions

There are use cases where a deployment is not needed, as predictions can be made asynchronously in a regular cadence via a custom trigger or a one-time event. This is where the **batch predictions** feature from DataRobot comes in. Batch predictions simply allow you to upload your data, compute predictions optionally with prediction explanations, and, once this is done, download the results.

This step can be done in both DataRobot Classic and Workbench. For the DataRobot Classic graphical UI experience, navigate to the **Make Predictions** sub-tab under the **Predict** tab of a leaderboard model. You will then be able to upload the dataset you want to generate one-off predictions with. This interface is shown in *Figure 18.23*:

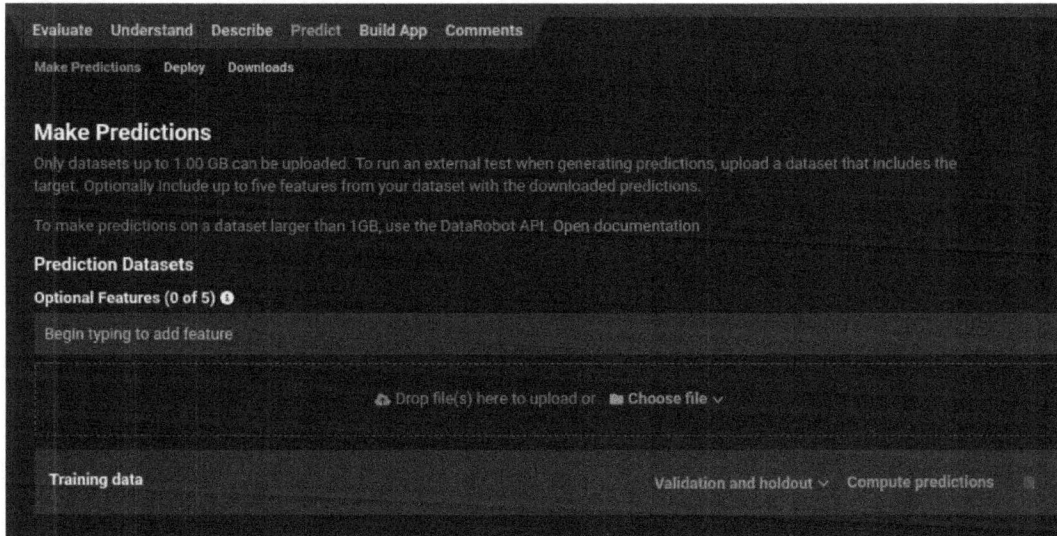

Figure 18.23 – Batch predictions functionality interface

There will still be use cases where real-time predictions are needed by single samples. This brings us to the next section—discussing how DataRobot manages the deployment of a blueprint.

Deploying a deep learning blueprint

DataRobot allows the deployment of a model directly through a trained blueprint in an experiment or a project, which we will explore in the next practical section. However, for more advanced users, the platform also allows the deployment of custom models through the **Custom Model Workshop** feature. Custom inference models are user-created, pre-trained models that can be uploaded to DataRobot as a collection of files coupled with either a drop-in environment or by a `requirements.txt` file. Once uploaded, users can create, test, and deploy custom inference models to DataRobot's centralized deployment hub. These custom models support different model types, which include regression, classification, and unstructured types where the input and output can be of various types.

To ensure the reliability and compatibility of your custom models, DataRobot provides a comprehensive testing suite in the Custom Model Workshop. The custom model testing suite encompasses a comprehensive range of evaluations, including the following:

- **Startup check**: This ensures that the custom model can be built and the custom model service can be launched without errors
- **Null imputation check**: This validates the handling of missing values

- **Side effects check**: This makes sure that a row of data predicted as part of a batch of data produces the same predictions as the same row of data predicted with a single row of data

- **Prediction verification**: This confirms the correctness of predictions

- **Performance check**: This gauges the efficiency and speed of the model

- **Stability check**: This evaluates the model's consistency and reliability

- **Duration check**: This measures the time taken for various tasks

By running these tests, you can verify the performance, stability, and prediction accuracy of your custom models before deployment. A bonus here with a supervised custom model is that it can be linked to training data, which will allow prediction explanations to be computed and data drift to be measured and monitored. Once your custom model is assembled and tested, you can deploy it alongside other blueprints in DataRobot, making it a versatile and powerful tool for advanced users. To deploy a model, either a custom or a local DataRobot model, a prerequisite is that you'd need to choose the reliability of the deployment that you want. Levels of reliability include the following:

- **Low**: This is suitable for non-critical, experimental, or low-priority use cases where occasional downtime or reduced performance is acceptable. This option provides minimal resources and infrastructure redundancy.

- **Medium**: This is ideal for moderately important use cases that require a balance between cost and performance. This level offers better resources and redundancy than the low option, but you may still experience some downtime or reduced performance during peak loads.

- **High**: This is recommended for important use cases that demand high availability and performance. This level provides increased resources, infrastructure redundancy, and faster response times to ensure consistent performance, even during heavy loads.

- **Critical**: This is designed for mission-critical applications where maximum availability and performance are essential. This option offers the highest level of resources, redundancy, and response times to ensure near-zero downtime and optimal performance under any conditions.

Choosing an appropriate reliability level allows DataRobot to configure an appropriate server machine type and infrastructure to host your model according to your requirements. As there are limitations to the deployment your organization signed up with, choosing an appropriate reliability level will make sure you don't overpay after passing the organization deployment limits, or it just makes sure you stay under the deployment limits. In other words, you must manage costs incurred when it comes to deployment. Let's continue through the previous tutorial and deploy the best-performing blueprint that wasn't trained into the validation or holdout partition.

Practically deploying a blueprint in DataRobot

Deploying a blueprint is as simple as going to the **Deploy** sub-tab under the **Predict** tab under a blueprint and then clicking the **Deploy model** button. You then need to choose the deployment reliabilitythe default is low. Then, click on the **Deploy model** button again in the same location. The interface for the first **Deploy model** button is shown in *Figure 18.24*:

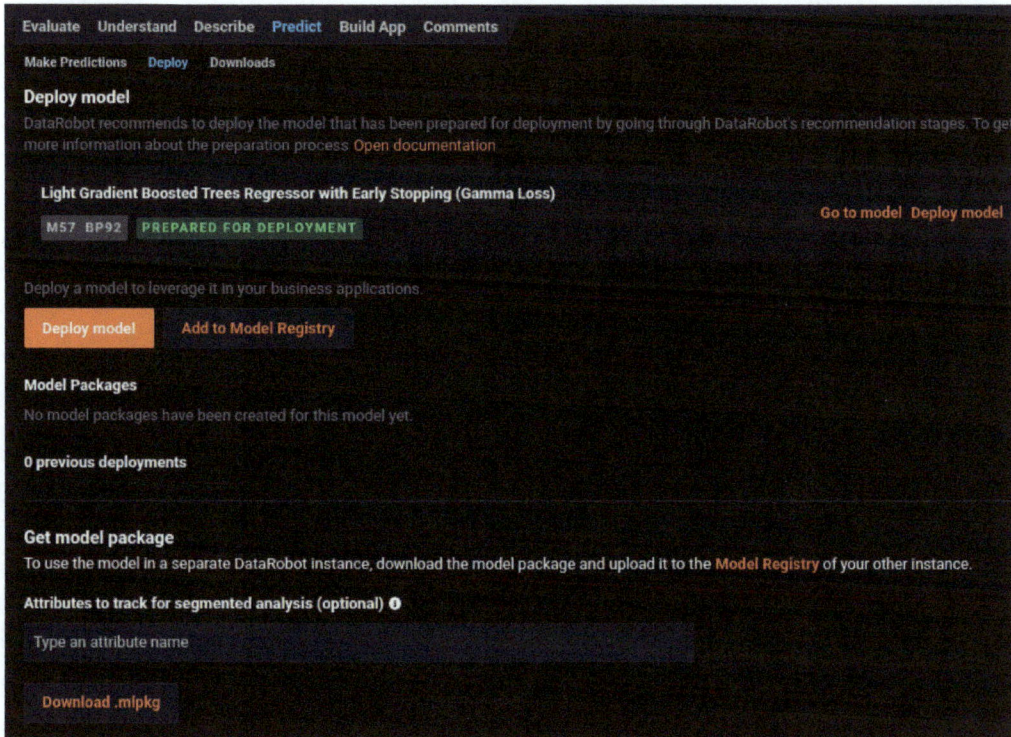

Figure 18.24 – Deploying a blueprint

And that's it! We have successfully deployed a model in DataRobot with the click of two buttons. Next, we will discuss how DataRobot governs its deployed blueprint.

Governing a deployed deep learning blueprint

In this section, we will discuss how DataRobot enables users to govern their deep-learning models effectively by providing comprehensive tools for model utilization, monitoring, and maintenance. With a focus on seamless integration, DataRobot allows users to deploy AI applications on cloud-based or on-premises infrastructure, manage prediction outputs, and monitor model performance using custom metrics and alerts. Furthermore, the platform supports data drift detection and offers retraining capabilities for continuous model improvement. We will explore these features in detail, demonstrating how DataRobot empowers users to efficiently manage their deep learning models and ensure optimal performance throughout their life cycle.

Governing through model utilization in DataRobot

Users can access their models through various means, such as API calls, Python interfaces, or DataRobot-made applications called **AI Apps**. The platform supports different output formats, such as JSON and CSV, allowing users to easily consume the prediction results. For image-based models, the input data needs to be encoded in the base64 format. DataRobot also facilitates the direct storage of predictions into databases, streamlining the process of incorporating model outputs into existing workflows or applications. Additionally, DataRobot allows for the scheduling of batch predictions with a deployed model to be executed regularly to the specified frequency and time.

Overall, the model utilization component in DataRobot simplifies the process of leveraging deep learning models and machine learning models in general, making it more accessible and efficient for users across various domains.

Now, let's continue on from the previous tutorial to get predictions using the DataRobot HTTP client library.

Practically consuming predictions from a deployed blueprint in DataRobot

In this section, we will continue on from the previous tutorial and use the datarobotx Python client library to generate predictions with the deployed model. Under each deployment, DataRobot includes low-level example code to make real-time prediction-api requests to a deployment. However, in this tutorial, we will utilize an easy-to-use, high-level library called datarobotx that simplifies making prediction-api requests. Let's start the step-by-step process:

1. Let's start by importing the libraries:

    ```
    import datarobotx as drx
    import pandas as pd
    ```

2. Next, we will initialize the deployment instance based on the deployment ID. You'd have to replace deployment_id with your own deployment ID here:

    ```
    deployment_id = "YOUR_DEPLOYMENT_ID"
    d = drx.Deployment(deployment_id)
    ```

3. Next, we will set the API token and the endpoint URL in a DataRobot context class. You'd have to set your own token here:

    ```
    drx.Context(
        token="YOUR_API_TOKEN",
        endpoint="https://app.datarobot.com/api/v2",
    )
    ```

4. Now, we will load the `trulia_one_row.csv` house pricing DataFrame provided in the code repository and make a prediction using the initialized deployment instance. Finally, we display the predictions:

```
one_row_df = pd.read_csv("trulia_one_row.csv")
prediction_results = d.predict(one_row_df, max_explanations=3)
print(prediction_results.to_json(indent=2))
```

This will produce the following results:

```
{"prediction":{"0":1020288.625},"explanation_1_
feature_name":{"0":"Price Sqr Ft"},"explanation_1_
strength":{"0":0.4690563465},"explanation_1_
actual_value":{"0":"$359\/
sqft"},"explanation_1_qualitative_strength":{"0":"+++"},"explanation_2_
feature_name":{"0":"Description"},"explanation_2_
strength":{"0":0.3055447842},"explanation_2_actual_
value":{"0":" The table is set in this large Plan 4 with
its top two levels of interactive living spaces for
educating and engaging. Four bedrooms suit home-based
entrepreneurs, stepchildren and a widowed parent who can
come together in an expansive gathering room. An elongated
deck invites open-air grilling and entertaining. A side-
by-side garage adds valuable storage. "},"explanation_2_
qualitative_strength":{"0":"++"},"explanation_3_
feature_name":{"0":"Bath"},"explanation_3_
strength":{"0":0.2212744913},"explanation_3_actual_
value":{"0":3.0},"explanation_3_qualitative_
strength":{"0":"++"}}
```

And that concludes the tutorial. We will now dive into how DataRobot implements model monitoring for a deployed model.

Governing through model monitoring in DataRobot

Model monitoring in DataRobot is an essential component that allows users to track the performance and health of their deployed deep learning models. The platform provides several features to ensure models maintain optimal performance over time:

- **Data drift detection**: DataRobot continuously monitors changes in the distribution of input data, identifying any deviations from the original training data. This feature helps users understand when their models might be at risk of becoming less accurate due to shifts in the underlying data.

- **Model performance monitoring**: Users can track the performance of their models over time. This includes the following:

  - **Accuracy**: This can be monitored by comparing the actual target values with the predicted values. Actual target values can be sent to the deployment any time after a prediction has been made with the prerequisite that an association ID has to be set and returned to connect the actual targets to the historical prediction requests.

- **Fairness**: This ensures that AI models continue to provide fair and unbiased predictions in a production environment. An association ID is similarly required here, as it is tied to the accuracy-related performance metric. Key aspects of the fairness functionality for deployed models include the following:

  - **Fairness metrics tracking**: The platform tracks various fairness metrics, such as disparate impact, demographic parity, and equal opportunity, enabling users to assess the fairness of their models across different subgroups within the protected attributes.

  - **Alerts and notifications**: DataRobot can be configured to send alerts and notifications if biases or disparities are detected, ensuring that users are promptly informed about any fairness issues that may arise during the model's life cycle.

- **Humility rules**: Users can set actions to execute based on undesirable conditions. Supported conditions are defined as uncertainties in predictions, outlier input values or ranges, and low observation regions. Supported actions are recording the trigger, overriding the prediction with a defined prediction, and throwing an error. This enhances the user's overall confidence in the model's predictions and mitigates the risk of it making incorrect decisions based on low-confidence predictions, out-of-distribution data, or low observation data points.

- **Deployment service health monitoring**: This includes total predictions made, total requests made, requests made over a defined time interval, aggregated response time (such as median), aggregated execution time (median), median peak load at calls per minute, data error rate, system error rate, number of consumers, and cache hit rate.

- **Deployment notifications**: Be notified about changes in the deployment, either for all changes or just critical changes.

- **Custom metrics tracking and alerting**: DataRobot enables users to define and monitor custom performance metrics specific to their use cases. Users can set up alerts to notify them when certain thresholds are reached, ensuring prompt response to any performance-related issues.

Now, let's practically explore the interface that DataRobot provides for model monitoring.

Practically monitoring a deployed blueprint in DataRobot

By clicking on the **Service Health** tab of the deployed model, you will be able to see the general service health monitoring. This is shown in *Figure 18.25*. Additionally, you can check out the dashboards for data drift, accuracy, humility, fairness, and custom metrics each in their own tab under the deployed model. Notifications, on the other hand, are by default configured to be sent to all deployment activities and can be configured based on preferences.

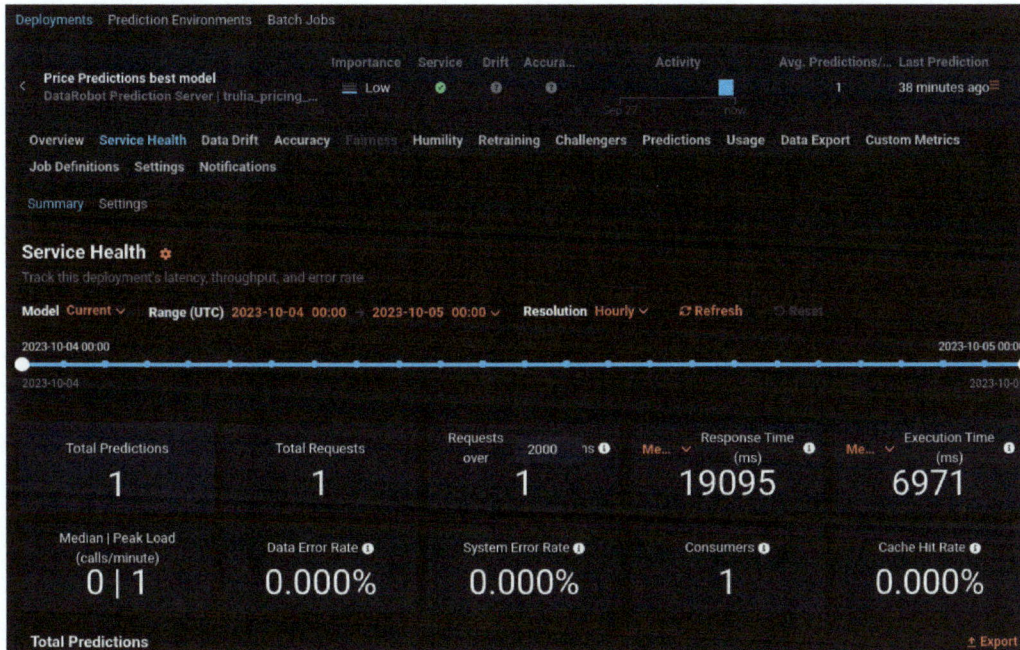

Figure 18.25 –The service health of the deployed model

Next, we will discuss how a user can maintain the performance of the deployed model and ensure that the model can continuously deliver value.

Governing through model maintenance in DataRobot

Model maintenance in DataRobot is a crucial aspect of managing deep learning models, ensuring that they continue to deliver accurate and reliable results throughout their life cycle. The platform provides several features to facilitate effective model maintenance:

- **Challenger models and model replacement**: DataRobot allows users to create and compare alternative models, known as challengers, against the currently deployed model. By evaluating the performance of these challenger models, users can identify potential improvements and decide if it's necessary to replace the existing model with a better-performing alternative. Once a better model has been identified, it is referred to as the **Champion model**. It will then replace the previous model for the existing deployment. This maintains the same deployment ID and ensures a seamless transition to a better model.

- **Model versioning**: DataRobot tracks and manages different versions of a model, allowing users to easily revert to previous versions if needed. This feature ensures that users can maintain a history of their models and compare their performance across different versions.

- **Retraining and retraining policies**: Users can register data received by the deployed model, including input data and delayed target values, into the AI Catalog. This enables the models to be retrained with the most recent data. Additionally, DataRobot's retraining policies provide a way to manage and automate the model updating process, ensuring that deployed models stay relevant and maintain optimal performance. Key aspects of retraining policies include the following:

  - **Customizable triggers**: Users can define specific triggers for retraining, such as data drift, performance degradation, or scheduled intervals, to initiate the retraining process automatically when certain conditions are met.

  - **Data integration**: Retraining policies facilitate the seamless integration of new data into the model updating process, ensuring that models are trained on the most recent and relevant information.

  - **Automated retraining and deployment workflow**: DataRobot automates the entire retraining process, from data ingestion and preprocessing to model building and validation, streamlining the model update workflow and reducing manual effort. You can choose to either maintain the same model with the same parameters, maintain the same model with hyperparameter optimization, or just choose the best model from autopilot. Retrained models are automatically evaluated against the new data using performance metrics, enabling users to assess the updated model's performance and determine if it's ready for redeployment. Once a retrained model meets the desired performance criteria you'd wish to achieve, DataRobot facilitates its seamless deployment, replacing the existing model with minimal interruption. By employing retraining policies, DataRobot simplifies and automates the model updating process, helping users ensure that their deployed AI models remain accurate, relevant, and high-performing as new data and insights become available.

By delving into the features supported by DataRobot and going through hands-on tutorials, we've gained significant insights into how DataRobot employs deep learning methods to process and analyze data, including unstructured and structured data. Now, let's examine some real-world examples that demonstrate the capabilities of this technology, as shared by customers who were enthusiastic about their experiences with DataRobot.

Exploring some customer success stories

DataRobot has empowered numerous organizations to achieve remarkable success through the implementation of deep learning solutions, particularly in handling unstructured data such as text and images. While most of these success stories remain confidential, we are fortunate to have a few customers who have enthusiastically shared their inspiring experiences, showcasing the transformative potential of deep learning in various industries. Some of these notable successes include the following:

- Lenovo, a leading technology company, successfully implemented DataRobot's Visual AI in its Brazilian laptop manufacturing facility to improve quality control and increase productivity. The Visual AI system helped increase label verification accuracy from 93% to 98% by automating the comparison of identification labels on laptops with their respective bill of materials. This implementation not only reduced errors in the manual labeling process but also had a positive impact on delivery times, customer satisfaction, and legal risk reduction for the manufacturer.

- OYAK Cement, a leading Turkish cement maker, successfully utilized DataRobot's AI solutions to optimize their manufacturing processes, resulting in reduced costs and CO_2 emissions. By implementing AI-assisted process control, OYAK increased alternative fuel usage by seven times, cutting almost 2% of total CO_2 emissions and reducing costs by approximately $39 million. The company was also able to predict and prevent mechanical failures more efficiently, improving overall operational efficiency and environmental sustainability.

- AUTOproff successfully implemented the DataRobot AI Platform, which included visual AI capabilities for processing image data of vehicles, to develop their Pricing Robot to make automated car value estimations. The AI-driven solution automated 55–60% of all estimates, leading to improved pricing accuracy and a significant reduction in the time required to generate quotes. As a result, the estimators could focus on rarer vehicles, enhancing their efficiency. The Pricing Robot's success has played a pivotal role in AUTOproff's European expansion, enabling the company to swiftly adapt to new markets. This has ultimately led to increased customer satisfaction and streamlined business operations.

For more information on the latest success stories, check out `https://www.datarobot.com/customers/`.

As we reach the end of this chapter, if you are interested in trying out the DataRobot AI Platform for yourself and don't already have access, you can subscribe for a free trial. And that's it! This will allow you to experience first-hand the powerful tools and automation features that DataRobot offers for 30 days (as of 28 September 2023), enabling you to focus on extracting significant value from your deep learning applications.

Summary

This chapter explored the DataRobot AI Platform and showcased the benefits an AI platform can provide to you in general. DataRobot streamlines the complex stages of the machine learning life cycle, providing an intuitive interface for data scientists, engineers, and researchers. By harnessing the potential of AI platforms such as DataRobot, users can accelerate the creation, training, deployment, and governance of intricate deep learning models, focusing on extracting significant value from their machine learning applications.

DataRobot offers automation, collaboration, and scalability for machine learning use cases. DataRobot provides support for various data types and advanced features such as bias and fairness mitigation, Composable ML, custom tasks, advanced tuning, and time-series modeling. DataRobot also enables users to deploy AI applications on cloud-based or on-premises infrastructure, manage prediction outputs, monitor model performance, and maintain models implemented in features such as **Challenger Models**, **Model Versioning**, **Retraining**, and **Retraining policies**.

While this chapter showcased the various features and capabilities of the DataRobot AI platform, it is not a comprehensive coverage of what the platform provides. Additionally, the company constantly evolves to attend to real-world data science needs, so any unsupported features may be added in the future. For a more detailed understanding, you can refer to the official documentation at `https://docs.datarobot.com/`.

In summary, AI platforms such as DataRobot offer a powerful solution for deep learning applications, streamlining and accelerating the deep learning life cycle. However, they are not a replacement for the expertise and creativity of data scientists, engineers, analysts, or researchers; instead, they serve as tools to assist practitioners in solving complex problems quickly and accurately.

As we move forward to the next chapter, we will delve deeper into the world of large language models, exploring their potential, challenges, and ways to create effective solutions. Building upon the foundation from all the previous chapters, we'll uncover how to harness the power of LLMs to tackle complex language-related tasks and create advanced, contextually-aware applications.

19

Architecting LLM Solutions

Large language models (**LLMs**) have revolutionized the field of **natural language processing** (**NLP**) and **artificial intelligence** (**AI**), offering remarkable versatility in tackling a variety of tasks. However, realizing their full potential requires addressing certain challenges and developing effective LLM solutions. In this chapter, we'll demystify the process of architecting LLM solutions, focusing on essential aspects such as memory, problem-solving capabilities, autonomous agents, and advanced tools for enhanced performance. We will be focusing on retrieval-augmented language models, which provide contextually relevant information, their practical applications, and methods to improve them further. Additionally, we'll uncover the challenges, best practices, and evaluation methods to ensure the success of an LLM solution.

Building upon these foundational concepts, this chapter will equip you with the knowledge and techniques necessary to create powerful LLM solutions tailored to your specific needs. By mastering the art of architecting LLM solutions, you will be better prepared to tackle complex challenges, optimize performance, and unlock the true potential of these versatile models in a wide range of real-world applications.

Specifically, we will cover the following topics:

- Overview of LLM solutions
- Handling knowledge for LLM solutions
- Evaluating LLM solutions
- Identifying challenges with LLM solutions
- Tackling challenges with LLM solutions
- Leveraging LLMs to build autonomous agents
- Exploring LLM solution use cases

Overview of LLM solutions

LLMs excel in diverse tasks such as answering questions, machine translation, language modeling, sentiment analysis, and text summarization. They generate unstructured text but can be guided to produce structured output. LLM solutions harness this ability and leverage custom data from knowledge bases to create targeted, valuable outcomes for organizations and individuals. By properly streamlining processes and enhancing output quality objectively, an LLM solution can unlock the true potential of LLM-generated content, making it more powerful and practical across applications.

The increasing accessibility of LLMs with pre-trained world knowledge has played a significant role in making these benefits more attainable for a broader audience. Thanks to various LLM providers and open source platforms, organizations and developers can now more easily adopt and integrate LLMs into their workflows. Prominent LLM providers, such as OpenAI (GPT-4 or GPT-3.5), Microsoft Azure, Google, and Amazon Bedrock, offer pre-trained models and APIs that can be seamlessly integrated into diverse applications. Additionally, the Hugging Face platform has made LLMs even more accessible by offering an extensive collection of open source models. Hugging Face provides a wide selection of pre-trained models and fine-tuning techniques while fostering an active community that continually contributes to enhancing LLMs.

As organizations and individuals harness the power of LLMs for their typical tasks and use cases, it is crucial to determine how to leverage custom knowledge effectively. This consideration ensures that LLMs are optimally utilized to address specific needs; this will be explored further in the *Handling knowledge for LLM solutions* section. By taking advantage of the increased accessibility and versatility of LLMs, organizations and individuals can unlock the full potential of these powerful models to drive innovation and improve outcomes.

Despite their impressive capabilities, LLMs face some limitations when it comes to solving more complex problems that they were not made to account for. Some of these limitations are as follows:

- Inability to access up-to-date information on recent events
- Tendency to hallucinate facts or generate imitative falsehoods
- Difficulties in understanding low-resource languages
- Lack of mathematical skills for precise calculations
- Unawareness of the progression of time

To overcome these limitations and enhance LLMs' problem-solving capabilities, advanced solutions can be developed by incorporating the following components:

- **Real-time data integration**: By connecting LLMs to real-time data sources such as APIs, databases, or web services, the model can access up-to-date information and provide more accurate responses.

- **Existing tool integration**: Incorporating existing tools and APIs into the LLM architecture can extend its capabilities, allowing it to perform tasks that would otherwise be impossible or challenging for a standalone model.

- **Multiple agents with different personas and contexts**: Developing a multi-agent system where each agent possesses a unique persona and context can help address the challenges of diverse problem-solving scenarios. These agents can collaborate, share information, and provide more comprehensive and reliable solutions.

Figure 19.1 shows an architecture that depicts the different approaches and methods that can be applied in an LLM solution and will be introduced in this chapter:

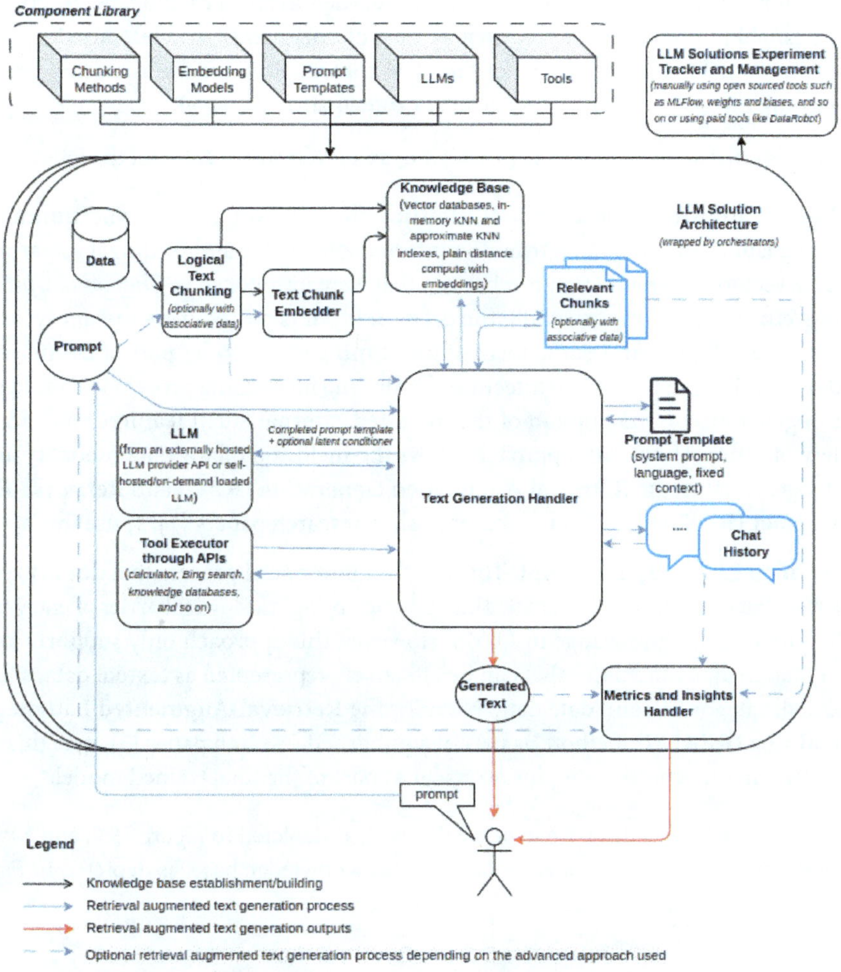

Figure 19.1 – LLM solution architecture

In the next few sections, we will dive into the individual components listed in this LLM solution architecture and LLM solutions in general more comprehensively. We will start with how knowledge is handled for LLM solutions.

Handling knowledge for LLM solutions

Domain knowledge is key in creating LLM solutions as it provides the background information and understanding they need to solve specific problems. This ultimately ensures the answers or actions the solutions come up with are on point and helpful. Domain knowledge needs to be included as context either as part of parametric memory, non-parametric memory, or a combination of both. Parametric memory refers to the parameters that are learned in an LLM. Non-parametric memory refers to an external library of knowledge, such as a list of documents, articles, or excerpts, that can be selectively chosen to be injected as part of the LLM context. This process is also referred to as an in-context learning method, knowledge retrieval, or information retrieval.

Non-parametric external knowledge can be provided to an LLM through either of the following options:

- **As a latent conditioner in the cross-attention mechanism**: Latent conditioning involves generating latent feature vectors from the external knowledge and feeding it as part of the key and value vectors in the attention mechanism that were introduced in *Chapter 6, Understanding Neural Network Transformers*, while the original input is passed in as the query vector. This approach typically requires some form of fine-tuning the decoder part of the network in an encoder-decoder transformer architecture. Ideally, the fine-tuning process will build a decoder that can generalize to the domain of the intended external latent features and can attend to a variety of information. This approach allows the inclusion of any data modality as external knowledge. Notably, the **Retrieval Augmented Generation (RAG)** and **Retrieval-Enhanced Transformer (RETRO)** methods from published research papers [1][2] use this approach.

- **As part of an LLM's input prompt**: This is a straightforward process that doesn't require any fine-tuning but can still benefit from it. This approach brings the lowest barrier of entry to leverage any custom domain knowledge in LLMs. However, this approach only supports knowledge represented in data modalities that can be effectively represented as textual data, such as text, numerical, categorical, and date data. Notably, the **Retrieval-Augmented Language Model Pre-Training (REALM)** method, as part of a published research paper [3], uses this approach for pre-training specifically and doesn't use it as part of the final trained model.

Both methods require a knowledge base to be established, as depicted in *Figure 19.2*, and a knowledge retrieval component that retrieves information from the knowledge base, as depicted in *Figure 19.1*:

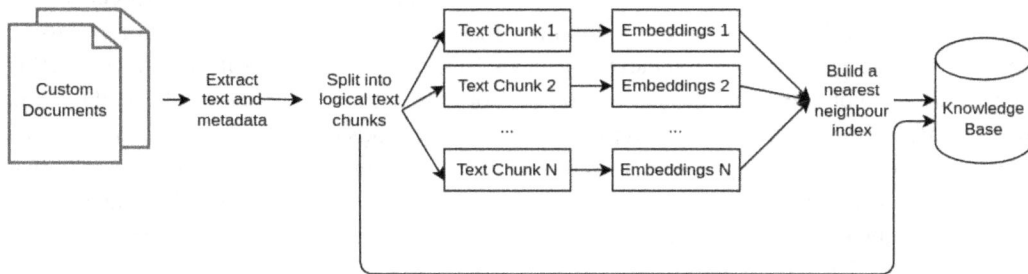

Figure 19.2 – Establishing a knowledge base

A short tabular summary of the REALM, RETRO, and RAG methods is presented in *Table 19.1*:

Knowledge retrieval methods	Retriever training	Retrieval integration
RAG	Fine-tune with a frozen base network	Latent conditioner with cross-attention
REALM	Full end-to-end training	Prepend to prompt specifically without a template
RETRO	Fine-tune with a frozen base network	Latent conditioner with cross-attention

Table 19.1 – Short overview of retrieval integration with LLM methods

A knowledge base requires that text data is pre-processed into appropriate logical chunks or segments. These segments are then transformed into embedding vectors using trained transformer models. Finally, a nearest neighbor index is constructed, enabling efficient retrieval of relevant information from the knowledge base. The nearest neighbor index can either be a simple KNN algorithm that computes raw distances between the prompt embedding vector, or an approximate KNN algorithm that approximates the distance computations. Both the index and the logical text chunks will then serve as the knowledge base, which can be used for retrieval. The method to perform retrieval can vary with different strategies, but the simplest form involves simply generating embedding from the prompt and returning the top k closest text chunks from the knowledge base using the index. These top k closest text chunks can then be included as part of the LLM prompt or as a latent conditioner.

For the approach of including the most relevant text chunks as part of the prompt, crafting a prompt template that can allow a specific spot to be inserted is the standard and can help organize information in a prompt properly. This can be as simple as using leading text such as `Context:`, following up with the retrieved relevant text chunks, and having new line separation before and after the context part of the prompt template.

While research papers often present published methods that encompass various aspects of the retrieval process in a single method, it is helpful to consider each component of building and using the knowledge base as separate, interchangeable parts. This allows for greater flexibility in selecting the most suitable components for specific situations. Moreover, although there is a published method known as RAG, it is worth noting that, in practice, the term RAG is commonly used to describe the general approach of integrating knowledge retrieval with LLMs, rather than referring solely to that specific method. Let's briefly go through the three key method-based components that can be freely modified according to the use case. We will also choose orchestrator tools that help streamline the implementation of these components.

Exploring chunking methods

The text chunking process affects the efficiency of LLM context utilization and the quality of the resulting LLM generation. Choosing an appropriate chunking method depends on the following factors:

- **The embedding model used for embedding vector generation**: Different pre-trained embedding models may have different requirements or limitations when it comes to text chunking. Two such requirements are the supported context size and the typical text context size that was used during pre-training.

- **The granularity of information needed for the expected prompts that will be made**: The level of detail or granularity required for the prompts can impact the choice of text chunking method. Depending on the specific use case, the method should be able to chunk the text into appropriate and concise segments that provide the necessary information for the LLM to generate accurate, concise, and relevant responses.

- **The nature of the text data used to build a knowledge base**: The characteristics of the text data itself can also influence the choice of text chunking method. For example, if the text data consists of long paragraphs or documents, a method that breaks the text into smaller chunks or sections may be more suitable. On the other hand, if the text data is already organized into logical segments, a method that preserves these segments may be preferred. Also, if the text data is Python code, it can be suitable to chunk the text by code methods.

There are several methods for chunking text, including sentence, paragraph, entity, topic, and section chunking. These methods help organize the text into meaningful units that can be processed by the LLM. One notable method that is useful and readily available is the recursive chunking method from the LangChain library. This method allows you to adjust the granularity of the chunks by recursively splitting the text using an ordered list of text separators, a maximum chunk size, and the percent of overlap between chunks. The maximum chunk size should be tailored to the context size supported by the embedding model, ensuring that the generated chunks can be effectively processed. Meanwhile, incorporating an overlap percentage helps minimize the risk of missing critical information that could be located at the boundaries of chunks.

Many document-specific chunking methods are created based on this recursive chunking method by specifying the appropriate ordered list of text separators. Specifically, as of `langchain==0.0.314`, recursive methods have been created for Python code with `PythonCodeTextSplitter`, markdown documents with the `MarkdownTextSplitter` class, and LaTeX-formatted text with the `LatexTextSplitter` class.

Next, let's dive into embedding model choices.

Exploring embedding models

Embedding models play a crucial role in generating a knowledge base for LLM solutions. These models are responsible for encoding the semantic information of text into vector representations, which are then used to retrieve relevant information from the knowledge base.

One benchmark that provides insights into the performance of text embedding models is the **Massive Text Embedding Benchmark (MTEB)**. MTEB evaluates the effectiveness of various embedding models on diverse embedding tasks. By assessing the leaderboard results, users can identify the top-performing models for their specific use cases. The MTEB leaderboard is available at `https://huggingface.co/spaces/mteb/leaderboard`. MTEB even includes the embedding models of paid providers such as the embedding model from OpenAI called `text-embedding-ada-002`.

When selecting an embedding model for knowledge base generation, it is essential to consider factors such as model size, embedding dimensions, and sequence length. Traditional embedding models such as GloVe offer high speed but may lack context awareness, resulting in lower average scores. On the other hand, models such as `all-mpnet-base-v2` and `all-MiniLM-L6-v2` strike a balance between speed and performance, providing satisfactory results. For maximum performance, larger models such as `bge-large-en-v1.5`, `ember-v1`, and `e5-large-v2` dominate the MTEB leaderboard, all with a 1.34 GB model size.

It's important to note that the choice of embedding model depends on the specific task and dataset being used. Therefore, thoroughly exploring the various tabs of the MTEB leaderboard and considering the requirements of the knowledge base generation process can help in selecting the most suitable embedding model.

MTEB, with its extensive collection of datasets and evaluation metrics, serves as a valuable resource for researchers and practitioners in the field of NLP. By leveraging the insights provided by MTEB, developers can make informed decisions when choosing an embedding model for knowledge base generation in LLM solutions.

Before we delve into exploring the knowledge base index types, it's essential to remember that the choice of chunking method and embedding model shapes the construction of your knowledge base. Both components play a crucial part in how effectively the LLM can retrieve and utilize knowledge. Now, let's dive deeper into the world of knowledge base index types and learn how they contribute to the efficiency of LLM solutions.

Exploring the knowledge base index types

The knowledge base index is the backbone of retrieval mechanisms in LLM solutions. It is the component that facilitates the efficient lookup of relevant information. While there are several ways of implementing this index, they all aim to provide a fast and efficient way of retrieving the most relevant text chunks from the knowledge base based on the input prompt.

Many options are available for building a knowledge base index. They range from manual code implementations to using various vector database libraries, service providers, and plugins. Some of these options are listed here:

- **Manual code**: You can manually implement an in-memory vector database using libraries such as `faiss`, a library for efficient similarity search of dense vectors, and `scipy`, a library for pairwise distance computations. This allows for customization but may require more effort and expertise while requiring bigger RAM allocations.

- **Service providers**: Various cloud providers offer vector database services. These include Pinecone, Chroma, Vespa, and Weaviate. These services handle the complexities of managing a vector database, providing scalable and robust solutions that can be easily integrated into your LLM architecture and solution.

- **Database tools with vector computation support**: Traditional database tools such as MongoDB, Neo4j, Redis, and PostgreSQL provide vector computation support through plugins. This can be a good option if you're already using these tools in your tech stack and want to leverage their capabilities for your knowledge base.

- **Plugins**: There are also plugins available, directly from LLM service providers, such as ChatGPT, that can help with the construction and maintenance of a knowledge base.

Choosing the right knowledge base index type depends on your specific requirements, such as the size of your knowledge base, the complexity of your retrieval needs, and the resources you have available. Consider factors such as scalability, ease of integration, cost, and the level of control you need over your knowledge base when making your choice. A recommendation is to only consider vector databases that are using the actual database technology or claim to do so when your knowledge base is big enough to matter. If your knowledge base is small, let's say in the six-digit range, raw distance computations take less than 1 second if you make one prompt per compute in Python! Next, we will briefly discover orchestrator libraries for LLM solutions.

Exploring orchestrator tools for LLM solutions

The process of architecting LLM solutions can be streamlined with the use of specific tools. Open sourced orchestrator libraries such as LangChain and LlamaIndex play a pivotal role in this context. Both tools simplify tasks such as setting up the knowledge base, integrating an LLM, and managing retrieval mechanisms. In general, an orchestrator significantly reduces the complexity and development time of LLM solutions.

In addition to open sourced orchestrator tools, there are also paid options available that provide advanced features and support. Some of these DataRobot, Microsoft Azure, IBM Watson, LangSmith, OpenAI, and Google Vertex AI. These platforms offer a wide range of pre-built models, integrations, and tools that streamline the entire pipeline, from data ingestion to model deployment and monitoring.

As you continue to explore these tools and methods, it's crucial to establish robust evaluation methods to measure the impact of these components on your solution, ensuring it meets its intended objectives. We'll delve deeper into these methods in the upcoming section.

Evaluating LLM solutions

Evaluating LLM solutions is a crucial step in harnessing their full potential and ensuring their effectiveness in various applications. By implementing a comprehensive set of evaluation approaches, organizations can better assess the performance, accuracy, and overall quality of the results from an LLM solution, while also considering the associated costs, adherence to safety standards, and potential negative impact on users. In other words, doing this provides you with valuable insights to help make any informed decisions. To achieve a comprehensive evaluation, we can view evaluation methods as part of either a quantitative measure or a qualitative measure. Let's dive into evaluation methods by these groups.

Evaluating LLM solutions through quantitative metrics

Quantitative metrics can be aggregated throughout a provided evaluation dataset and can provide a more quick, comprehensive, and objective measure to compare multiple LLM solution setups. Here are some examples of quantitative metrics:

- **Comprehension and fluency-based metrics**: Flesch Reading Ease, Coleman Liau Index, and SMOG readability.

- **Facts-based metrics**: Any metrics that use the facts provided by a knowledge base for inference:

  - **Factual consistency**: This refers to comparing generated text with facts stated in the knowledge base. It is important to note that relevant facts might not always be available in the knowledge base. This metric is also known as the extractiveness metric. To measure factual consistency, you can use either semantic similarity, which focuses on differences in the meaning of the text, or lexical similarity, which emphasizes matching words in the text.

  - **Factual relevance**: This is about how relevant the provided facts are, without considering the LLM generation. This is possible when you have ranked relevant document labels.

- **Generated text relevance/accuracy metric**: This metric evaluates the relevance and accuracy of the text generated by an LLM in comparison to an ideal ground truth. It can be computed using similarity metrics or self-evaluation techniques. Self-evaluation can be further broken down into the following areas:

- **With access to token probabilities:** The average of log probabilities is used to assess the quality of the generated text. Higher log probabilities indicate that the model is more confident in its output, suggesting greater relevance and accuracy.

- **Without access to token probabilities:** SelfCheckGPT is a method that can be employed to evaluate the generated text without relying on token probabilities. This approach leverages the LLM's capabilities to assess the quality of its generated content, providing an alternative measure of relevance and accuracy.

- **Runtime metrics:** The time taken to generate text, the number of tokens processed, and so on.

- **Cost metrics:** The number of tokens generated, API call costs, hosting costs, and so on.

- **Guardrail violations metrics:** The percentage of outputs that violate predefined standards. Examples of guardrails are toxicity levels and hate speech degree.

- **Adversarial performance metrics:** The performance measures in handling adversarial inputs. These were introduced more comprehensively in *Chapter 14, Analyzing Adversarial Performance*.

- **Bias and fairness metrics:** Quantitative measures for assessing biases in the generated text. These were introduced more comprehensively in *Chapter 13, Exploring Bias and Fairness*.

- **Any supervised classification or regression metrics:** This can be applied to the results or resulting actions from an LLM solution.

Note that quantitative methods such as generated text relevance and factual consistency metrics, which use similarity metrics to compare two sets of text, are not as reliable as supervised model metrics such as accuracy. These metrics should be taken with a grain of salt. Additionally, a nice bonus with quantitative metrics is that they can be used for monitoring a deployed model programmatically. Next, we will dive into qualitative manual evaluations.

Evaluating LLM solutions through qualitative evaluation methods

Qualitative methods, which involve human feedback and manual assessments, complement quantitative measures and provide a comprehensive understanding of LLM performance. It can also sometimes be the only way for evaluation when there are no reference ground truth datasets. Here are some examples of qualitative LLM solution evaluation methods:

- **Human feedback scores:** These are the users' ratings or rankings of generated responses to gauge effectiveness and relevance. Examples include grammar and coherence of text.

- **Generated text relevance evaluation:** This involves manually assessing the generated text's relevance to the given context or prompt.

- **Prediction explanations:** These assess the reasoning behind the generated text or predictions, which can help identify potential biases or faulty logic in the LLM solution.

- **Ethical and legal compliance**: This ensures that the generated text adheres to ethical and legal guidelines through manual review.

By employing a combination of quantitative metrics and qualitative manual evaluations, organizations can gain a deeper understanding of LLM performance and identify potential areas for improvement. In general, try to treat LLM solutions as no different from any supervised machine learning projects and evaluate them vigorously, similarly to how you would in a supervised machine learning project. This holistic approach to evaluating LLM solutions not only ensures consistent performance and compliance but also helps in aligning these powerful models with specific needs and objectives, driving innovation and improving outcomes in various applications.

Identifying challenges with LLM solutions

Despite their impressive capabilities, LLMs face challenges when solving complex real-world problems. In this section, we will explore some of the challenges faced by LLM solutions and discuss possible ways to tackle them. We will explore challenges by high-level groups, as follows:

- **Output and input limitations**:

  - **LLMs just produce text**: Text output can help provide value for a lot of businesses. However, many other use cases require predictions and recommendations in entirely different formats.

  - **The context size of an LLM is limited**: The issue is that with a large input size, you need exponentially more compute resources to train and predict. So, context size usually stays in a token range of one to three thousand. This issue should be prevalent only for use cases that require long context, as a few thousand context sizes should be enough for most use cases.

  - **An LLM is a text-specific model**: Other data modalities are not supported by default.

  - **Repetitive retrieved information**: The information that's retrieved from a knowledge base can be highly relevant but repetitive and numerous. As the context size of an LLM is limited, a risk arises when multiple pieces of information are placed as the context is repetitive and takes up most of the context limit quota.

- **Knowledge and information-related challenges**:

  - **Inability to access up-to-date information**: LLMs may not know about recent events or developments, leading to outdated or inaccurate information being provided in their responses.

  - **Handling low-resource languages**: LLMs can struggle with understanding and processing languages with limited data or resources.

  - **Unawareness of the progression of time**: LLMs may not understand the concept of time, leading to confusion when dealing with time-sensitive information.

  - **Information loss**: LLMs are shown to look at the beginning and end of sentences, but not so much the middle, and thus lose the most information placed in the middle.

- **Single index failures**: This challenge arises when an LLM lacks sufficient knowledge about a specific topic or area due to limitations in its training data. For instance, if you ask an LLM about a newly opened local restaurant that wasn't covered in its training data, the LLM may provide limited or irrelevant information.

- **Incomplete content retrieval from documents:** When retrieval of a chunked sentence gets the right document, but the actual content needed is below the retrieved chunk in the same document, LLMs may not provide the complete or accurate information required by the user.

  Example: In a documentation search for a software's installation process, the LLM retrieves a section mentioning the installation, but the actual step-by-step instructions are located in the following section of the document. Consequently, the user only receives an overview without the necessary details for proper installation.

- **The use irrelevant information in the context:** LLMs may use irrelevant information from their context as a basis for their output, essentially mimicking or echoing opinions found in the context even if they are not applicable or appropriate for the given situation. This phenomenon, referred to as sycophancy, can lead to misleading or unhelpful responses.

- **The global knowledge base summarization task can't be executed accurately**: A retrieval process is unaware of the type of knowledge base it requests and thus can't execute the global summarization task effectively.

- **Accuracy and reliability**:

  - **Hallucinations**: LLMs can generate false or misleading information that may appear plausible but is not based on facts. This phenomenon is known as hallucination.

  - **Lack of mathematical skills**: LLMs often cannot perform precise calculations or solve complex mathematical problems. This issue is more widely known as it is slightly controversial, depending on how you look at it.

  - **Imitative falsehoods**: These are false statements that LLMs generate because they mimic common misconceptions found in the training data. Since the model learns from the data it's trained on, it might inadvertently reproduce widely held but incorrect beliefs. For example, if many people believe that a specific food causes a particular illness, an LLM might generate a similar statement, even if it's not scientifically accurate.

  - **Non-imitative falsehoods**: These are false statements that arise due to the model's inability to fully achieve its training objective. This includes hallucinations, which are statements that seem plausible but are incorrect. For instance, an LLM might generate a statement about a historical event that never occurred, but the statement may sound convincing to someone who is not knowledgeable about that specific event.

- **Runtime performance issues**: An LLM's runtime can be slow. Additionally, by adding a knowledge base to it, the entire process can become slower than it already is.

- **Ethical implications and societal impacts**: The widespread adoption and deployment of LLMs comes with several ethical implications and societal impacts. As these models learn from vast amounts of data, they may inadvertently inherit biases present in the training data, leading to biased outputs, perpetuating stereotypes, or promoting misinformation. Furthermore, LLMs can generate content that may inadvertently promote harmful behavior, hate speech, or violate privacy concerns. The following ethical challenges are involved in the usage of an LLM solution:

  - **Bias and fairness**: Ensuring that the LLM does not exhibit biased behavior or discriminate against specific user groups based on their race, gender, age, or other protected attributes. Consider the case that a bank uses an LLM to analyze loan applications and determine creditworthiness. The LLM has been trained on historical data, which may contain biases against certain ethnic groups. As a result, the LLM might reject loan applications from these groups at a higher rate, even when the applicants have good credit scores.

  - **Privacy concerns**: LLMs may inadvertently generate **personally identifiable information** (**PII**) or sensitive data in their outputs, which raises privacy concerns and potential legal issues. Consider the case where a healthcare organization uses an LLM to generate personalized health recommendations for its clients. The LLM can inadvertently include specific patient names and medical conditions in the generated advice, which then gets shared publicly, violating patient privacy.

  - **Misinformation and disinformation**: LLMs can potentially generate misleading or false information, which can contribute to the spread of misinformation and disinformation. Consider the case where an LLM is used by a news agency to automatically summarize and publish news articles. The model unintentionally generates a summary that misrepresents the original story, leading to the spread of misinformation about a crucial business merger.

  - **Safety**: Ensuring that the content generated by LLMs adheres to ethical guidelines, legal regulations, and community standards while avoiding promoting harmful or offensive content. Consider the case where an e-commerce platform uses an LLM to generate product descriptions for sellers. The LLM can create a description that promotes a potentially harmful product, such as a recalled item or an item that violates safety regulations, exposing the platform to legal and ethical issues.

  - **Transparency and explainability**: Ensuring that the decisions made by LLMs are transparent, understandable, and justifiable to users and stakeholders. Consider the case where an insurance company uses an LLM to assess risk and determine premiums for customers. A customer receives a significantly higher premium and requests an explanation for the increase. The LLM's decision-making process is, by itself, opaque and difficult to understand, making it challenging for the company to provide a clear and justifiable explanation.

Now that we have identified these challenges, let's move on to the next section, where we will explore potential solutions and strategies to overcome these limitations.

Tackling challenges with LLM solutions

Tackling the pesky challenges that LLMs face is key to unlocking their full potential and making them our trusty tools or sidekicks in solving real-world problems. Only by tackling these challenges can an LLM solution be formed objectively and effectively. In this section, we'll dive into various complementary strategies that can help us tackle these challenges and boost the performance of LLMs by its high-level issue type. We will start with output and input limitations.

Tackling the output and input limitation challenge

Navigating the output and input limitation challenges is vital for unlocking the full potential of LLMs, allowing them to efficiently process diverse data types, formats, and context sizes while delivering accurate and reliable results. The solutions are as follows:

- **Customized pre-processing**: Design tailored pre-processing techniques to transform non-text data into a format that can be efficiently processed by LLMs. For example, design a structure that places structured tabular data as the LLM prompt.

- **Use context limit expansion neural network components**: Implement advanced neural network components such as LongLORA, which requires you to fine-tune an existing model, to expand the context window size, allowing LLMs to process larger amounts of information. However, it is essential to note that this option might not be available for external LLM providers and might only be feasible if you are considering hosting your own LLM model.

- **LLM context optimization**: Any wasted space or repetitive content limits the depth and breadth of the answers we can extract and generate. There are three possible methods here:

  - Select only the most relevant and unique information to be included in the LLM's context window. The maximal marginal relevance algorithm can be used to find a set of both relevant and unique sets of information from the distance scores.

  - Consider compressing and summarizing the information provided, which can also be done by an LLM, and then use the summarized information as context in the main LLM prompt.

  - Apply knowledge retrieval on demand instead of by default. This on-demand behavior can be enforced by treating the RAG as a tool and either teaching the LLM to use it via fine-tuning or in-context learning.

Next, we will tackle the challenges with knowledge and information.

Tackling the knowledge- and information-related challenge

Addressing the output and input limitation challenges is crucial for enhancing the versatility and effectiveness of LLMs in solving a wider range of real-world problems across various data modalities and context sizes. The solutions are as follows:

- **Real-time data integration**: Connecting LLMs to real-time data sources such as APIs, databases, or web services can help them access up-to-date information and provide more accurate responses. Incorporating relevant information from knowledge bases, using the RAG approach, is part of this solution. RAG can also help reduce hallucinations compared to if a model is fine-tuned with custom data if a rigorous strict prompt is made to instruct the LLM to not deviate from the context provided in the prompt.

- **Tool integration**: Enhancing LLM architecture by integrating existing tools, APIs, and specialized algorithms can significantly extend their capabilities, allowing them to tackle tasks that are challenging or impossible for standalone models. Tools can be used to retrieve extra input context needed for the generation process. Alternatively, they can be used to accomplish specific tasks that the generated text tells them to do. Examples include leveraging external search engines, domain-specific APIs, and computational libraries to provide accurate responses, solve complex mathematical problems, or address queries related to real-time data. For LLMs such as GPT-3.5, which have API access, this can be achieved through effective few-shot prompting, while advanced models such as **Toolformer** and **WebGPT** by OpenAI showcase the potential of integrating external tools seamlessly into the LLM's parametric memory and framework. WebGPT can browse the internet by detecting the Bing search engine identifier it generates and subsequently execute the search before continuing the generation it's appended. Toolformer, on the other hand, is an LLM that can autonomously select and utilize APIs, integrating tools such as calculators, Q&A systems, search engines, translators, and calendars for improved generation. This is a key functionality of transforming an LLM into an agent that can accomplish real-world tasks.

- **Reordering the relevant context position in the LLM context**: This solution involves reordering the input text to distribute important information more evenly throughout the context. By following a specific pattern, such as [1, 3, 5, 7, 9, 10, 8, 6, 4, 2], the LLM is encouraged to pay equal attention to all parts of the text, reducing the likelihood of missing valuable information placed in the middle.

- **Utilizing the surrounding information from the same document in LLM context:** This solution enhances the LLM's understanding by incorporating additional information from the source document. Expanding the scope of retrieval to include surrounding text or metadata helps the LLM generate more accurate and comprehensive responses, ensuring it considers the broader context. This approach improves the LLM's ability to address complex questions and provide well-informed responses, which effectively solves the documentation search use case issue.

- **Filtering out irrelevant context using the LLM:** Before proceeding with the generation task, the LLM is employed to identify and remove any irrelevant context. This refined context is then used for generating responses. This seemingly simple and logical method has demonstrated its effectiveness in most cases as introduced in the paper https://arxiv.org/abs/2311.11829v1. Moreover, the black box nature of this technique allows for easy implementation, contributing to more intuitive and natural LLM-generated content.

- **Regularly building an up-to-date knowledge base**: To address the issue of single index failures, it is essential to maintain and update the LLM's knowledge base regularly. This ensures that the LLM stays current with recent developments and can provide accurate information across a wide range of subjects, ultimately enhancing its reliability and effectiveness in solving real-world problems.

- **Treat RAG as a tool an LLM can use dynamically based on its generation**: This will help solve the problem of not being able to perform summarization at the global level of a knowledge base. Similar to how Deadpool is aware of being a comic book character, we need the retrieval process to be aware of the type of knowledge base it is retrieving from, along with a special handler for summarization tasks. A bonus here is to allow the LLM to configure how many rows of relevant text to return to the scope of summarization that can be expanded and shrunk as required.

- **Multi-index retrieval**: To address the issue of single-index failures, a multi-index retrieval approach can be employed. This solution involves decomposing – or in other words, chunking – the user's query into multiple components and retrieving information from various sources or knowledge indexes. This multi-faceted search strategy helps gather more diverse and comprehensive information, reducing the likelihood of overlooking relevant details due to a single index's limitations. Consider a user asking about a rare bird species. Using a single index might yield limited information. With a multi-index retrieval approach, the LLM would do the following:

    I.   Decompose the query into components (for example, habitat, diet, and appearance).

    II. Retrieve data from various sources (for example, ornithology databases, nature websites, and social media).

    III. Aggregate and synthesize the data to generate a comprehensive response.

- **Set up directed acyclic graph (DAG) workflows**: Setting up DAG workflows involves organizing a series of tasks or processes in a structured, non-circular sequence to efficiently process multiple sources of information and extend an LLM's functionality. In the context of LLMs, a DAG workflow can be manually designed to connect various tools, APIs, and algorithms while addressing the challenges related to real-time data integration, tool integration, and multi-index retrieval. Let's consider a use case where a user wants to plan a trip and needs information on various aspects of the destination, such as weather, attractions, and local cuisine. An LLM could use the DAG workflow to address this complex query efficiently. Here's an example of a DAG workflow for LLMs:

    I.   Decompose the user's query into sub-queries or components, specifically, weather forecast, top attractions, and cuisines topics.

    II. For each subquery, identify relevant tools, APIs, or data sources. For weather forecast, we will retrieve data from a weather API. For top attractions, we will extract information from travel website knowledge base. For local cuisine, we will gather data from restaurant review website APIs.

III.   Apply summarization to each fact separately before using it as part of the LLM's input context.

IV.   Execute the LLM generation process with the user query and the summarized facts.

V.   Publish the results on a website through an API.

This DAG is depicted in *Figure 19.3*:

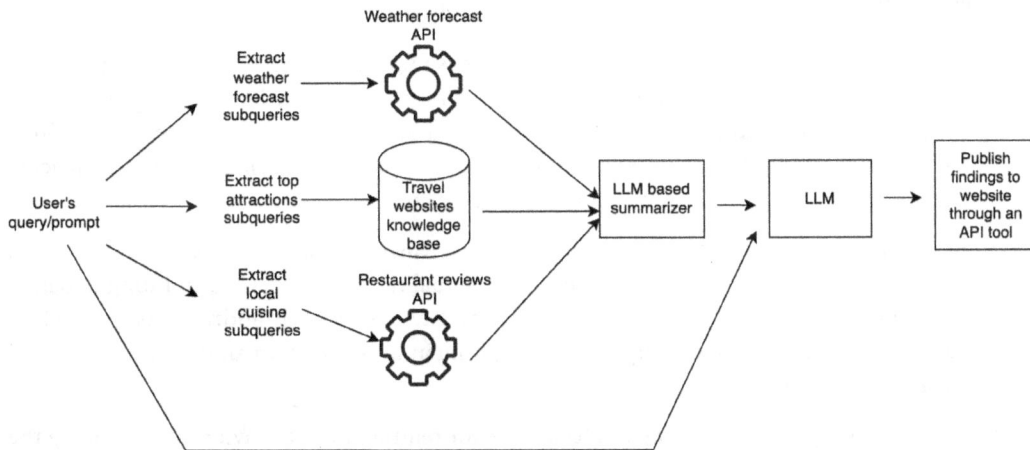

Figure 19.3 – An example LLM DAG workflow

By setting up a manual DAG workflow, an LLM can efficiently process information from multiple sources, leverage external tools and APIs, and provide accurate and reliable responses to a wide range of real-world problems.

This strategy helps the LLM provide a more accurate and detailed response, even when information is scarce or not readily available in a single index. The type of problem this solves is more commonly known as multi-hop question answering.

Next, we will tackle the challenges of accuracy and reliability.

Tackling the challenges of accuracy and reliability

Ensuring the accuracy and reliability of LLMs is crucial for building trust in their abilities and making them effective problem-solving tools. The solutions that can help solve accuracy and reliability-related challenges are as follows:

- **Treat the LLM solution as any modeling experiment**: Pair the LLM with a knowledge base, evaluate its performance using relevant metrics, and gather insights to fine-tune its capabilities iteratively according to the deep learning life cycle. This will help you choose a model that at least produces fewer hallucinations and can help you understand its effectiveness for your use case.

- **Fine-tune retriever embedding models**: This is instead of just depending on pre-trained embedding models or embedding model providers. This can improve the retrieval accuracy, thereby boosting the quality of LLM-generated responses.

- **Prompt engineering**: Prompt engineering is the process of crafting effective and targeted prompts to guide a language model's response, thereby improving its accuracy, relevance, and overall performance. Consider implementing the following techniques:

  - **Chain-of-thought (CoT)**: The method encourages LLMs to generate step-by-step reasoning traces, leading to more accurate and structured responses for tasks involving arithmetic, commonsense reasoning, and other problem-solving scenarios. By guiding the LLM through a series of reasoning steps, CoT helps reduce issues such as fact hallucination while enhancing the overall quality and coherence of the generated content.

  - **ReAct**: This method is a framework that interleaves reasoning traces and task-specific actions, enabling LLMs to generate more reliable and factual responses. By incorporating dynamic reasoning and interaction with external sources, ReAct effectively addresses issues such as fact hallucination and error propagation, resulting in improved human interpretability and trustworthiness of LLMs.

  - **Prompt tuning**: Prompt tuning is a technique for refining LLM behavior by optimizing the input prompts using gradient-based methods, which allows for better control over the model's responses, and leads to improved accuracy and relevance in various problem-solving tasks. By fine-tuning prompts, users can effectively guide the LLM to generate more desirable and context-specific outputs. This only applies to LLMs you can host yourself, however.

- **Relying on well-engineered prompts**: Leverage published well-engineered prompts instead of crafting one of your own. This is a technique used by Langchain and Auto-GPT. AutoGPT is an open source Python application based on GPT-4. It automates the execution of tasks without requiring multiple prompts, using AI agents to access the web and perform actions with minimal guidance. Unlike ChatGPT, AutoGPT can execute larger tasks such as creating websites and developing marketing strategies without needing step-by-step instructions. It has various applications, such as generating content, designing logos, and developing chatbots.

- **Rejection sampling (best-of-n) reference**: Use rejection sampling techniques to improve the quality of generated responses by selecting the best response from multiple attempts. The best response can be evaluated through a chosen metric.

- **Re-ranking relevance distance scores from knowledge retrieval**: Knowledge retrieval is in the domain of recommendation systems. A common technique that's used is to implement proper regression-based recommendation models to re-rank relevance distance scores. This can help provide more accurate and potentially more personalized relevant information with more contextual data. This technique is used by most real-world large-scale recommendation-based products, such as YouTube.

- **Iterative retrieval and generation**: Use techniques such as self-ask, Active RAG, and ITER-RETGEN to generate temporary responses, evaluate their quality, and iteratively refine them using retrieved knowledge. This approach can reduce hallucinations and improve the quality of LLM-generated content.

- **Multi-agent systems**: Develop a multi-agent system composed of agents with unique personas and contexts to address diverse problem-solving scenarios. These agents can collaborate, share information, and provide more comprehensive and reliable solutions. An example of this is **AutoAgents**. AutoAgents is an innovative framework that adaptively generates and coordinates multiple specialized agents to build an AI team according to different tasks. The framework consists of two stages: drafting and execution. In the drafting stage, an agent team and execution plan are generated based on the input task, while the execution stage refines the plan through inter-agent collaboration and feedback to produce the outcome. AutoAgents can dynamically synthesize and coordinate multiple expert agents to form customized AI teams for diverse tasks. Experiments on open-ended question-answering and trivia creative writing tasks demonstrate the effectiveness of AutoAgents compared to existing methods. AutoAgents offers new perspectives for tackling complex tasks by assigning different roles to different tasks and promoting team cooperation.

Next, we will dive into solutions for the runtime performance challenge.

Tackling the runtime performance challenge

The runtime performance challenge is a critical issue that can significantly impact the efficiency and effectiveness of language models. As LLMs continue to grow in complexity and scale, optimizing their runtime performance becomes more crucial than ever. The solutions to solve this issue are as follows:

- **Caching outputs**: Temporarily store results to avoid recomputing information, enabling faster response times and improved performance. This approach is particularly useful when dealing with repetitive or similar queries.

- **GPUs and GPU inference accelerators**: This is only applicable for LLMs you host yourself. LLMs need to run with these components to run in a reasonable time. These were introduced in more detail in *Chapter 15, Deploying Deep Learning Models to Production*.

- **Use approximate KNN indexes:** Approximate KNN indexes take a much longer time than basic KNN indexes to be set up. However, after setting them up, the inference time can be 1,000 times faster while maintaining a reasonable retrieval accuracy. Go to `https://ann-benchmarks.com/` to understand the implications of different approximate KNN algorithms. The `scann` algorithm and the `faiss` *IVFPQFS* algorithm provide a good balance between index build time, index size, retrieval recall, and retrieval runtime. However, an approximate KNN algorithm is only required with large knowledge bases since the retrieval speed of a small knowledge base is already fast, which is less than 1 second. Typically, suitable data dimensions lie in the range of three-digit vector column sizes, and seven-digit vector row sizes.

Tackling the challenge of ethical implications and societal impacts

Addressing the ethical implications and societal impacts of LLM solutions is crucial for ensuring their responsible and sustainable deployment across various applications. By considering the ethical and societal consequences of LLM-generated content, developers can create models that respect user values, adhere to legal guidelines, and contribute positively to society.

The strategies to tackle these challenges are as follows:

- **Bias and fairness mitigation**: Consider the following strategy in the context of the methods that were introduced in *Chapter 13*, *Exploring Bias and Fairness*:

  - **Data collection and preparation**: Ensure a diverse and representative dataset for fine-tuning LLM models. Balance sensitive attributes in the data, and eliminate or control potential biases that may arise from these attributes. Additionally, you can instruct the LLM to specifically not perpetuate bias in natural language as part of the input context. Better yet, empower users to define their preferences, values, and ethical guidelines, enabling LLMs to generate content that aligns with individual user needs and values.

  - **Bias mitigation during fine-tuning**: Implement techniques such as counterfactual data augmentation, adversarial training, or re-sampling during the fine-tuning process to reduce the impact of biased features and improve fairness.

  - **Post-processing**: Modify LLM-generated content using techniques such as equalized odds post-processing to ensure fairness in the outputs. This can be applied when using LLM providers such as OpenAI GPT-4 or fine-tuned open source models.

  - **Monitoring and evaluation**: Continuously monitor LLM-generated content for potential biases using bias and fairness metrics, and adjust the model as needed to ensure compliance with ethical guidelines and fairness requirements.

- **Privacy-preserving techniques**: Adopt privacy-preserving approaches, such as differential privacy, federated learning, and homomorphic encryption, to protect sensitive information in the training data and generated content. Implement policies and guidelines to prevent the inadvertent disclosure of **Personal Identifiable Information** (**PII**) in LLM-generated content.

- **Fact-checking and credibility assessment**: Incorporate fact-checking and credibility assessment mechanisms into LLM solutions to reduce the risk of generating misleading or false information. This can involve integrating LLMs with external knowledge sources, such as knowledge databases, to verify the accuracy of generated content. You can also instruct the LLM to be humble and not return a statement if no facts can be used for verification.

- **Content moderation and guardrails**: Implement content moderation techniques, such as keyword filtering, machine learning-based classifiers, and human-in-the-loop review processes, to prevent the generation of harmful or offensive content. Establish guardrails, such as toxicity thresholds or ethical guidelines, to ensure that LLM-generated content adheres to community standards and legal regulations.

- **Transparency and explainability**: Develop methods for enhancing the transparency and explainability of LLM-generated content, such as providing reasoning traces, saliency maps, or counterfactual explanations. The concepts that were introduced in *Chapter 11, Explaining Neural Network Predictions*, and *Chapter 12, Interpreting Neural Networks*, can be applied to an LLM.

By implementing these strategies, developers can create LLM solutions that not only respect user values and adhere to legal guidelines but also contribute positively to society. Addressing the ethical implications and societal impacts of LLMs is an essential step toward building trust in the technology and ensuring its responsible and sustainable deployment across various applications.

With a comprehensive understanding of the challenges associated with LLMs mentioned, as well as their potential solutions, we can now turn our attention to addressing the overarching challenge of LLM solution adoption across organizations and industries.

Tackling the overarching challenge of LLM solution adoption

One overarching challenge in realizing the full potential of LLM solutions lies in their adoption across organizations and industries. Similar to the adoption of any machine learning or deep learning solution, the key factor driving the adoption of LLMs is confidence. Confidence in the technology's capabilities, its effectiveness in addressing specific use cases, and its ability to deliver tangible results are essential for widespread adoption.

To overcome this challenge, it is crucial to systematically educate organizations about the power of LLMs, their diverse applications, and how they can be tailored to meet specific needs. This includes demonstrating the benefits of LLMs through real-world success stories, providing practical guidance on implementing LLM solutions, and offering support for organizations navigating the complexities of integrating LLMs into their workflows.

Building confidence in LLM solutions involves thoroughly evaluating their performance, addressing the challenges discussed earlier in this chapter, and ensuring that the solutions meet their intended objectives. By implementing a comprehensive set of evaluation approaches, including quantitative metrics and qualitative manual evaluations, organizations can better assess the performance, accuracy, and overall quality of the results from an LLM solution. These evaluations should be conducted iteratively, allowing for ongoing refinement and improvement of the LLM solution.

Additionally, addressing the challenges identified in this chapter, such as output and input limitations, knowledge and information-related challenges, accuracy and reliability issues, and runtime performance challenges, is essential for building confidence in LLM solutions. By leveraging the strategies and techniques discussed in this chapter, organizations can optimize the performance of LLMs and ensure their effectiveness in various applications.

Another essential aspect of building confidence in LLM solutions is effective communication and collaboration with stakeholders. This includes sharing evaluation results, discussing the benefits and potential limitations of LLMs, and addressing any concerns that stakeholders may have regarding the adoption of LLM solutions.

In conclusion, adopting LLM solutions successfully requires a combination of rigorous evaluation, addressing challenges, and effective communication with stakeholders. By treating the adoption of LLM solutions with the same level of care as any machine learning or deep learning solution, organizations can build confidence in the capabilities and performance of LLMs, unlocking their full potential in a wide range of real-world applications. And with that, we have explored the challenges that plague LLMs and their solutions in detail.

In the following section, we will dive deeper into leveraging LLMs to build autonomous agents, which can significantly expand and improve our problem-solving skills.

Leveraging LLM to build autonomous agents

One promising area in which LLMs can be harnessed is the development of autonomous agents that can efficiently solve complex problems and interact with their environment. This section will focus on leveraging LLMs to build such agents and discuss the key aspects that contribute to their effectiveness.

Autonomous agents are AI-driven entities that can perform tasks, make decisions, and interact with their environment independently. By incorporating LLMs into these agents, developers can create versatile and adaptive systems that can tackle a wide range of challenges. Here are some essential components of LLM-powered autonomous agents:

- **Planning and decision making**: LLMs can be utilized to generate plans and strategies that guide the agent's actions, taking into account the context and goals.

- **Observing and learning from the environment**: LLMs can be trained to observe and interpret the environment, learning from past experiences and adjusting their behavior accordingly.

- **Collaborative problem-solving**: Multi-agent systems can be developed, where each agent has a unique persona and context. These agents can collaborate, share information, and provide more comprehensive and reliable solutions.

- **Self-refinement**: Autonomous agents can use LLMs to analyze their performance, identify areas for improvement, and refine their strategies and behaviors over time.

Agents encompass parts and pieces of the solutions to the challenges identified in the *Identifying challenges with LLM solutions* section of this chapter. Additionally, the solutions that were introduced here can be combined to expand the scope of problems the overall architected LLM solution can cover. Examples of published agent methods that were also introduced earlier in this chapter are WebGPT, Toolformer, Auto-GPT, and AutoAgents. Autonomous agents that leverage LLMs are the key to making powerful LLM solutions. By combining the strengths of LLMs with the adaptability and decision-making capabilities of agents, developers can create groundbreaking systems that can revolutionize various domains and industries.

With a comprehensive understanding of LLM solutions and their potential applications, let's explore some specific use cases where they can be employed effectively.

Exploring LLM solution use cases

In this section, we will explore some fascinating real-world applications where LLM solutions can truly shine. This will give you a sense of how revolutionary LLM solutions are. The use cases are as follows:

- **Travel itinerary planner**: LLMs can be employed to develop advanced travel itinerary planners that generate personalized trip plans based on user preferences and constraints. By integrating LLMs with travel APIs, such as flight, hotel, and attraction databases, as well as real-time data sources such as weather and traffic information, these planners can provide context-aware recommendations tailored to individual traveler needs. Notably, companies such as Booking.com and Expedia integrated this into their products, and Agoda announced that they will be working on it.

- **Intelligent tutoring systems**: LLMs can be used to develop intelligent tutoring systems that offer personalized learning experiences for students. By integrating LLMs with educational content, assessment tools, and learner data, these systems can generate targeted learning materials, provide real-time feedback, and adapt to individual learning needs. This enables a more efficient and engaging learning experience for students. Notably, Duolingo is a company that implemented such a solution in their gamified language learning product.

- **Automated email responses**: LLM solutions can be employed to develop automated email response systems that handle various types of inquiries, such as customer support, sales inquiries, or general information requests. By integrating LLMs with email APIs, CRM systems, and relevant knowledge bases, email responses can be personalized, accurate, and contextually relevant. This helps businesses streamline their customer communication and provide efficient support. Notably, Nanonets AI is a company that implemented this as part of their product.

- **Code generation**: LLMs can be used to generate code snippets, algorithms, or entire software programs based on user input or specific requirements. Solutions such as GitHub Copilot leverage LLMs to assist developers in writing code, suggesting relevant code snippets, and completing sections of code based on context. By integrating LLMs with code repositories, programming language APIs, and domain-specific knowledge bases, code generation can be tailored to specific programming languages, frameworks, and use cases, improving developer productivity.

- **Customer support chatbots**: LLM solutions can be employed to develop advanced, context-aware chatbots that can handle customer inquiries and support requests more effectively. By integrating LLMs with **customer relationship management (CRM)** systems and knowledge bases, chatbots can provide personalized and accurate responses to customer queries. This helps businesses improve their customer support services, reduce response times, and increase customer satisfaction. Companies such as forethought.ai, Ada, and EBI.AI provide such a solution in their product.

- **Medical diagnosis and treatment suggestions**: LLMs can be employed to develop advanced diagnostic tools that analyze patient symptoms, medical history, and relevant medical literature to suggest potential diagnoses and treatment options. By integrating LLMs with **electronic health record (EHR)** systems, medical databases, and domain-specific knowledge bases, these tools can help healthcare professionals make more informed decisions and improve patient outcomes. Notably, Harman, a Samsung company, implemented and offered such a solution as part of their offered services.

- **Personal finance management**: LLMs can be used to develop intelligent personal finance management applications that provide tailored financial advice, budgeting suggestions, and investment recommendations based on user-specific financial goals and risk tolerance. By integrating LLMs with banking APIs, stock market data, and financial knowledge bases, these applications can offer context-aware financial planning and guidance to users. Although not exactly a service or product, Bloomberg has developed BloombergGPT, a 50-billion parameter large language model designed specifically for finance, which showcases the potential of LLMs in the financial domain.

- **Creative content generation**: LLMs can be employed to generate creative content, such as stories, poetry, or music, based on user inputs, preferences, or inspirations. By integrating LLMs with databases of literary works, music libraries, and knowledge bases on creative techniques and styles, these applications can produce unique and engaging content that caters to individual artistic tastes and needs. Notably, Jasper built a platform to account for this use case.

- **Legal document analysis and drafting**: LLMs can be used to develop advanced legal document analysis and drafting tools that assist legal professionals in reviewing contracts, identifying potential issues, and generating legal documents based on specific requirements. By integrating LLMs with legal databases, contract templates, and domain-specific knowledge bases, these tools can help streamline legal work and improve efficiency in the legal industry. Notably, netdocuments implemented this use case with their product.

- **Smart home automation**: LLMs can be employed to develop intelligent home automation systems that understand natural language commands and adapt to user preferences and routines. By integrating LLMs with smart home devices, APIs, and user behavior data, these systems can provide a more intuitive and personalized home automation experience, enabling users to control their home environment with ease and convenience. Amazon Alexa is a prime example of this use case.

In each of these use cases, integrating LLMs with the relevant tools, APIs, and data sources ensures that the generated content, recommendations, and responses are accurate, contextually relevant, and tailored to specific needs, enhancing user experiences and providing valuable support in various domains.

Summary

In this chapter, we explored LLMs and their potential to address real-world problems and create value across various applications. We discussed the key aspects of architecting LLM solutions, such as handling knowledge, interacting with real-time data and tools, evaluating LLM solutions, identifying and addressing challenges, and leveraging LLMs to build autonomous agents. We also emphasized the importance of retrieval-augmented language models for providing contextually relevant information and examined various techniques and libraries to improve LLM solutions.

We also discussed the limitations of LLMs, such as output and input limitations, knowledge and information-related challenges, accuracy and reliability issues, runtime performance challenges, ethical implications and societal impacts, and the overarching challenge of LLM solution adoption. To tackle these limitations, we presented various complementary strategies, such as real-time data integration, tool integration, prompt engineering, rejection sampling, multi-agent systems, runtime optimization techniques, bias and fairness mitigation, content moderation, and enabling LLM transparency and explainability. Lastly, we discussed leveraging LLMs to build autonomous agents, which can significantly expand and improve problem-solving abilities in diverse applications.

By understanding the intricacies of LLM solutions and applying the strategies and techniques discussed in this chapter, organizations and individuals can harness the full potential of LLMs to drive innovation and improve outcomes across various applications.

By reading *The Deep Learning Architect Handbook*, you have gained invaluable insights into the various stages of the deep learning life cycle, exploring critical aspects from planning and data preparation to model deployment and governance. By reaching the end of this enlightening journey, you are now armed with the knowledge and skills to design, develop, and deploy effective deep learning solutions. To build upon this strong foundation, consider taking the following next steps:

- Apply your newfound knowledge to real-world projects, either in your professional field or through open source contributions, to gain practical experience and deepen your understanding

- Stay up-to-date with the latest research, trends, and breakthroughs in deep learning by attending conferences, following influential researchers, and reading research papers

- Explore specialized areas within deep learning that interest you, such as reinforcement learning, generative adversarial networks, or few-shot learning, to further expand your expertise

- Collaborate with fellow deep learning enthusiasts and professionals, joining communities, discussion forums, and social media groups to exchange ideas, share experiences, and learn from each other

- Consider pursuing advanced courses, certifications, or even a degree in deep learning or a related field to enhance your education and qualifications

Embrace the challenges and triumphs that lie ahead, for with the mastery of building deep learning models with intricate deep learning architectures, a keen understanding of bias and fairness, and the ability to monitor and maintain model performance, you are well-prepared to unleash the full potential of deep learning and drive innovation across a vast array of applications. Here's to your continued success and growth in the world of deep learning!

Further reading

1. *RAG*: https://doi.org/10.48550/arXiv.2005.11401
2. *RETRO*: https://arxiv.org/pdf/2112.04426.pdf
3. *REALM*: https://doi.org/10.48550/arXiv.2002.08909

Index

A

X

Y

Z

‹packt›

www.packtpub.com

Subscribe to our online digital library for full access to over 7,000 books and videos, as well as industry leading tools to help you plan your personal development and advance your career. For more information, please visit our website.

Why subscribe?

- Spend less time learning and more time coding with practical eBooks and Videos from over 4,000 industry professionals
- Improve your learning with Skill Plans built especially for you
- Get a free eBook or video every month
- Fully searchable for easy access to vital information
- Copy and paste, print, and bookmark content

Did you know that Packt offers eBook versions of every book published, with PDF and ePub files available? You can upgrade to the eBook version at packtpub.com and as a print book customer, you are entitled to a discount on the eBook copy. Get in touch with us at customercare@packtpub.com for more details.

At www.packtpub.com, you can also read a collection of free technical articles, sign up for a range of free newsletters, and receive exclusive discounts and offers on Packt books and eBooks.

Other Books You May Enjoy

If you enjoyed this book, you may be interested in these other books by Packt:

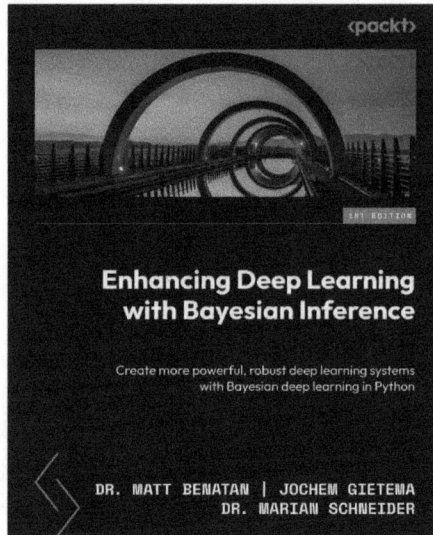

Enhancing Deep Learning with Bayesian Inference

Matt Benatan, Jochem Gietema, Marian Schneider

ISBN: 978-1-80324-688-8

- Understand advantages and disadvantages of Bayesian inference and deep learning

- Understand the fundamentals of Bayesian Neural Networks

- Understand the differences between key BNN implementations/approximations

- Understand the advantages of probabilistic DNNs in production contexts

- How to implement a variety of BDL methods in Python code

- How to apply BDL methods to real-world problems

- Understand how to evaluate BDL methods and choose the best method for a given task

- Learn how to deal with unexpected data in real-world deep learning applications

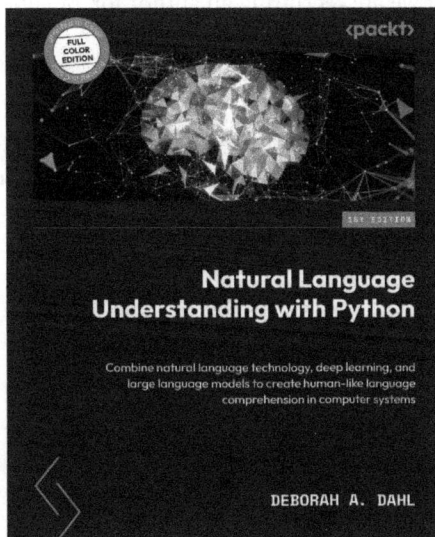

Natural Language Understanding with Python

Deborah A. Dahl

ISBN: 978-1-80461-342-9

- Explore the uses and applications of different NLP techniques
- Understand practical data acquisition and system evaluation workflows
- Build cutting-edge and practical NLP applications to solve problems
- Master NLP development from selecting an application to deployment
- Optimize NLP application maintenance after deployment
- Build a strong foundation in neural networks and deep learning for NLU

Packt is searching for authors like you

If you're interested in becoming an author for Packt, please visit authors.packtpub.com and apply today. We have worked with thousands of developers and tech professionals, just like you, to help them share their insight with the global tech community. You can make a general application, apply for a specific hot topic that we are recruiting an author for, or submit your own idea.

Share Your Thoughts

Now you've finished *The Deep Learning Architect's Handbook*, we'd love to hear your thoughts! Scan the QR code below to go straight to the Amazon review page for this book and share your feedback or leave a review on the site that you purchased it from.

https://packt.link/r/1-803-24379-1

Your review is important to us and the tech community and will help us make sure we're delivering excellent quality content.

Download a free PDF copy of this book

Thanks for purchasing this book!

Do you like to read on the go but are unable to carry your print books everywhere?

Is your eBook purchase not compatible with the device of your choice?

Don't worry, now with every Packt book you get a DRM-free PDF version of that book at no cost.

Read anywhere, any place, on any device. Search, copy, and paste code from your favorite technical books directly into your application.

The perks don't stop there, you can get exclusive access to discounts, newsletters, and great free content in your inbox daily

Follow these simple steps to get the benefits:

1. Scan the QR code or visit the link below

https://packt.link/free-ebook/9781803243795

2. Submit your proof of purchase
3. That's it! We'll send your free PDF and other benefits to your email directly

9 781803 243795